LITHIUM ION BATTERIES – NEW DEVELOPMENTS

Edited by **Ilias Belharouak**

Lithium Ion Batteries – New Developments
Edited by Ilias Belharouak

Published by InTech
Janeza Trdine 9, 51000 Rijeka, Croatia

Publishing Process Manager Marija Radja

Technical Editor Teodora Smiljanic

Cover Designer InTech Design Team

Additional hard copies can be obtained from orders@intechopen.com

Lithium Ion Batteries – New Developments, Edited by Ilias Belharouak
p. cm.
ISBN 978-953-51-0077-5

Meet the editor

Dr. Ilias Belharouak is a Materials Scientist Leader and Energy Storage Expert in the Chemical Sciences and Engineering Division at Argonne National Laboratory, Illinois, USA. He published over 200 peer-reviewed publications, papers at professional society meetings, and patents. He is a reviewer for several international journals in the field of electrochemistry, batteries, and materials physics and chemistry. Dr. Belharouak was recognized with several awards including R&D-100 awards, Pacesetter awards, and Federal and State Laboratory Consortium Awards. Dr. Belharouak holds Ph.D. and Master's Degrees in Materials Science from the Institute for Solid State Chemistry, National Center for Scientific Research, University Bordeaux 1, France; and a Bachelor's Degree in Inorganic Chemistry, University Cadi Ayyad, Morocco.

Contents

LITHIUM ION BATTERIES – NEW DEVELOPMENTS

Edited by **Ilias Belharouak**

Lithium Ion Batteries – New Developments
Edited by Ilias Belharouak

Published by InTech

Janeza Trdine 9, 51000 Rijeka, Croatia

Publishing Process Manager Marija Radja

Technical Editor Teodora Smiljanic

Cover Designer InTech Design Team

Additional hard copies can be obtained from orders@intechopen.com

Lithium Ion Batteries – New Developments, Edited by Ilias Belharouak
p. cm.
ISBN 978-953-51-0077-5

Preface

This book is intended as a collection of essays covering the most recent advances in science and engineering related to lithium-ion batteries. Rechargeable lithium-ion batteries were first commercialized for the consumer electronics industry by Sony in the early 1990s. A typical commercial lithium-ion cell consists of a carbonaceous anode, a transition metal oxide cathode such as $LiCoO_2$, and an organic electrolyte soaked in a separator that separates the cathode and anode. During the last two decades, research groups in Asia, Europe, and the United States have invented and developed cathode and anode materials and electrolytes that can be used in electrochemical couples that store more and more electricity per mass and volume. As a result of these advances, lithium-ion battery technology has evolved beyond the consumer electronics industry and is now making serious inroads in the transportation industry; still, challenges at the battery system level remain with regard to energy, power, cost, life, and safety. Even though lithium-ion batteries have higher safety risk and cost compared to traditional lead-acid batteries, they are slowly making inroads in other industries such as stationary storage systems for wind farms and solar plants. Reducing battery cost and improving battery energy and safety will always be critical goals. The ultimate benefits of this energy storage technology would be reduced fossil fuel consumption and substantially lower greenhouse gas emissions, lessening the threat of global warming.

The eight chapters in this book cover topics on advanced anode and cathode materials, materials design, materials screening, electrode architectures, diagnostics and materials characterization, and electrode/electrolyte interface characterization for lithium batteries. All these topics were carefully chosen to reflect the most recent advances in the science and technology of rechargeable Li-ion batteries, to provide wide readership with a platform of subjects that will help in the understanding of current technologies, and to shed light on areas of deficiency and to energize prospects for future advances.

Finally, I wish to thank colleagues and authors from renowned scientific institutions who have suggested the inclusion of a wide range of topics covering all aspects of modern lithium-ion batteries. I am especially grateful to colleagues at Argonne National Laboratory who read chapters of this book, provided invaluable comments, and contributed with a chapter on redox shuttle additives for lithium-ion battery.

To conclude, I must express my thanks to Mrs. Marija Radja, who uncomplainingly worked diligently in sorting out the chapters of this book in the final version that is now between the hands of scientists and other interested readers. Even with all this help, I must confess that some mistakes will remain; for these I ask pardon.

Ilias Belharouak
Materials Scientist Leader and Energy Storage Expert,
Chemical Sciences and Engineering Division at Argonne National Laboratory,
Illinois,
USA

1

Synthesis Processes for Li-Ion Battery Electrodes – From Solid State Reaction to Solvothermal Self-Assembly Methods

Verónica Palomares[1] and Teófilo Rojo[1,2]
[1]*Universidad del País Vasco/Euskal Herriko Unibertsitatea,*
[2]*CIC Energigune,*
Spain

1. Introduction

Since 1990, Li-ion batteries became essential for our daily life, and the scope of their applications is currently expanding from mobile electronic devices to electric vehicles, power tools and stationary power grid storage. The ever-enlarging market of portable electronic products and the new demands of the transportation market and stationary storage require cells with enhanced energy density, power density, cyclability and safety. In short, to get better performance. These new needs have boosted research and optimization of new materials for Li-ion batteries.

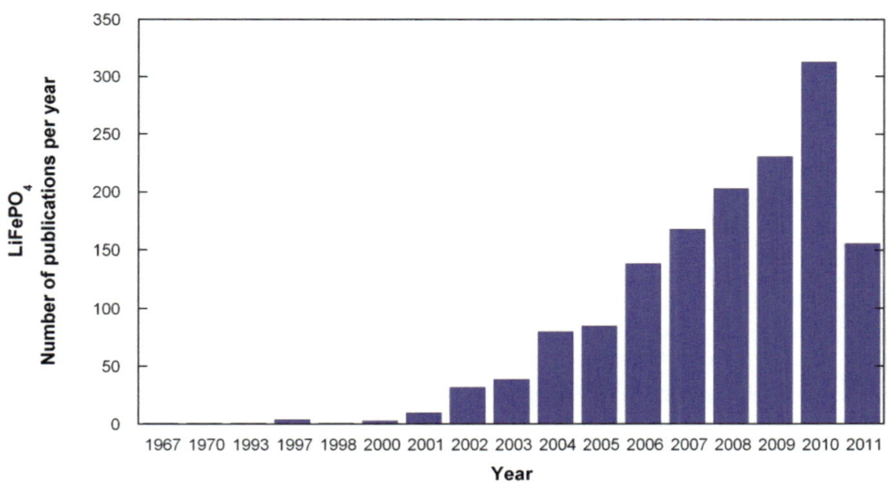

Fig. 1. Number of scientific publications about $LiFePO_4$ material in the last 40 years. Source: Scifinder Scholar™ 2007.

The aim of this work is to show the evolution of chemical preparative methods used to synthesize new electroactive materials or to ameliorate electrochemical performance of the existing ones, and to compare the improvement of performance achieved by the new materials processing. This way, the synthesis methods of several electrodic materials for Li-ion batteries will be analyzed. Mainly cathode materials, such as layered oxides derived from $LiCoO_2$ or $LiMn_2O_4$ spinel derivatives will be described. Olivine $LiFePO_4$ phase, a material that, besides having the right voltage to present safety attributes is made of low cost and abundant elements, will be specially remarked because of its extraordinary importance in the last years (figure 1).

In recent years, nanoscience has irrupted strongly in the battery materials field. Not only the performance of previously known materials was improved significantly by nanodispersion and nanostructuring, but also new materials and electrochemical reactions have emerged. Thus, the fabrication of nanostructured electrodes has become one of the main goals in battery materials.

First, the small size and large surface area of nanomaterials provide greater contact area between the electrode material and the electrolyte. Second, the distance the Li ions have to diffuse across the electrode is shortened. Therefore, faster charge/discharge ability, that is, a higher rate capability, can be expected for nanostructured electrodes. For very small particles, the chemical potentials for lithium ions and electrons may be modified, resulting in a change of electrode potential. Moreover, the range of composition over which solid solutions exist is often more extensive for nanoparticles, and the strain associated with intercalation is often better accommodated. Furthermore, even new electrochemical reactions, such as conversion reactions for anodes have appeared in nanostructured electrodes. Thus, morphology and size of electrode materials have become a key factor for their performance and the synthesis processes have been evolved toward nanoarchitectured materials.

This chapter will provide an overview on most used synthesis methods from the beginning of Li-ion batteries major research up to the newest ones. Materials performance evolution due to new processing systems will be discussed.

2. Conventional synthesis methods

Classical synthesis methods can be classified in solid reactions and solution methods, according to the precursors used (Figure 2).

Ceramic process is the simplest and most traditional synthesis method because of its easy procedure and easy scale-up. It consists on manual grinding of the reactants and their subsequent heating in air, oxidative, reducing or inert atmosphere, depending on the targeted compound. The great disadvantage of this method is the need for high calcination temperatures, from 700 to 1500° C, which provokes the growth and sinterization of the crystals, leading to micrometer-sized particles (>1 μm) [Eom, J. et al. (2008); Cho, Y. & Cho, J. (2010); Mi, C.H. et al. (2005); Yamada, A. et al. (2001)]. The macroscopic dimensions of as synthesized particles leads to limited kinetics of Li insertion/extraction and makes difficult the proper carbon coating of phosphate particles [Song, H-K. et al. (2010)]. For this reason it was necessary to add carbon during or after the grinding process, which implies the use of an extra grinding step [Liao, X.Z. et al. (2005); Zhang, S.S. et al. (2005); Nakamura, T. et al. (2006); Mi, C.H. et al. (2005)]. Mechanochemical activation can be considered as a variant of the ceramic method, but the final calcination temperature is lower, of about 600° C [Kwon, S.J. et al. (2004); Kim, C.W. et al. (2005); Kim, J-K. et al. (2007)]. This way, grain size is slightly lower due to mechanical milling.

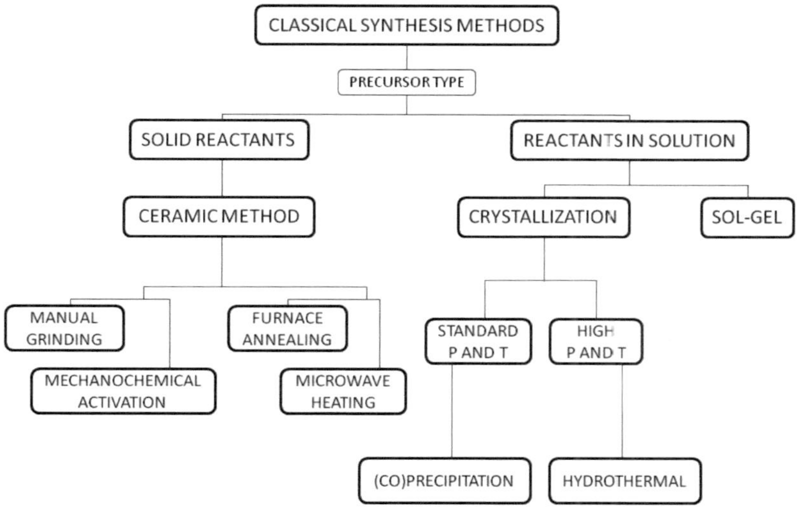

Fig. 2. Schematic of the classical synthesis methods used to prepare electrode materials for Li-ion batteries.

Hand-milled precursors can also be activated by microwave radiation [Song, M-S. et al. (2007)]. If at least one of the reactants is microwave sensitive, the mixture can get sufficiently high temperatures so as to achieve the reaction and obtain the targeted compound in very short heating times, between 2 and 20 minutes. This factor makes this synthesis method an economic way to obtain desired phases. Sometimes, when a carbonaceous composite is desired, active carbon can be used to absorb microwave radiation and to heat the sample [Park, K.S. et al. (2003)]. Organic additives such as sucrose [Li, W. et al. (2007)], glucose [Beninati, S. et al. (2008)] or citric acid [Wang, L. et al. (2007)] can be used in the initial mixture in order to get in situ carbon formation. Oxide-type impurity generation is not usually indicated in literature, but, sometimes, the reaction atmosphere is so reducing that iron carbide (Fe_7C_3) or iron phosphide (Fe_2P) are generated as secondary phases [Song, M-S. et al (2008)]. Particle size of phosphates obtained by this synthesis method ranges between 1 and 2 μm, but two effects have been reported with regard to this parameter. The growth of particles was correlated with the increase of microwave exposure times. However, in the presence of greater amounts of carbon precursor the particles decrese in size leading to 10-20 nm particles.

Synthesis methods that comprise the dissolution of all reactants promote greater homogeneity in final samples. Both coprecipitation and hydrothermal processes consist on the precipitation and crystallization of the targeted compound under normal (coprecipitation) or high (hydrothermal) temperature and pressure conditions. Usually coprecipitation involves a subsequent heating process, which enhances particle growth [Park, K.S. et al. (2004); Yang, M-R. et al. (2005)]. Nevertheless, recent advances in direct precipitation method have produced narrow particle size materials, of about 140 nm, with enhanced electrochemical properties in terms of specific capacity (147 mAh g^{-1} at 5C rate) as well as in terms of cyclability (no significant capacity fade after more than 400 cycles) with no carbon coating [Delacourt, C. et al. (2006)]. On the other hand, hydrothermal synthesis is an effective method to obtain well-crystallized materials with well-defined morphologies, where no additional high-temperature treatment is needed, but no small size particles can be obtained. Tryphilite crystals of about 1x3 μm have been produced by this method without carbonaceous coating [Yang, S. et al. (2001); Tajimi, S. et al. (2004); Dokko, K. et al. (2007); Kanamura, K. and Koizumi, S. (2008)]. Conductive carbon coating can be produced by using diverse additives that also act as reductive agents, such as sucrose, ascorbic acid [Jin, B. and Gu, H-B. (2008)] or carbon nanotubes [Chen, J. and Whittingham, M.S. (2006)]. The preparation of $LiFePO_4$ samples by hydrothermal method using heating temperatures below 190° C has been demonstrated to create olivine phases with some inversion between Fe and Li sites, with 7% of the iron atoms in lithium sites, and also the presence of small amounts of Fe(III) in the material. Lithium ion diffusion in $LiFePO_4$ is one-dimensional, because the tunnels where Li ions are located run along the b axis are not connected, so lithium ions residing in the channels cannot readily jump from one tunnel to another if Fe(III) ions are present. Thus, any blockage in the tunnel will block the movement of the lithium ions. This way, the presence of iron atoms on the lithium sites prevents the diffusion of Li ions down the channels in the structure and jeopardizes electrochemical performance. For this reason, materials synthesized under hydrothermal conditions at 120° C did not reach 100 $mAh \cdot g^{-1}$ [Yang, S. et al. (2001)]. The use of higher temperatures, the addition of L-ascorbic acid, carbon nanotubes or a subsequent annealing process (500-700° C) under nitrogen atmosphere can produce ordered $LiFePO_4$ phases that are able to deliver sustainable capacities of 145 $mAh \cdot g^{-1}$ [Whittingham, M.S. et al. (2005); Chen, J. et al. (2007)].

A study by Nazar et al. on the different variables that influence the hydrothermal processes concludes that, in the first place, crystal size can be controlled by reaction temperature and precursors concentration inside the reactor, because higher precursor concentration creates higher quantity of nucleation sites, thus leading to smaller particle sizes. In the second place, decrease of synthesis temperature also entails smaller particle size, but shorter reaction times do not have remarkable influence on the product morphology, once the minimum reaction time is surpassed [Ellis, B. et al. (2007a)].

Among the solution methods, sol-gel process is a classical method used to obtain different types of inorganic materials [Kim, D.H. and Kim, J. (2007); Pechini, P. Patent; Baythoun, M.S.G. and Sale, F.R. (1982)]. Apart from the homogeneity promoted by the starting reactants solution, this method allows the introduction of a carbon source that can act as particle size control factor, leaves a carbon that can be useful to create carbon composites, and, finally, allows the use of lower heating temperatures than in solid state reaction methods [Hsu, K-F. et al. (2004); Chung, H-T. et al. (2004); Choi, D. and Kumta, P.N. (2007)]. This way, synthesizing one phase by ceramic or sol-gel method under the same thermal treatments allows getting lower particle size for sol-gel samples [Piana, M. et al. (2004)].

3. New synthetic methods directed towards nanostructured materials

Apart from the classical preparative methods, a wide variety of synthetic approaches has been developed to improve the rate capabilities of the materials. The rate determining step in the electrodes of Li-ion batteries is supposed to be solid state diffusion. Faster kinetics is expected with smaller particle size because the diffusion length is shorter. For this purpose, Li-ion battery electrode materials have been built in very different nanoarchitectures, such as nanotubes, nanobelts, nanowires, nanospheres, nanoflowers and, nanoparticles. These synthesis methods have been focused to obtaining nanostructured electrode materials (figure 3).

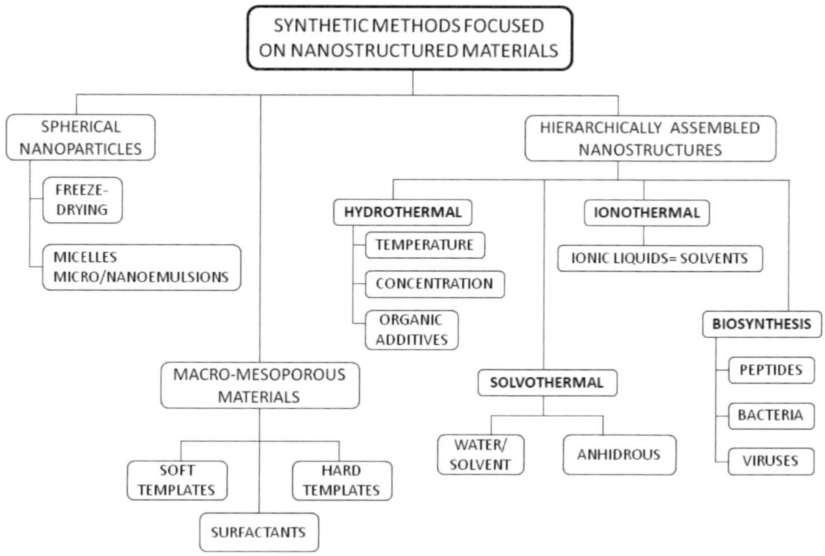

Fig. 3. Schematic of the synthesis methods used to prepare nanostructured electrode materials for Li-ion batteries.

3.1 Synthesis of spherical nanoparticles

Freeze-drying synthesis method presents advantages such as homogeneity of reactants, the possibility of introducing a carbon source and the use of lower calcinations temperatures [Palomares, V. et al. (2009a)]. Rojo et al. applied this synthesis process to prepare $LiFePO_4/C$ composites for the first time, getting nanosized phosphate particles of 40 nm completely surrounded by a carbonaceous web with 141 $mAh{\cdot}g^{-1}$ specific capacity at 1C rate [Palomares, V. et al. (2007)].

Freeze-drying process consists on solvent elimination from a frozen solution by sublimation. Sublimation process is thermodynamically favoured versus fusion or evaporation below solvent triple point pressure and temperature conditions (figure 4). First, the reactant solution needs to be frozen (from A to B point), and under low temperature, and low pressure conditions a direct sublimation process can be feasible (from C to E points).

However, the presence of any solute does alter triple point location. Freeze-drying technique allows maintaining stoichiometry and homogeneity of a multicomponent solution in the final dried product [Paulus, M. (1980)], and also provides promotes small size particles.

The starting solution is frozen so millimeter sized droplets with high specific area are formed. These droplets are dried under low temperature and vacuum conditions in order to get a spongy solid that is calcined at low temperature to obtain the targeted compound. Optimization of this synthesis method has led to 10 nm sized $LiFePO_4$ particles embedded in a carbonaceous web that enhances the electrochemical performance due to greater surface area of nanosized particles and to homogeneous carbon coating that connects the active material [Palomares et al, (2011)].

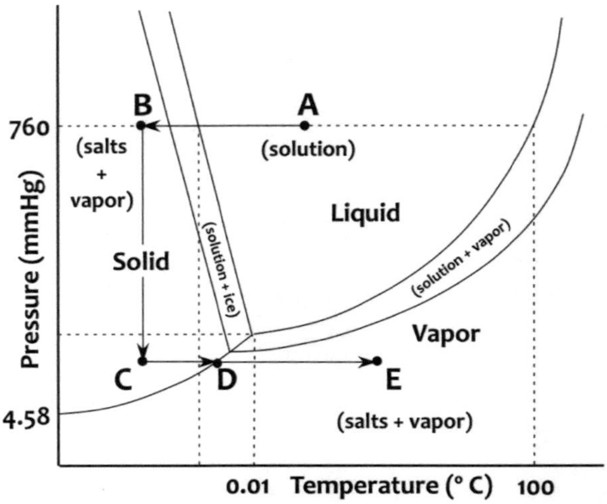

Fig. 4. Water phase diagram. Freeze-drying process is marked by arrows.

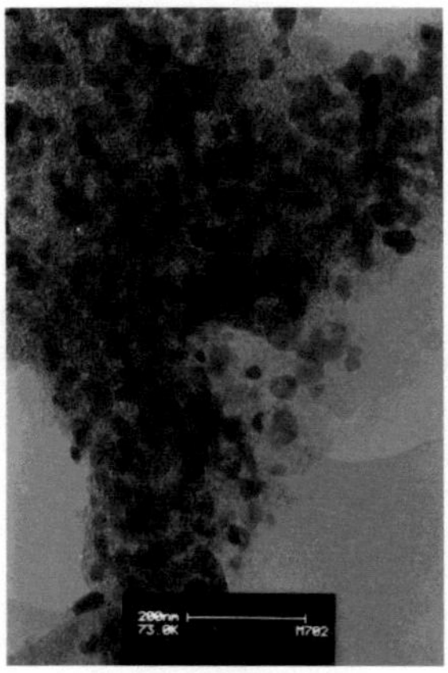

Fig. 5. $LiFePO_4/C$ nanocomposites prepared by freeze-drying. [Palomares et al.(2007)]

Although carbonaceous coating for these freeze-dried materials is very homogeneous, it has been demonstrated that it can only replace a small proportion of the conductive carbon additives used to prepare positive electrodes based on $LiFePO_4$ compound [Palomares, V. et al. (2009b)]. Deep characterization of *in situ* produced carbon showed that, in spite of its high specific surface, it presents high disorder, which is not favourable to a good electrochemical performance, and does not have enough conductivity to act as conductive additive in these cathodes.

Swollen micelles and microemulsions make up another synthesis method that lead to discrete nanoparticles with controlled chemical composition and size distribution [Li, M. et al. (1999)]. In this synthesis method, chemical reactions are carried out in an aqueous media within a restricted volume, limited by the array of surfactant and co-surfactant molecules. The versatility of this technique allows its use in the preparation of different electrode materials for lithium ion batteries. The obtained solid products exhibit a controlled size and shape, remaining well dispersed due to their isolation from other particles by the surfactant molecules during the synthesis [Aragón, M.J. et al. (2010)]. There are three different processes to obtain nanoparticles by the reverse micelles methods. The first one consists on mixing different emulsions which contain the necessary reagents in aqueous solution, so coalescence of pairs of droplets results in the formation of the solids in a confined volume. The second one involves reacting by diffusion of one of the reagents through the oil phase and the surfactant molecular layer. The last one requires thermolysis within individual droplets to get the target compound of a controlled size. $LiCoO_2$ cathode material has been prepared by the last process, providing 140 mAh·g^{-1}. Thermal decomposition of the micelles was achieved by putting the emulsion in contact with a hot organic solvent, such as kerosene at 180° C. $LiMn_2O_4$ was also obtained by the same method, leading to 200 nm diameter particles with good electrochemical performance.

Rod-like $LiFePO_4$/C composite cathodes have also been synthesized by reverse micelles method, using kerosene with Tween#80 surfactant as oil phase, and annealing the obtained precursor at 650° C in N_2 atmosphere [Hwang, B-J. et al. (2009)]. Morphology of this composite consisted on rod-like porous aggregates made of tiny primary nanoparticles. This special arrangement of primary particles provided better accommodation of volume changes during cycling, better electrical connection with the current collector and efficient electron transport. Galvanostatic cycling of this composite showed very good results for this rod-like composite, with a specific capacity of 150 and 95 mAh·g^{-1} at C/30 and 5C, respectively.

3.2 Synthesis of macro or mesoporous materials

One approach to new positive electrode materials for high rate applications is to synthesize three dimensionally ordered macroporous or mesoporous solids. Such materials are composed of micrometer-sized particles within which identical ordered pores of diameter 2-50 nm exist with walls of 2-8 nm thickness. Unlike nanoparticles, which can become disconnected one from another as they expand or contract on cycling, mesoporous materials, since they have the same dimensions as the intercalation cathodes in conventional lithium-cells, suffer less from the problem of disconnection. Furthermore, they may be fabricated in the same way as conventional materials, yet the internal porosity permits the electrolyte to flood the particles ensuring a high contact area and hence a facile lithium transfer across the interface, as well as short diffusion distances for Li^+ transport within the walls, where intercalation takes place [Bruce, P.G. (2008a)].

Ordered mesoporous solids can be built with silica structures [Bruce, P.G. et al. (2008b)]. The first example of an ordered mesoporous lithium transition-metal oxide, the low temperature polymorph of $LiCoO_2$, has been synthesized and shown to exhibit superior properties as a cathode compared with the same compound in nanoparticle form. This material showed 40 Å size pores and a wall thickness of 70 Å. Synthesis of this sample comprised the use of KIT-6 silica as a template. Impregnation of the silica in Co precursor solution, subsequent annealing and silica template dissolution produced mesostructured Co_3O_4. This porous oxide reacted with LiOH by solid state reaction to get $LiCoO_2$. The ordered mesoporous material demonstrates superior lithium cycling during continuous intercalation/removal for 50 cycles [Jiao, F. Et al. (2005)].

Mesoporous structures can also be prepared by using soft colloidal crystals as templates. In 1997, Velev first reported the use of colloidal latex spheres, in the range of 150 nm to 1 μm as templates to produce silica macroporous structures [Velev, O.D. et al. (1997)].A colloidal crystal consists on an ordered array of colloid particle that is analogous to a standard crystal whose repeating subunits are atoms or molecules [Pieranski, P. (1983)]. They are usually formed from closed-packed spheres such as latex, poly(styrene) (PS), silica or PMMA (poly(methyl methacrylate)) microbeads. After infiltration of the precursors solution into the opal structure, the assembly is usually calcined in air at temperatures between 500 and 700° C. This way, void spaces between particles are filled by the fluid precursors, and these latter are converted into a solid before removal of the template material.

Colloidal crystal templates were first reported as additive to form electrode materials for Li-ion batteries in 2002 [Sakamoto, J.S., Dunn, B. (2002)], and has also been used for the preparation of 3-D ordered macroporous $LiMn_2O_4$ spinel [Tonti, D. et al. (2008)]. Lithium iron phosphate has been successfully templated using colloidal crystal templates of PMMA of 100, 140 and 270 nm diameter spheres to produce porous, open lattice electrode materials, which featured pores in the mesoporous (10-50 nm), meso-macroporous (20-80 nm), and macroporous (50-120 nm) ranges, respectively [Doherty, C.M. et al. (2009)]. The well-stacked PMMA colloidal crystals provided robust scaffolding in which the $LiFePO_4$ precursor solution was infiltrated and then condensed. Once the PMMA spheres were removed through the calcinations process at different annealing temperatures ranging from 320 to 800° C, the $LiFePO_4$ featured an open lattice structure with residual carbon left over from the decomposed colloidal crystal template. Figure 6 shows the crystal colloidal systems used for this research, with well-organized, stacked homogeneous diameter spheres, and also the open porous structures of the template $LiFePO_4$, with a continuous open lattice structure with long-range order.

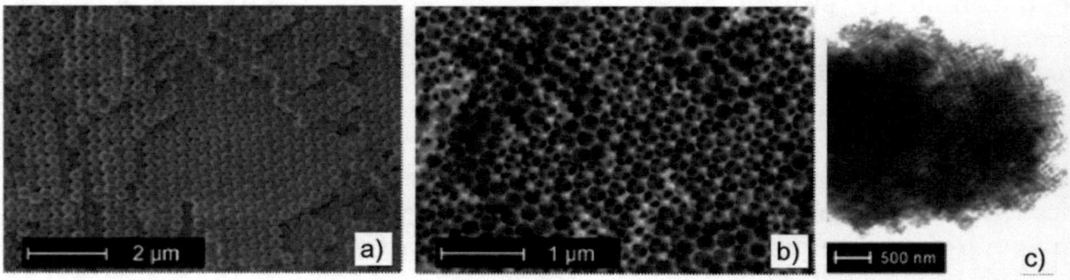

Fig. 6. a) Micrograph of the used colloidal crystal system; b) $LiFePO_4$ templated with 270 nm sized PMMA spheres; and c) Regular channels formed from the closed-packed beads, which allow for good electrolyte access to the $LiFePO_4$ surfaces. (Doherty, C.M. et al. (2009)).

All produced materials suffered from shrinkage of the porous structure, being the pore diameters for each of the samples approximately 40% of the initial bead diameter. Growing crystallite sizes were recorded when the calcination temperature was increased from 500 to 800° C. Despite the higher surface area of the material made from the smallest bead diameter (100 nm), its electrochemical performance was the poorest of the three of them. This can be due to restriction of the electrolyte access caused by poor interconnectivity between the pores that leaves some $LiFePO_4$ surface electrochemically inactive. Micrographs of the sample showed some areas where the small pores had collapsed and become blocked during thermal treatment and, thus, open lattice was not present. For this reason, it can be said that interconnectivity of the pore structure is essential for good electrolyte penetration as well as efficient charge transfer. This way, the templated samples prepared with the larger spheres (270 nm) would offer both good interconnectivity and better electrolyte access to surfaces within large micrometer-sized $LiFePO_4$ particles. The advantage of using colloidal crystal templating to produce high power $LiFePO_4$ electrodes is that it allows the pore sizes to be tailored while controlling the synthesis conditions. It increases the surface area and decreases the diffusion distance while maintaining an interconnected porous structure to provide efficient charge transfer and reduced impedance.

Mesoporous electrode materials have also been fabricated by using a cationic surfactant in fluoride medium, such as $Li_3Fe_2(PO_4)_3$ [Zhu, S. et al. (2004)]. This material showed an average pore diameter of 3.2 nm and a wall thickness of 2.2 nm. In this case, self-assembly process that leaded to mesoporous material was based on Coulombic interactions between the head groups of the surfactant (cetyltrimethylammonium $CTMA^+$) and F^- ions, that encapsulate Fe^{2+} species. Fe^{2+} ions are located between ion pairs [$LiPO_4Fe^{2+}$] and [F^-CTMA^+]. Cathode performance of this self-assembled $Li_3Fe_2(PO_4)_3$ material was better than the observed in other studies described in literature, with a specific capacity above 100 $mAh \cdot g^{-1}$ at 200 $mA \cdot g^{-1}$.

3.3 Synthesis of nanostructured materials by hydrothermal/solvothermal methods

While the hunt for high performance electrode materials for Li-ion batteries remains the main research objective, cost associated with producing these materials is now becoming another overriding factor. Sustainability, renewability and green chemistry concepts must be also taken into consideration when selecting electrode materials processing methods for the next generation of batteries, especially for high volume applications. Thus, turning to low cost and green systems has led to rediscover hydrothermal and solvothermal approaches for battery materials. Furthermore, solvothermal and hydrothermal processes can be easily tuned to obtain nanostructures using different solvent/cosolvent/surfactant systems.

Obtaining $LiFePO_4$ small sized particles by hydrothermal synthesis method is possible by controlling several factors, as it was proposed in L.F. Nazar's work (2007). In this synthesis method, the main parameters that have influence on particle size in the absence of organic molecules (size modifiers) are reaction temperature and precursor concentration. As is the case for processing methods that require annealing treatments, e.g. sol-gel or ceramic methods, the use of lower temperatures induce the formation of smaller particles. On the other hand, an increase of the reactant concentration creates more nucleation sites and therefore produces much smaller particles. For example, 1-5 μm sized $LiFePO_4$ crystals were obtained by using low concentrations of precursors, but 250 nm size crystals were synthesized when concentration was increased threefold. The effect of both parameters in particle structure and size must be taken into account when trying to produce small sized materials. Although lower reaction temperatures favour smaller particle size, in low temperature hydrothermally synthesized $LiFePO_4$ Fe disorder in the structure was observed, and this structural defect was detrimental to electrochemical performance. Thus, a balance between both parameters must be seeked.

The use of organic compounds in hydrothermal medium can have two effects. First, some specific molecules can control particle morphology, and, second, the organic compounds in the medium serve as reductive agents and carbon precursor in order to get a carbonaceous coating around the particles. When the main purpose of the organic molecule is the attainment of a carbon coating, there exist two possibilities: organic product decomposition during hydrothermal process, as is the case of ascorbic acid; or the use of a further annealing treatment, as it is necessary for citric acid. Water soluble polymers can also be used as carbon precursors to get nanosized particles. For example, polyacrilic acid leads to 300-500 nm $LiFePO_4$ aggregates made of 75-100 nm size particles. These polymers coordinate strongly to crystal faces, and inhibit the nucleation and growth.

Non ionic surfactants are also widely used to control particle size and shape. For example, Pluronics P123, FC4 and Jeffamine compounds, make possible the access to smaller and more homogeneous particle size (150-300 nm) than when reductive additives are used. Materials prepared by using these surfactants present more homogeneous and smaller particles than those obtained with reductive agents. These surfactants can control particle size but do not decompose under reaction conditions, thus mild thermal treatments are needed in order to decompose them to carbon [Nazar, L.F. et al. (2007)].

On the other hand, the use of mixed water/solvent systems can also lead to the formation of diverse nanostructures. The solvent usually employed for this purpose is an alcohol. For example, an ethyleneglycol/water (EG/W) system has been developed to synthesize $LiFePO_4$ nanodendrites under hydrothermal conditions using dodecylbenzenesulphonicacid sodium salt (SDBS) as soft template to control crystal growth [Teng, F. et al. (2010)]. The advantages of this kind of synthetic approach are: one-step synthesis, environmental friendliness and low cost, due to the inexpensive solvent, as well as the short heating time used. This EG/W system is a promising reaction medium to provide well controlled crystallization, but the optimal EG/W proportion must be fixed for each desired compound because a too high proportion of ethyleneglycol should provoke only a partial solubility of the initial reactants. The role of the EG/W mixture is matching the reactivity of the precursors, which facilitates the formation of the desired olivine $LiFePO_4$ structure instead of impurity crystals, such as Li_3PO_4 and $Fe_4(PO_4)_3(OH)_3$. Moreover, the solubility of the precursors in EG/W will be smaller than the solubility in deionized water. Thus, the precursors will have a higher degree of supersaturation in the EG/W system, favouring the nucleation and growth of the desired crystals.

Surfactant in the system is used to adjust the size and morphology of the particle. SDBS molecules strongly coordinate onto the newly formed surfaces during crystallization and accordingly inhibit the crystal growth along these surfaces [Xiang, J. et al. (2008); Zhou, G. et al. (2007); Huang, Y. et al. (2009); Leem, G. et al. (2009)]. The proposed formation mechanism is depicted in figure 7.

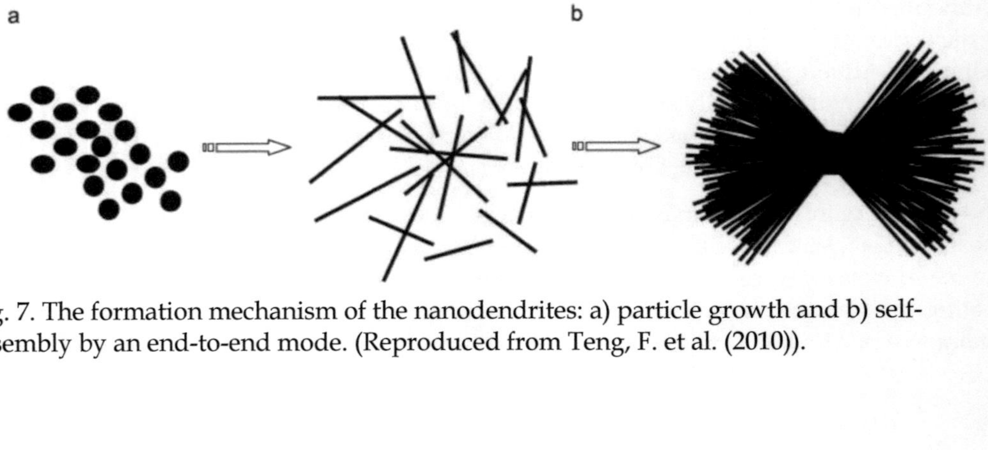

Fig. 7. The formation mechanism of the nanodendrites: a) particle growth and b) self-assembly by an end-to-end mode. (Reproduced from Teng, F. et al. (2010)).

During crystal growth, the surfactant acts as strong coordinating agent by binding to some crystal faces, and accordingly inhibits the crystal growth along the other defined crystal plane. As a result, at the early stages of the process, SDBS directs crystals to grow along the crystal direction whose crystal plane weakly bonds with the surfactant molecules. As a result, nanorods are formed. SDBS molecules can also act as soft template to assemble the nanorod building blocks into the final hierarchical structures, in which the nanorods are tightly attached together by their ends. This end-to end self-assembly results from the Van der Waals attraction of hydrophobic interaction on the surfactant molecules bonded to the end of the nanorods. Since surfactant molecules are preferentially and strongly adsorbed on the nanorod side, the stronger electrostatic force exists on the nanorod side than that on the end. It seems that there is a balance between electrostatic repulsion interaction and hydrophobic attraction interaction. Hence, the electrostatic repulsion interaction on the side is stronger than that on the end between nanorods, which refrains the side-by-side attachment. The long hydrophobic chains of the SDBS molecules bond on the nanorods will be attracted to one another through hydrophobic interaction. As a result, the nanorods are attached with each other by their ends to form hierarchical structures.

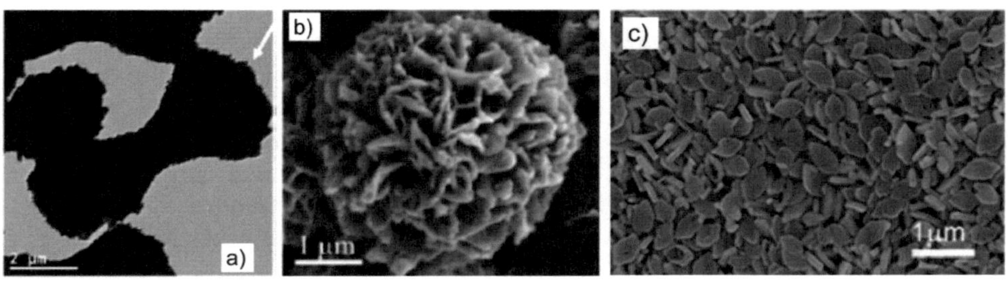

Fig. 8. Micrographs of different hierarchical structures prepared by solvothermal method, a) from ref. [Teng, F. et al. (2010), b) from Sun, C. et al. (2011), and c) is from Yang, S. et al. (2010).

An EG/W system has also been recently employed for the synthesis of porous $LiFePO_4$ microspheres by Goodenough's group in the presence of ethylenediamine [Sun, C. et al. (2011)]. In this case, ethyleneglycol/water ratio was 1/14, thus water was the main reaction medium. The role of EG in this solvothermal process consisted on limiting particle growth and preventing the agglomeration by means of chelates of EG (complexing agent) and on reducing Fe^{3+} into Fe^{2+} ions in solution (reductive agent). The presence of ethylenediamine as cosolvent in low concentration demonstrated to be critical for the formation of hierarchical flowerlike microspheres. Due to its strong chelating ability for some transition-metal ions, ethylenediamine has a great influence on the release of isolated iron ions, and, thus, on crystal growth. As-prepared hierarchical structures showed very good electrochemical properties, with a specific capacity value of 90 $mAh \cdot g^{-1}$ at 5C.

Polyethylene glycol/water (PEG/W) is another binary system that has been used to prepare electrode materials for Li ion batteries [Yang, S. et al. (2010)]. The use of different PEG/water proportions, heating temperatures (140° C and 180° C) and synthesis times (9-24 hours) leads to several morphologies, such as rod-like nanoparticles (50x200nm), nanoplates (100 nm thick and 800 nm long) and microplates (300 nm thick and 3μm long). Electrochemical evaluation of nanoplates showed very good results, with a specific capacity of 120-110 $mAh \cdot g^{-1}$ at 20C for 1000 cycles.

Anhydrous solvothermal processes also play an important role on the growth of hierarchically assembled nanostructures. Most of solvothermal systems are based on organic alcohols with different carbon chains. For example, benzyl alcohol has been reported in the last years to achieve nanostructures by solvothermal synthesis [Jia, F. et al. (2008); Pinna, N. et al. (2005)]. Benzyl alcohol and poly(vinyl pyrrolidone) (PVP) have been used to prepare $LiFePO_4$ dumb-bell structures, starting from an Fe^{3+} precursor [Yang, H. et al. (2009)]. Although benzyl alcohol is a reduced solvent that can partly reduce transition-metal ions [Niederberger, M. (2007)], the addition of LiI to the starting reactants is needed to accomplish complete reduction of Fe^{3+} to Fe^{2+} by I^- species. This way, benzyl alcohol provides a reductive environment and ensures reduction started with LiI. The role of PVP is related to the construction of the hierarchically self-assembled microstructures, and the adjustment of the amount of PVP is crucial to control both the morphology and size of the $LiFePO_4$ microstructures. The proposed mechanism for the formation of these dumb-bell structures is based on a dissolution-recrystallization process and it involves five steps (figure 9). The first one is the massive precipitation of Li_3PO_4 nuclei during the mixing of the reactants and the growth of the crystals in the shape of a rectangle. Second, the crystals start to aggregate in a system energy minimization driven process, and with the guide of PVP molecules, that bond to some crystal faces. Third, aggregation continues in an oriented way, to form pseudocubic 3C structure to minimize surface energy. Fourth, with increasing temperature and pressure, solubility of Li_3PO_4 increases and this phase starts to dissolve in order to crystallize $LiFePO_4$ nuclei. Driven by energy minimization principle, initial $LiFePO_4$ nanoplates assemble in edge to edge and layer-by-layer mode, and then, piled-up nanoplates tend to tilt at both ends, thus forming notched structures. This specific growth fashion can be due to lattice tension or surface interaction in the edge areas. Fifth, dissolution and recrystallization process complete by assembling more nanoplates onto the edges of the notched structures; these edges thicken and dumb-bell structures are formed. These nanostructures have not been electrochemically tested at high rates, and a 110 $mAh \cdot g^{-1}$ specific capacity value is obtained when cycling at C/30.

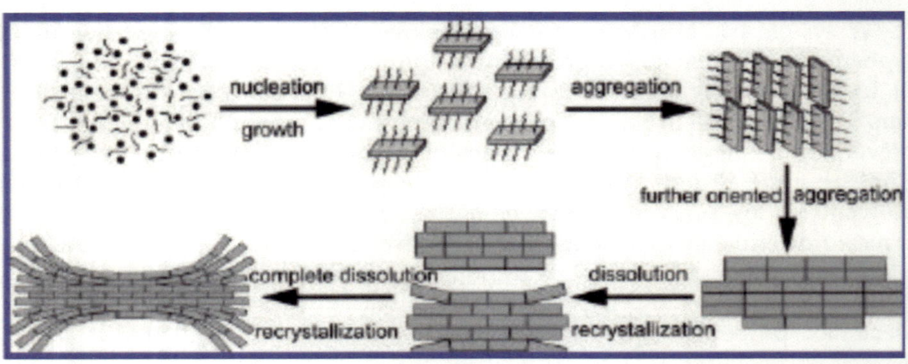

Fig. 9. Schematic illustration for the formation of hierarchically dumbbell-like $LiFePO_4$ microstructures. (Reproduced from ref Yang, H. et al. (2009)).

Ethylene glycol is the other alcohol largely used in solvothermal processes, not only combined with water, but also in anhydrous media, mixed with different surfactants or capping agents. $LiFePO_4$ nanorods and nanoflowers have been built by using this product as solvent, with short heating times (4-15 minutes) and relatively low temperature of 300° C

[Rangappa, D. et al. (2010)]. Three solvent combinations were tested, which induced different morphologies. When only EG was used as solvent, rectangular nanoplates of 50-100 nm width were produced. When EG and hexane were combined as solvent and cosolvent, long nanorods with 150 nm diameter and 700 nm long were obtained. In the case of using EG and oleic acid as solvent and surfactant, short nanorods were produced. In this last case, an increase in the reaction time provoked hierarchically self-assembly of the nanorods to form flower-like structures. Time-dependent experiments on EG-oleic acid system showed reaction sequence (figure 10). At first, $LiFePO_4$ spheres are formed. These spheres grow oriented to form rod-like structures, due to ethylene glycol, that directs the one dimensional growth by forming hydrogen bonds with certain faces of the crystals. With increasing reaction time, oleic acid molecules adsorb onto the surface of the nanorods, acting as a capping agent and also as a surfactant. The interaction of these oleic acid molecules leads to the decrease in the rods length and forms hierarchically flower like microstructure with prolonged reaction time. Electrochemical tests of these nanostructures showed a specific capacity of 154 mAh·g^{-1} at C/10.

A wide sloping region is usually observed in most hierarchically nanostructured materials. This can be attributed to a pseudo-capacitive effect, which consists on charge storage of Li ions from faradaic processes occurring at the surface of the materials. The presence of this kind of effect in hierarchically nanostructured materials makes sense in samples with great specific surface, such as nanoflowers.

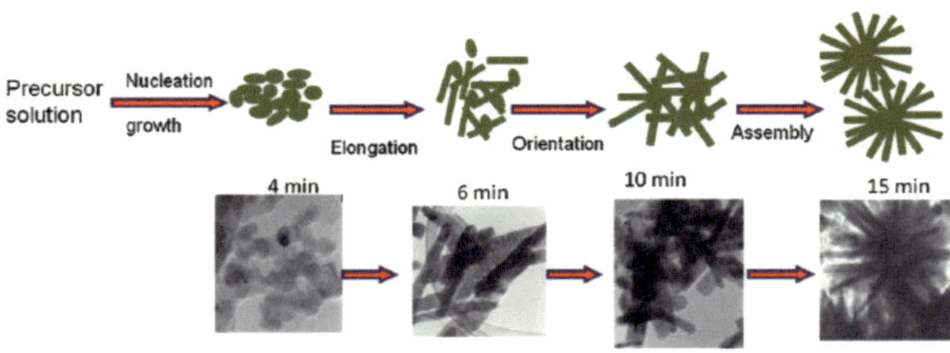

Fig. 10. Schematic illustration of the directed growth of $LiFePO_4$ nanorods and 3D hierarchical nanoflower by nanorods self-assembly. (Reproduced from Rangappa, D. et al. (2010))

3.4 Synthesis of nanostructured materials by ionothermal process

New synthetic methods derived from solvothermal approach, such as ionothermal processes have been used to obtain nanopowders of $LiMPO_4$ (M= Mn, Co and Ni), Li_xMSiO_4 [Nytén, A. et al. (2005)] and Li and Na fluorophosphates battery materials [DiSalvo, F.J. et al. (1971); Ellis, B.L. et al. (2007b)] using low heating temperature. Ionothermal synthesis has emerged when a great amount of research work is aimed at new low-cost processes to make highly electrochemically optimized electrode materials. This alternative route is considered as a new low cost synthesis process because it demands much less energy than high temperature ceramic routes. In spite of the higher cost of ionic liquids compared to water, it has been proved that these solvents can be reused without purification when used to prepare the same material, what leads to a significant cost decrease and minimizes waste production [Tarascon, J-M. et al. (2010)]. Ionothermal synthesis has also been carried out successfully by using microwave rather than traditional heating, which reduces reaction time and required energy for the synthesis.

Ionothermal synthesis is based on the use of an ionic liquid as reacting medium instead of water in solvothermal conditions. Ionic liquids are a class of organic solvents with high polarity and a preorganized solvent structure [Del Popolo, M. G. and Voth, G. A. (2004)]. Room temperature (or near-room-temperature) ionic liquids are classically defined as liquids at ambient temperatures (or <100 °C) that are made of organic cations and anions. They have excellent solvating properties, little measurable vapor pressure, and high thermal stability. Solvating properties and fusion temperatures will depend on the combination of cations and anions chosen. In the area of materials science, there have been several reports of ionic liquids being used as solvents with very little or controlled amounts of water involved in the synthesis [Antonietti, M et al. (2004)]. Most of these studies concentrated on amorphous materials and nanomaterials.

Like water, ionic liquids resulting from compatible cationic/anionic pairs have excellent solvent properties. In addition, they possess high thermal stability and negligible volatility so the use of autoclave is not mandatory. Moreover, because of the flexible nature of the cationic/anionic pairs, they present, as solvents, great opportunities to purposely direct nucleation. Over the past decade, ionothermal synthesis has developed into an advantageous synthetic technique for the preparation of zeotypes [Lin, Z-J. et al. (2008)] and other porous materials such as metal organic framework compounds (MOFs), but there has been very limited use made of this technique in the synthesis of inorganic compounds.

The unique feature of ionothermal synthesis is that the ionic liquid acts as both the solvent and the template provider. Many ionic liquid cations are chemically very similar to species that are already known as good templates (alkylimidazolium-based, pyridinium-based ionic liquids). Many are relatively polar solvents, making them suitable for the dissolution of the inorganic components required for the synthesis. One of the defining properties of ionic liquids is their lack of a detectable vapor pressure which effectively results in the elimination of the safety concerns associated with high hydrothermal pressures and has also led to their use in microwave synthesis.

In the process of ionothermal route, since there are no other solvents added to the reaction mixture, the theory holds that no other molecules are present to act as space fillers during the synthesis. This means that ionothermal method ideally removes the competition between template–framework and solvent–framework interactions that are present in hydrothermal preparations. This, however, is the idealistic scenario, which is not always attainable due to the possible decomposition of a small fraction of the ionic liquid cations, resulting in smaller template cations which may preferentially act as the structure directing agent in the ionic liquid solvent [Parnham, E. R. and Morris, R. E. (2006)].

Recent molecular modeling studies indicate that the structures of ionic liquids are characterized by long range correlations and distributions that reflect the asymmetric structures of the cations. Long-range asymmetric effects of this kind potentially increase the likelihood of transferring chemical information from the template cation to the framework, a situation that is desirable if full control over the templating process is to be achieved [Parnham, E.R. and Morris, R.E. (2007)].

Compound	Precursors	Ionic Liquids	Particle size (nm)
$LiFePO_4$	LiH_2PO_4 + $FeC_2O_4 \cdot 2H_2O$	EMI-TFSI 1-ethyl-3-methylimidazolium Bis(trifluoromethanesulfonyl)imide	150-300
$LiMnPO_4$	LiH_2PO_4 + $MnC_2O_4 \cdot 2H_2O$	EMI-TFSI 1-ethyl-3-methylimidazolium Bis(trifluoromethanesulfonyl)imide	100-400
$LiMPO_4$ (M= Ni, Co)	LiH_2PO_4 + $MC_2O_4 \cdot 2H_2O$	EMI-TFSI 1-ethyl-3-methylimidazolium Bis(trifluoromethanesulfonyl)imide	800-1000
Na_2FePO_4F	Na_3PO_4+ $FeF_2/FeCl_2$	C2 1-butyl-2,3-dimethylimidazolium Bis(trifluoromethanesulfonyl)imide	<50
Na_2MnPO_4F	Na_3PO_4+ $MnF_2/MnCl_2$	C2 1-butyl-2,3-dimethylimidazolium Bis(trifluoromethanesulfonyl)imide	<50
$Na_2Fe_{1-x}Mn_xPO_4F$ ($0 \leq x \leq 0.15$)	Na_3PO_4+ $MnF_2/MnCl_2$	C2 1-butyl-2,3-dimethylimidazolium Bis(trifluoromethanesulfonyl)imide	<50
$Na_2Fe_{1-x}Mn_xPO_4F$ ($0.25 \leq x \leq 1$)	Na_3PO_4+ $MnF_2/MnCl_2$	C2 1-butyl-2,3-dimethylimidazolium Bis(trifluoromethanesulfonyl)imide	<50
$LiFePO_4F$	Li_3PO_4+FeF_3	Triflate 1-butyl-3-methylimidazolium Trifluoromethanesulfonate	<50
$LiTiPO_4F$	Li_3PO_4+TiF_3	C2-OH 1,2-dimethyl-3-(3-hdroxypropyl)-imidazolium Bis(trifluoromethanesulfonyl)imide	<50
$LiFeSO_4F$	$FeSO_4 \cdot H_2O$+LiF	EMI-TFSI 1-ethyl-3-methylimidazolium Bis(trifluoromethanesulfonyl)imide	600-1200
$LiMnSO_4F$	$MnSO_4 \cdot H_2O$+ LiF	EMI-TFSI 1-ethyl-3-methylimidazolium Bis(trifluoromethanesulfonyl)imide	600-1200

Table 1. List of compounds prepared by ionothermal process with different ionic liquids [Tarascon, J-M. et al. (2010)].

Tarascon et al. were the first to apply ionothermal synthesis to battery electrode materials. They tested several ionic liquids based on different cationic and anionic species as reacting medium for the preparation of $LiFePO_4$ phase [Recham, N. et al. (2009a)]. The ionic liquids used demonstrated to have an impact on $LiFePO_4$ nucleation/growth and to behave as a structural directing agent. For example, the use of a CN functionalized EMI-TFSI ionic liquid [1-ethyl-3-methylimidazolium bis(trifluoromethanesulfonyl)imide] produced needle-like powders along [010] direction that perfectly piled up to form a larger needle. This change on the ending group modified the polar character of the reacting medium and its solvating properties, thus, influencing crystal growth. Platelet-like particles along [020] direction were created by decreasing the polarity via the use of a C_{18}-based EMI cation. Formation of these two morphological kinds can be explained in terms of competing energy surfaces, and it is directly connected with the nature of the ionic liquid, solvating power, polarity and aptitude to specifically absorb on one of the surfaces. Depending on the ionic liquid properties, surface energy minimization of the system will take place via a different mechanism. All $LiFePO_4$ produced materials were electrochemically active, but those with the best performance were of 300 and 500 nm size, showing 150 mAh·g^{-1} specific capacity at C/10 without carbon coating.

Ionothermal process was extended to the synthesis of size-controlled Na-based fluorophosphates [Na_2MPO_4F (M= Fe, Mn)] [Recham, N. et al. (2009b)]. These phases are attractive electrode materials because they are based on economic metals and can be used in both Li- and Na-based batteries. Furthermore, the fluorides possess higher electronegativity which increases the ionicity of the bonds and, thus, their redox potentials due to inductive effect. Nanosized samples of Na_2FePO_4F and Na_2MnPO_4F of about 25 nm diameter were prepared in a 1,2-dimethyl-3-butylimidazolium bis(trifluoromethanesulfonyl imide) ionic liquid, in contrast with coarse powders obtained by ceramic method. Electrochemical performance of the iron compound reached better results than that of ceramic material, with 115 mAh·g^{-1}, better initial capacity, lower irreversible capacity, lower polarization and better capacity retention. Manganese phase did not show electrochemical activity, which follows the general tendency of Mn-based compounds having worse electrochemical performance than their Fe counterpart, for example in $LiMPO_4$ and Li_2MSiO_4 families. There are two factors related to this phenomenon. First, the strong Jahn-Teller distortion on Mn^{3+}, that affects its coordination sphere; and, second, the poor electronic-ionic conductivity of the materials, owing to the greater ionicity of the M-O bonding.

Ionothermal process has been successfully used for preparing new electroactive materials that had not been achieved before, such as $LiFeSO_4F$. This material possesses an adequate structure to favor Li ions migration along channels. This electroactive phase had not been synthesized before because it is not accessible, neither by ceramic process nor in water medium. It decomposes at temperatures beyond 375° C and in water medium. This new cathode material showed electrochemical activity at 3.6 V, and a reversible specific capacity of 140 mAh·g^{-1}, very close to theoretical specific capacity vale of 151 mAh·g^{-1}. Thus, this preparative process has demonstrated to be a useful tool to synthesize nanosized new and known electroactive materials.

3.5 Synthesis of nanostructured materials with biological agents

The latest trend in battery materials processing is using biomineralization process in order to build controlled nanoarchitectured compounds under ambient conditions [Ryu, J. et al. (2010)]. Biomimetic chemistry involves the utilization of actual biomolecular entities such as proteins, bacteria and viruses to act either as a growth medium or as a spatially constrained nanoscale reactor for the generation of nanoparticles. Biosystems have the inherent capabilities of molecular recognition and self-assembly, and thus are an attractive template for constructing and organizing the nanostructure. Ryu et al. synthesized nanostructured transition metal phosphate via biomimetic mineralization of peptide nanofibers (figure 11). Peptides self-assembled into nanofibers displaying numerous acidic and polar moieties on their surface and readily mineralized with transition metal phosphate by sequential treatment with aqueous solutions containing transition metal cations and phosphate anions. $FePO_4$-mineralized peptide nanofibers were thermally treated at 350° C to fabricate $FePO_4$ nanotubes with inner walls coated with a thin layer of conductive carbon by carbonization of the peptide core. As formed carbon coated $FePO_4$ nanotubes showed high reversible capacity (150 mAh·g^{-1} at C/17) and good capacity retention during cycling.

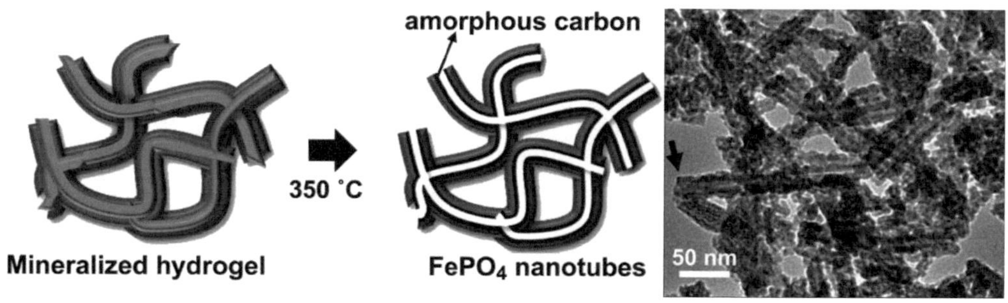

Fig. 11. a) Schematic of $FePO_4$ nanotubes synthesis by heat treatment of peptide/$FePO_4$ hybrid nanofibers; and b) transmission micrograph of tubular structures. [Reproduced from Ryu et al. (2010)].

Bacillus pasteurii bacterium has been extensively used to provoke calcite precipitation and it can generate a basic medium from urea hydrolysis that helps growing of $LiFePO_4$ nanofilaments at 65° C. Beer yeast has also been reported as a biomimetic template that has been used to prepare $LiFePO_4$ with enhanced surface area and conductivity [Li, P. et al. (2009)]. Engineered viruses have also been reported as templates to synthesize various electrode materials [Mao, Y. et al. (2007)], such as gold-cobalt oxide nanowires that consisted on 2-3 nm diameter nanocrystals prepared with modified bacteria M13 virus, with enhanced capacity retention [Tam, K.T. et al. (2006)]. Tobacco mosaic virus has also been used as a template for the synthesis of nickel and cobalt surfaces. This virus was genetically engineered to express a novel coat protein cysteine residue, and to vertically pattern virus particles into gold surfaces via gold-thiol interactions. Gold-supported vertically aligned virion particles served as vertical templates for reductive deposition of Ni and Co at room temperature via electroless deposition, and thus produced high surface area electrodes [Royston, E. et al. (2008)].

4. Conclusions

Nanostructure and hierarchical growth of electroactive materials has led to enhanced performances of lithium ion rechargeable cells such as higher capacity, improved rate capability, and sustained capacity retention for longer cycles. The key feature for the achievement of this goal has been the evolution of synthetic methods. From the classical ceramic to more advanced solvothermal processes, or to interdisciplinary views such as the use of biological agents to produce electrode materials, it is clear that, at present, synthetic approaches must go beyond the limits of traditional methods and set up a continuous knowledge feedback with surrounding disciplines in order to find new synthetic routes that lead to better performing materials and, thus, to a better quality of life.

The control of reaction conditions by using nano-shaped precursors, hard or soft templates or agents to limit the growth of particles allows the creation of useful featured structures that are able to overcome typical problems in battery materials.

5. Acknowledgements

This work was financially supported by the Ministerio de Educación y Ciencia (MAT2010-19442), the Gobierno Vasco/Eusko Jaurlaritza (GIU06-11 and ETORTEK CIC ENERGIGUNE 10), which we gratefully acknowledge. V.P. thanks the Universidad del País Vasco/Euskal Herriko Unibertsitatea for a postdoctoral fellowship.

6. References

Antonietti, M; Kuang, D.; Smarsly, B. & Zhou, Y. (2004) Ionic liquids for the convenient synthesis of functional nanoparticles and other inorganic nanostructures. *Angew. Chem.*, Vol. 43, pp. 4988– 4992, ISSN 1521-3773.

Aragón, M.J.; Lavela, P.; León, B.; Pérez-Vicente, C.; Tirado, J. L. & Vidal-Abarca, C. (2010) On the use of the reverse micelles synthesis of nanomaterials for lithium-ion batteries. *J. Solid State Electrochem.*, Vol. 14, pp. 1749-1753, ISSN 1432-8488.

Baythoun, M.S.G. & Sale, F.R. (1982) Production of strontium-substituted lanthanum manganite perovskite powder by the amorphous citrate process. *J. Mater. Sci.*, Vol. 17, pp. 2757-2769, ISSN 1573-4803.

Beninati, S.; Damen, L. & Mastragostino, M. (2008) MW-assisted synthesis of $LiFePO_4$ for high power applications. *J. Power Sources*, Vol. 180, pp. 875-879, ISSN 0378-7753.

Bruce, P.G. (2008a) Energy storage beyond the horizon: Rechargeable lithium batteries. *Solid State Ionics*, Vol. 179, pp. 752-760, ISSN 0167-2738.

Bruce, P.G.; Scrosati, B. & Tarascon, J-M. (2008b) Nanomaterials for Rechargeable Lithium Batteries. *Angew. Chem. Int. Ed.*, Vol. 47, pp. 2930-2946, ISSN 1521-3773.

Chen, J. and Whittingham, M.S. (2006) Hydrothermal synthesis of lithium iron phosphate. *Electrochem. Commun.*, Vol. 8, pp. 855-858, ISSN 1388-2481.

Chen, J.; Wang, S. & Whittingham, M.S. (2007) Hydrothermal synthesis of cathode materials. *J. Power Sources*. Vol. 174, pp. 442-448, ISSN 0378-7753.

Cho, Y. & Cho, J. (2010) Significant Improvement of $LiNi_{0.8}Co_{0.15}Al_{0.05}O_2$ Cathodes at 60°C by SiO_2 Dry Coating for Li-Ion Batteries. *J. Electrochem. Soc.*, Vol. 157, pp. A625-A629, ISSN 0013-4651.

Choi, D. & Kumta, P.N. (2007) Surfactant based sol–gel approach to nanostructured $LiFePO_4$ for high rate Li-ion batteries. *J. Power Sources*, Vol. 163, pp. 1064-1069, ISSN 0378-7753.

Chung, H-T.; Jang, S-K.; Ryu, H. W. & Shim, K-B. (2004) Effects of nano-carbon webs on the electrochemical properties in $LiFePO_4/C$ composite. *Solid State Commun.*, Vol. 131, pp. 549-554, ISSN 0038-1098.

Del Popolo, M. G. & Voth, G. A. (2004) On the structure and dynamics of ionic liquids. *J. Phys. Chem. B*, Vol. 108, pp.1744–1752, ISSN 1520-5207.

Delacourt, C.; Poizot, P.; Levasseur, S. & Masquelier, C. (2006) Size Effects on Carbon-Free $LiFePO_4$ Powders: The Key to Superior Energy Density. *Electrochem. Solid State Lett.*, Vol. 9, pp. A352-A355, ISSN 1099-0062.

DiSalvo, F.J.; Schwall, R.; Geballe, T. H.; Gamble, F. R. & Osiecki, J. H. (1971) Superconductivity in layered compounds with variable interlayer spacings. *Phys. Rev. Lett.*, Vol. 27, pp. 310-313, ISSN 0031-9007.

Doherty, C.M.; Caruso, R. A.; Smarsly, B. M. & Drummond, C. J. (2009) Colloidal Crystal Templating to Produce Hierarchically Porous $LiFePO_4$ Electrode Materials for High Power Lithium Ion Batteries. *Chem. Mater.*, Vol. 21, pp. 2895-2903, ISSN 0897-4756.

Dokko, K.; Koizumi, S.; Nakano, H. & Kanamura, K. (2007) Particle morphology, crystal orientation, and electrochemical reactivity of $LiFePO_4$ synthesized by the hydrothermal method at 443 K. *J. Mater. Chem.*, Vol. 17, pp. 4803-4810, ISSN 0959-9428.

Ellis, B.; Kan, W. H.; Makahnouk, W. R. M. & Nazar, L. F. (2007a) Synthesis of nanocrystals and morphology control of hydrothermally prepared $LiFePO_4$. *J. Mater. Chem.* Vol. 17, pp. 3248-3254, ISSN: 0959-9428.

Ellis, B.L.; Makahnouk, W. R. M.; Makimura, Y.; Toghill, K. & Nazar, L. F. (2007b) A multifunctional 3.5 V iron-based phosphate cathode for rechargeable batteries. *Nat. Mater.*, Vol. 6, pp. 749-753, ISSN 1476-1122.

Eom, J.; Ryu, K.S. & Cho, J. (2008) Dependence of Electrochemical Behavior on Concentration and Annealing Temperature of Li_xCoPO_4 Phase-Grown $LiNi_{0.8}Co_{0.16}Al_{0.04}O_2$ Cathode Materials. *J. Electrochem. Soc.*, Vol. 155, pp. A228-A233, ISSN 0013-4651.

Hsu, K-F.; Tsay, S-Y. & Hwang, B-J. (2004) Synthesis and characterization of nano-sized $LiFePO_4$ cathode materials prepared by a citric acid-based sol–gel route. *J. Mater. Chem.*, Vol. 14, pp. 2690-2695, ISSN 0959-9428.

Huang, Y.; Wang, W.; Liang, H. & Xu, H. (2009) Surfactant-Promoted Reductive Synthesis of Shape-Controlled Gold Nanostructures. *Cryst. Growth Des.*, Vol. 9, pp. 858-862, ISSN 1528-7483.

Hwang, B-J.; Hsu, K-F.; Hu, S-K; Cheng, M-Y.; Chou, T-C.; Tsay, S-Y. & Santhanam, R. (2009) Template-free reverse micelle process for the synthesis of a rod-like $LiFePO_4$/C composite cathode material for lithium batteries. *J. Power Sources*, Vol. 194, pp. 515-519, ISSN 0378-7753.

Jia, F.; Zhang, L. Z.; Shang, X. Y. & Yang, Y. (2008) Non-Aqueous Sol–Gel Approach towards the Controllable Synthesis of Nickel Nanospheres, Nanowires, and Nanoflowers. *Adv. Mater.*, Vol. 20, pp.1050-1054, ISSN 1521-4095.

Jiao, F.; Shaju, K. M. & Bruce, P. G. (2005) Synthesis of Nanowire and Mesoporous Low-Temperature $LiCoO_2$ by a Post-Templating Reaction. *Angew. Chem. Int. Ed.*, Vol. 44, pp. 6550-6553, ISSN 1521-3773.

Jin, B. & Gu, H-B. (2008) Preparation and characterization of $LiFePO_4$ cathode materials by hydrothermal method. *Solid State Ionics*, Vol. 178, pp. 1907-1914, ISSN 0167-2738.

Kanamura, K. & Koizumi, S. (2008) Hydrothermal synthesis of $LiFePO_4$ as a cathode material for lithium batteries. *J. Mater. Sci.*, Vol. 43, pp. 2138-2142, ISSN 1573-4803.

Kim, C.W.; Lee, M.H.; Jeong, W.T. & Lee, K.S. (2005) Synthesis of olivine $LiFePO_4$ cathode materials by mechanical alloying using iron(III) raw material. *J. Power Sources*, Vol. 146, pp. 534-538, ISSN 0378-7753.

Kim, J-K.; Choi, J-W.; Cheruvally, G.; Kim, J-U.; Ahn, J-H.; Cho, G-B.; Kim, K-W. & Ahn, H-J. (2007) A modified mechanical activation synthesis for carbon-coated $LiFePO_4$ cathode in lithium batteries. *Matter. Lett.*, Vol. 61, pp. 3822-3825, ISSN 0167-577X.

Kim, D.H. & Kim, J. (2007) Synthesis of $LiFePO_4$ nanoparticles and their electrochemical properties. *J. Phys. Chem. Solids*, Vol. 68, pp. 734-737, ISSN 0022-3697.

Kwon, S.J.; Kim, C.W.; Jeong, W.T. & Lee, K.S. (2004) Synthesis and electrochemical properties of olivine $LiFePO_4$ as a cathode material prepared by mechanical alloying. *J. Power Sources*, Vol. 137, pp. 93-99, ISSN 0378-7753.

Leem, G.; Sarangi, S.; Zhang, S.; Rusakova, I.; Brazdeikis, A.; Litvinov, D. & Lee, T. R. (2009) Surfactant-Controlled Size and Shape Evolution of Magnetic Nanoparticles. *Cryst. Growth Des.*, Vol. 9, pp. 32-34, ISSN 1528-7483.

Li, M.; Schnablegger, H. & Mann, S. (1999) Coupled synthesis and self-assembly of nanoparticles to give structures with controlled organization. *Nature*, Vol. 402, pp. 393-395, ISSN 0028-0836.

Li, W.; Ying, J.; Wan, C.; Jiang, C.; Gao, J. & Tang, C. (2007) Preparation and characterization of $LiFePO_4$ from $NH_4FePO_4 \cdot H_2O$ under different microwave heating conditions. *J. Solid State Electrochem.*, Vol. 11, pp. 799-803, ISSN 1432-8488.

Li, P.; He, W.; Zhao, H. & Wang, S. (2009) Biomimetic synthesis and characterization of the positive electrode material $LiFePO_4$. *J. Alloys and Comp.*, Vol. 471, pp. 536-538, ISSN 0925-8388.

Liao, X.Z.; Ma, Z-F.; He, Y-S.; Zhang, X-M.; Wang, L. & Jiang, Y. (2005) Electrochemical Behavior of $LiFePO_4$/C Cathode Material for Rechargeable Lithium Batteries. *J. Electrochem. Soc.*, Vol. 152, pp. A1969-A1973, ISSN 0013-4651.

Lin, Z-J.; Alexandra, Y. L.; Slawin, M. Z. & Morris, R. E. (2008) Hydrogen-bond-directing effect in the ionothermal synthesis of metal coordination polymers. *Dalton Trans.*, pp. 3989-3994, ISSN 1364-5447.

Mao, Y.; Park, T-J.; Zhang, F.; Zhou, H. & Wong, S. S. (2007) Environmentally Friendly Methodologies of Nanostructure Synthesis. *Small*, Vol. 3, pp.1122-1139, ISSN 1613-6810.

Mi, C.H.; Zhao, X. B.; Cao, G. S. & Tu, J. P. (2005) *In Situ* Synthesis and Properties of Carbon-Coated $LiFePO_4$ as Li-Ion Battery Cathodes. *J. Electrochem. Soc.*, Vol. 152, pp. A483-A487, ISSN 0013-4651.

Mi, C.H.; Cao, G.S. & Zhao, X.B. (2005) Low-cost, one-step process for synthesis of carbon-coated $LiFePO_4$ cathode. *Matter. Lett.*, Vol. 59, pp.127-130, ISSN 0167-577X.

Nakamura, T.; Miwa, Y.; Tabuchi, M. & Yamada, Y. (2006) Structural and Surface Modifications of $LiFePO_4$ Olivine Particles and Their Electrochemical Properties. *J. Electrochem. Soc.*, Vol. 153, pp. A1108-A1114, ISSN 0013-4651.

Niederberger, M. (2007) Nonaqueous Sol-Gel Routes to Metal Oxide Nanoparticles. *Acc. Chem. Res.*, Vol. 40, pp. 793-800, ISSN 1520-4898.

Nytén, A.; Abouimrane, A.; Armand, M.; Gustafsson, T. & Thomas, J. O. (2005) Electrochemical performance of Li_2FeSiO_4 as a new Li-battery cathode material. *Electrochem. Commun.*, Vol. 7, pp. 156-160, ISSN 1388-2481.

Palomares, V. Goñi, A.; Gil de Muro, I.; de Meatza, I.; Bengoechea, M.; Miguel, O. & Rojo, T. (2007) New freeze-drying method for $LiFePO_4$ synthesis. *J. Power Sources*, Vol. 171, pp. 879-885, ISSN 0378-7753.

Palomares, V.; Goñi, A.; Gil de Muro, I.; de Meatza, I.; Bengoechea, M.; Cantero, I. & Rojo, T. (2009a) Influence of Carbon Content on $LiFePO_4$/C Samples Synthesized by Freeze-Drying Process. *J. Electrochem. Soc.*, Vol. 156, pp. A817-A821, ISSN 0013-4651.

Palomares, V.; Goñi, A.; Gil de Muro, I.; de Meatza, I.; Bengoechea, M.; Cantero, I. & Rojo, T. (2009b) Conductive additive content balance in Li-ion battery cathodes: Commercial carbon blacks vs. *in situ* carbon from $LiFePO_4$/C composites. *J. Power Sources*, Vol. 195, pp. 7661-7668, ISSN 0378-7753.

Palomares, V.; Goñi, A.; Gil de Muro, I.; Lezama, L.; de Meatza, I.; Bengoechea, M.; Boyano, I. & Rojo, T. (2011) Near heterosite $Li_{0.1}FePO_4$ Phase Formation as Atmospheric Aging Product of $LiFePO_4$/C Composite. Electrochemical, Magnetic and EPR Study. *J. Electrochem. Soc.*, Vol. 158(9), pp. A1042-A1047, ISSN 0013-4651.

Park, K.S.; Son, J. T.; Chung, H. T.; Kim, S. J.; Lee, C. H.; & Kim, H. G. (2003) Synthesis of $LiFePO_4$ by co-precipitation and microwave heating. *Electrochem. Commun.*, Vol. 5, pp. 839-842, ISSN 1388-2481.

Park, K.S.; Kang, K.T.; Lee, S.B.; Kim, G.Y.; Park, Y.J. & Kim, H.G. (2004) Synthesis of $LiFePO_4$ with fine particle by co-precipitation method. *Mater. Res. Bull.*, Vol. 39, pp. 1803-1810, ISSN 0025-5408.

Parnham, E. R. & Morris, R. E. (2006) 1-Alkyl-3-methylimidazolium bromide ionic liquids in the ionothermal synthesis of aluminium phosphate molecular sieves. *Chem. Mater.*, Vol. 18, pp. 4882–4887, ISSN 0897-4756.

Parnham, E.R. & Morris, R.E. (2007) Ionothermal Synthesis of Zeolites, Metal–Organic Frameworks, and Inorganic- Organic Hybrids. *Acc. Chem. Res.*,Vol. 40, pp. 1005-1013, ISSN 1520-4898.

Paulus, M. (1980) Freeze-drying: a method for the preparation of fine sinterable powders and low temperature solid state reaction. *Fine Part. Process. Int. Symp.*, Vol. 1, pp. 27-50.

Pechini, P. Patent 3.330.697, July 11, 1967.

Piana, M.; Cushing, B.L.; Goodenough, J.B. & Penazzi, N. (2004) A new promising sol–gel synthesis of phospho-olivines as environmentally friendly cathode materials for Li-ion cells. *Solid State Ionics*, Vol. 175, pp. 233-237, ISSN 0167-2738.

Pieranski, P. (1983) Colloidal Crystals. *Contemp. Phys.*, Vol. 24, p.25-73, ISSN 0010-7514.

Pinna, N.; Grancharov, S.; Beato, P.; Bonville, P.; Antonietti, M. & Niederberge, M. (2005) Magnetite Nanocrystals: Nonaqueous Synthesis, Characterization, and Solubility. *Chem. Mater.*, Vol, 17, pp. 3044-3049, ISSN 0897-4756.

Rangappa, D.; Sone, K.; Kudo, T. & Honma, I. (2010) Directed growth of nanoarchitectured $LiFePO_4$ electrode by solvothermal synthesis and their cathode properties. *J. Power Sources*, Vol. 195, pp. 6167-6171, ISSN 0378-7753.

Recham, N.; Dupont, L.; Courty, M.; Djellab, K.; Larcher, D.; Armand, M. & Tarascon, J.-M. (2009a) Ionothermal Synthesis of Tailor-Made $LiFePO_4$ Powders for Li-Ion Battery Applications. *Chem. Mater.*, Vol. 21, pp. 1096-1107, ISSN 0897-4756.

Recham, N.; Chotard, J-N.; Dupont, L. ; Djellab, K.; Armand, M. & Tarascon, J-M. (2009b) Ionothermal Synthesis of Sodium-Based Fluorophosphate Cathode Materials. *J. Electrochem. Soc.*, Vol. 156, pp. A993-A999, ISSN 0013-4651.

Ryu, J.; Kim, S-W.; Kang, K. & Park, C. B. (2010) Mineralization of Self-assembled Peptide Nanofibers for Rechargeable Lithium Ion Batteries. *Adv. Mater.*, Vol 22, pp. 5537-5541, ISSN 0935-9648.

Royston, E.; Ghosh, A.; Kofinas, P.; Harris, M. T. & Culver, J. N. (2008) Self-Assembly of Virus-Structured High Surface Area Nanomaterials and Their Application as Battery Electrodes. *Langmuir*, Vol. 24, pp. 906-912, ISSN 1520-5827.

Sakamoto, J.S. & Dunn, B. (2002) Hierarchical battery electrodes based on inverted opal structures. *J. Mater. Chem.*, Vol. 12, pp. 2859-2861, ISSN 0959-9428.

Song, M-S.; Kang, Y.M.; Kim, J-H.; Kim, H-S.; Kim, D-Y.; Kwon, H-S. & Lee, J.Y. (2007) Simple and fast synthesis of $LiFePO_4$-C composite for lithium rechargeable batteries by ball-milling and microwave heating. *J. Power Sources*, Vol. 166, pp. 260-265, ISSN 0378-7753.

Song, M-S.; Kim, D-Y.; Kang, Y-M.; Kim, Y-I.; Lee, J-Y. & Kwon, H-S. (2008) Synthesis and characterization of Carbon Nano Fiber/$LiFePO_4$ composites for Li-ion batteries. *J. Power Sources*, Vol. 180, pp. 546-552, ISSN 0378-7753.

Song, H-K.; Lee, K.T.; Kim, M.G.; Nazar, L.F. & Cho, J. (2010) Recent Progress in Nanostructured Cathode Materials for Lithium Secondary Batteries. *Adv. Funct. Mater.*, Vol. 20, pp. 3818-3834, ISSN 1616-3028.

Sun, C.; Rajasekhara, S.; Goodenough, J. B. & Zhou, F. (2011) Monodisperse Porous $LiFePO_4$ Microspheres for a High Power Li-Ion Battery Cathode. *J. Am. Chem. Soc.*, Vol. 133, pp. 2132-2135, ISSN 0002-7863.

Tajimi, S.; Ikeda, Y.; Uematsu, K.; Toda, K. & Sato, M. (2004) Enhanced electrochemical performance of $LiFePO_4$ prepared by hydrothermal reaction. *Solid State Ionics*, Vol. 175, pp. 287-290, ISSN 0167-2738.

Tam, K.T.; Kim, D-W.; Yoo, P. J.; Chiang, C-Y.; Meethong, N.; Hammond, P. T.; Chiang, Y-M. & Belcher, A. M. (2006) Virus-Enabled Synthesis and Assembly of Nanowires for Lithium Ion battery Electrodes. *Science*, Vol. 312, pp. 885-888, ISSN 1095-9203.

Tarascon, J-M.; Recham, N.; Armand, M.; Chotard, J-N.; Barpanda, P.; Walker, W. & Dupont, L. (2010) Hunting for Better Li-Based Electrode Materials via Low Temperature Inorganic Synthesis. *Chem. Mater.*, Vol. 22, pp. 724-739, ISSN 0897-4756.

Teng, F.; Santhanagopalan, S.; Asthana, A.; Geng, X.; Mho, S-I.; Shahbazian-Yassar, R. & Meng, D. D. (2010) Self-assembly of $LiFePO_4$ nanodendrites in a novel system of ethyleneglycol–water. *J. Crystal Growth*, Vol. 312, pp. 3493-3502, ISSN 0022-0248.

Tonti, D.; Torralvo, M. J.; Enciso, E.; Sobrados, I. & Sanz, J. (2008) Three-Dimensionally Ordered Macroporous Lithium Manganese Oxide for Rechargeable Lithium Batteries. *Chem. Mater.*, Vol. 20, pp. 4783-4790, ISSN 0897-4756.

Velev, O.D.; Jede, T. A.; Lobo, R. F. & Lenhoff, A. M. (1997) Porous silica via colloidal crystallization. *Nature*, Vol. 389, pp. 447-448, ISSN 0028-0836.

Wang, L.; Huang, Y.; Jiang, R. & Jia, D. (2007) Preparation and characterization of nano-sized $LiFePO_4$ by low heating solid-state coordination method and microwave heating. *Electrochimica Acta*, Vol. 52, pp.6778-6783, ISSN 0013-4686.

Whittingham, M.S.; Song, Y.; Lutta, S.; Zavalij, P. Y. & Chernova, N. A. (2005) Some transition metal (oxy)phosphates and vanadium oxides for lithium batteries. *J. Mater. Chem.*, Vol. 15, pp. 3362-3379, ISSN 0959-9428.

Xiang, J.; Cao, H.; Warner, J. H. & Watt, A. A. R. (2008) Crystallization and Self-Assembly of Calcium Carbonate Architectures. *Cryst. Growth Des.*, Vol. 8, pp. 4583-4588, ISSN 1528-7483.

Yamada, A.; Chung, S. C. & Hinokuma K. (2001) Optimized $LiFePO_4$ for Lithium Battery Cathodes. *J. Electrochem. Soc.*, Vol. 148, pp. A224-A229, ISSN 0013-4651.

Yang, S. ; Zavalij, P.Y. & Whittingham, M. S. (2001) Hydrothermal Synthesis of Lithium Iron Phosphate Cathodes. *Electrochem. Commun.*, Vol. 3, pp. 505-508, ISSN 1388-2481.

Yang, M-R.; Ke, W-H. & Wu, S-H. (2005) Preparation of $LiFePO_4$ powders by co-precipitation *J. Power Sources*, Vol. 146, pp. 539-543, ISSN 0378-7753.

Yang, H.; Wu, X-L.; Cao, M-H. & Guo, Y-G. (2009) Solvothermal Synthesis of $LiFePO_4$ Hierarchically Dumbbell-Like Microstructures by Nanoplate Self-Assembly and Their Application as a Cathode Material in Lithium-Ion Batteries. *J. Phys. Chem. C*, Vol. 113, pp. 3345-3351, ISSN 1932-7447.

Yang, S.; Zhou, X.; Zhang, J. & Liu, Z. (2010) Morphology-controlled solvothermal synthesis of $LiFePO_4$ as cathode material for lithium-ion batteries. *J. Mater. Chem.*, Vol. 20, pp. 8086-8091, ISSN 0959-9428.

Zhang, S.S.; Allen, J.L.; Xu, K. & Jow, T.R. (2005) Optimization of reaction condition for solid-state synthesis of $LiFePO_4$-C composite cathodes *J. Power Sources*, Vol. 147, pp. 234-240, ISSN 0378-7753.

Zhou, G.; Lü, M.; Yang, Z.; Tian, F.; Zhou, Y. & Zhang, A. (2007) Fabrication of Novel Palladium Microstructures through Self-Assembly. *Cryst. Growth Des.*, Vol. 7, pp. 187-190, ISSN 1528-7483.

Zhu, S.; Zhou, H.; Miyoshi, T.; Hibino, M.; Honma, I. & Ichihara, M. (2004) Self-Assembly of the Mesoporous Electrode Material $Li_3Fe_2(PO_4)_3$ Using a Cationic Surfactant as the Template. *Adv. Mater.*, Vol. 16, pp. 2012-2017, ISSN 1521-4095.

2

Preparation and Electrochemical Properties of Cathode and Anode Materials for Lithium Ion Battery by Aerosol Process

Takashi Ogihara
University of Fukui,
Japan

1. Introduction

Lithium ion battery (LIB) has been used as energy storage devices for portable electronics since 1990 years. Recently, these are well noted as the power sources for the vehicles such as electric vehicles and hybrid electric vehicles. Both layered type $LiCoO_2$, $LiNiO_2$ and spinel type $LiMn_2O_4$ is the most important cathode materials because of their high operating voltage at 4 V (Mizushima, et.al, 1980, Guyomard, et.al, 1994). So far, $LiCoO_2$ has been mostly used as cathode material of commercial LIB. However, $LiCoO_2$ and $LiNiO_2$ have a problem related to capacity fading due to the instability in rechargeable process. Cobalt is also expensive and its resource is not sufficient. Therefore, $LiCoO_2$ cathode material is not suitable as a LIB for EV and HEV. On the other hand, $LiMn_2O_4$ is regarded as a promising cathode material for large type LIB due to their advantages such as low cost, non-toxicity and thermally stability (Pegeng, et.al, 2006). It was also known that Ni-substitute type $LiMn_2O_4$ ($LiNi_{0.5}Mn_{1.5}O_4$) was exhibited rechargeable behavior at about 5 V (Markovsky, et.al, 2004, Idemoto, et.al, 2004, Park, et.al, 2004). $LiNi_{0.5}Mn_{1.5}O_4$ has been considerably noticed as a cathode material with high power density which had an active potential at 5 V. The layered type $LiCo_{1/3}Ni_{1/3}Mn_{1/3}O_2$ was found to exhibit superior high potential cathode properties. This had rechargeable capacity with more than 150 mAh/g at higher rate and a milder thermal stability, but shows significantly capacity fading during the long rechargeable process. Recently, olivine type phosphate compound is noted as an alternative cathode material. $LiFePO_4$ and $LiMnPO_4$ were expected as next generation materials for large LIB because of low-cost, environmentally friendly, high thermally stability and electrochemical performance. On the other hand, the oxide type anode such as spinel type $Li_4Ti_5O_{12}$ is expected as the candidate for the replacement of carbon anodes because of better safety. LIB which is consisted of $LiFePO_4$ cathode and $Li_4Ti_5O_{12}$ anode offers to high safety and long life cycle. Therefore, it is expected as the application of HEV or power supply for load levelling in wind power generation and solar power generation. So far, we have been developed spray pyrolysis technique as a aerosol process to prepare $LiFePO_4$ and $Li_4Ti_5O_{12}$ powders for LIB. In this chapter, the powder processing and electrochemical properties of $LiFePO_4$ cathode and $Li_4Ti_5O_{12}$ anode materials by spray pyrolysis were described.

2. Spray pyrolysis process

Spray pyrolysis is a versatile process regarding the powder synthesis of inorganic and metal materials (Messing, et.al, 1993, Dubois, et.al, 1989, Pluym, et.al, 1993). An atomizer such as ultrasonic (Ishizawa, et.al, 1985) or two-fluid nozzle (Roy, et.al, 1977) is often used to generate the mist. The mist is droplet in which the inorganic salts or metal organic compound is dissolved in water or organic solvent. The droplets were dried and pyrolyzed to form oxide or metal powders at elevated temperature. The advantages of spray pyrolysis are that the control of particle size, particle size distribution and morphology are possible. Furthermore, the fine powders with homogeneous composition can be easily obtained because the component of starting solution is kept in the mist derived from an ultrasonic atomizer or two-fluid nozzle. Each metal ion was homogeneously blending in each mist. Each mist play a role as the chemical reactor at the microscale. The production time was very short (less than 1 min). In the other solution process such sol-gel, hydrothermal, precipitation, hydrolysis, the oxide powders were often prepared for few hours. In addition, the process such as the separation, the drying and the firing step must be done after the chemical reaction in the solution. The oxide powders are continuously obtained without these steps in the spray pyrolysis. So far, it has been reported that this process is effective in the multicomponent oxide powders such as $BaTiO_3$ (Ogihara, et.al, 1999) and alloy powders such as Ag-Pd (Iida, et.al, 2001). Recently, layered type of lithium transition metal oxides such as $LiCoO_2$ (Ogihara, et.al 1993), $LiNiO_2$ (Ogihara, et.al, 1998), $LiNi_{0.5}Mn_{1.5}O_4$ (Park, et.al, 2004), $LiNi_{1/3}Mn_{1/3}Co_{1/3}O_2$ (Park, et.al, 2004) and spinel type of lithium transition metal oxides such as $LiMn_2O_4$ (Aikiyo, et.al, 2001), which are used as the cathode materials for Li ion batteries also have been synthesized by spray pyrolysis. It has been clear that these cathode materials derived from spray pyrolysis showed excellent rechargeable performances. This revealed that the particle characteristics such as uniform particle morphology, narrow size distribution and homogeneous chemical composition led to higher rechargeable capacity, higher efficiency, long life cycle and higher thermal stability.

3. $LiFePO_4$/C cathode materials

The electrochemical properties of olivine-type $LiFePO_4$ cathode materials exhibit a relatively high theoretical capacity of 170 mAh/g and a stable cycle performance at high temperatures. However, in the past, the low electrical conductivity of $LiFePO_4$ prevented its application as a cathode material for the lithium-ion battery. Therefore, conductive materials such as carbon and foreign metals were added to $LiFePO_4$ in order to enhance its electrical conductivity (Padhi, et.al, 1997, Bewlay, et.al, 2004, Wang, et.al, 2005, Barker, et.al, 2003). So far, the composite materials of $LiFePO_4$ and carbon have been synthesized by various types of solution techniques such as sol-gel method, hydrothermal thermal, emulsion and spray drying and solid state reaction. On the other hand, the carbon coating on $LiFePO_4$ powders have been also carried out after the preparation of pure $LiFePO_4$ powders. The advantage of spray pyrolysis is that the precursor of $LiFePO_4$/C materials is obtained at one step for very short time. The various types of organic compounds such as white sugar, ascorbic acid and citric acid were used as a carbon source to enhance the electrical conductivity of $LiFePO_4$ in spray pyrolysis.

3.1 Preparation of $LiFePO_4$/C cathode materials

$LiNO_3$, $Fe(NO_3)_3 \bullet 9H_2O$, and H_3PO_4 were used as starting materials. They were weighted out to attain a molar ratio of Li:Fe:P = 1:1:1 and then dissolved in double distilled water to prepare the aqueous solution. Various types of organic compounds such as sucrose, fructose, sugar or citric acid were added to aqueous solutions up to 60 wt% as carbon source. Figure 1 shows the schematic diagram of the spray pyrolysis apparatus. It consisted of an ultrasonic transducer, electric furnace, and cyclone. The mist of aqueous solution was generated with ultrasonic transducer (2.4 MHz) with 0.08 dm^3/s of air carrier gas. The droplet size (D_P) of mist generated using an ultrasonic transducer were very small and can be determined by equation (1), where ρ is the density of water as a solvent, γ is the surface tension of water, f is the frequency of the transducer. The mist was introduced to the electric furnace. The pyrolysis temperature of electric furnace was 500 °C. As-prepared $LiFePO_4$/C powders were continuously collected by using the cyclone. As-prepared $LiFePO_4$/C powders were heat treated at 700 °C for 10 h in the electric furnace under argon (95 %)/hydrogen (5 %) atmosphere.

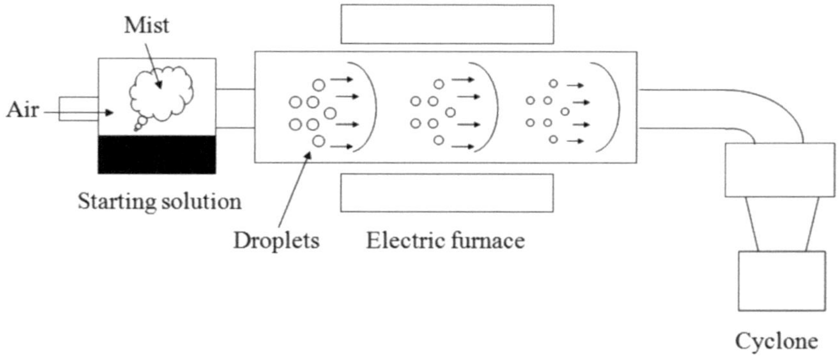

Fig. 1. Schematic diagram of spray pyrolysis with ultrasonic transducer

$$D_p = 0.34\left(\frac{8\pi\gamma}{\rho f^2}\right)^{1/3} \quad (1)$$

3.2 Particle characterizations of $LiFePO_4$/C cathode materials

Figure 2 shows SEM photographs of as-prepared $LiFePO_4$/C powders prepared by spray pyrolysis of an aqueous solution of sucrose and citric acid. The average particle size, morphology, and microstructure of the $LiFePO_4$/C powders were determined using a scanning electron microscope (SEM). The as-prepared $LiFePO_4$/C particles had a spherical morphology with a smooth surface and non-aggregation regardless of the type of carbon sources used. Figure 2 also shows that these have hollow particles. This resulted in the drastic decomposition of organic acid in the step of pyrolysis. The average particle sizes of as-prepared $LiFePO_4$/C powders obtained from sucrose and citric acid were approximately 1 μm. The particle size distribution of $LiFePO_4$/C powders ranged from 0.2 μm to 3 μm. It was found that these powders had a broad size distribution because of the broad size distribution of the mist generated by the two-fluid nozzle. The specific surface area of the powders was measured by the BET method using nitrogen adsorption.

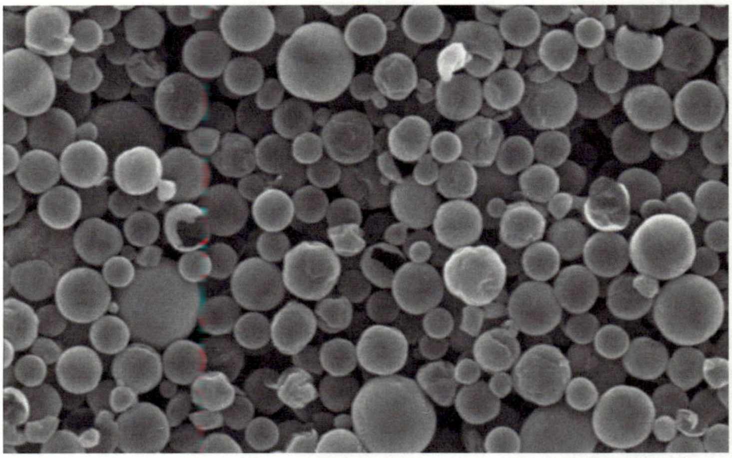

Fig. 2. SEM photograph of $LiFePO_4/C$ powders derived from spray pyrolysis

The specific surface area of the $LiFePO_4/C$ powders was approximately 10 m^2/g regardless of the type of carbon sources used; this suggests that the particle microstructure of $LiFePO_4/C$ powders was porous. The particle densities of the $LiFePO_4/C$ powders obtained from sucrose and citric acid were 3.5 kg/m^3 and 3.2 kg/m^3, respectively. It was considered that the hollow or porous microstructure led to a reduced particle density of the $LiFePO_4/C$ powders.

3.3 Electrochemical properties of $LiFePO_4/C$ cathode materials

Figure 3 shows the rechargeable curves of $LiFePO_4$ and $LiFePO_4/C$ cathodes at 1C. The long plateau was observed at about 3.5 V in the rechargeable curves. The discharge capacity of carbon-free $LiFePO_4$ cathode was about 20 mAh/g because of the poor electrical conductivity. It was found that the rechargeable capacity of $LiFePO_4$ was considerably improved by the addition of carbon. The discharge capacity of $LiFePO_4/C$ cathode derived from citric acid exhibited 150 mAh/g. That of $LiFePO_4/C$ cathode derived from sucrose exhibited 149 mAh/g.

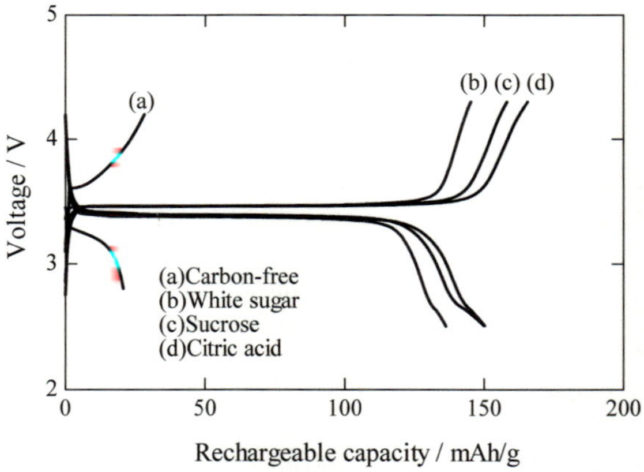

Fig. 3. Rechargeable curves of $LiFePO_4$ and $LiFePO_4/C$ cathodes at 1 C

The discharge capacity of $LiFePO_4/C$ cathode derived from other organic compound such as fructose, white sugar also exhibited 136 mAh/g. The rechargeable capacity of $LiFePO_4/C$ cathode derived from citric acid was higher than that derived from sucrose. The carbon content was 2.6 wt% in $LiFePO_4/C$ particles derived from citric acid. The carbon content was 7.1 wt% in $LiFePO_4/C$ particles derived from sucrose. Because the particle size of $C/LiFePO_4$ particles derived from citric acid is close to that of $LiFePO_4/C$ particles derived from sucrose, the excess carbon content (4.5 wt%) may be led to the loss for energy density of $LiFePO_4/C$ cathode derived from sucrose. Figure 4 shows the change of initial discharge capacity of $LiFePO_4/C$ cathode derived from citric acid.

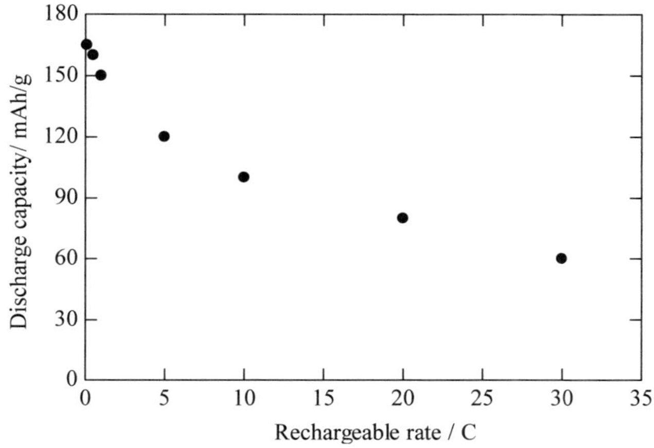

Fig. 4. Relation between rechargeable rate and discharge capacity

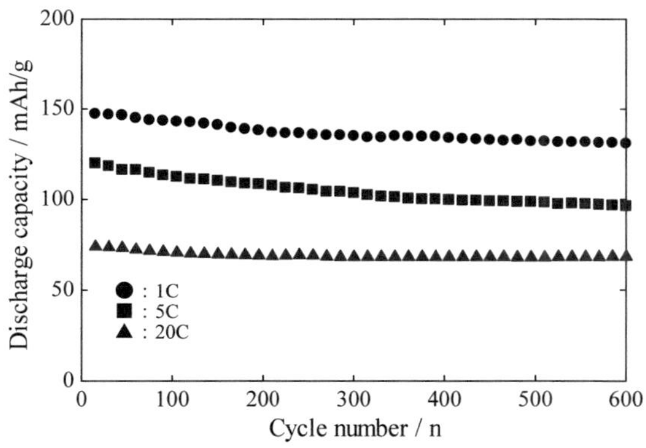

Fig. 5. Cycle performance of $LiFePO_4/C$ cathode at rechargeable rate indicated

The initial discharge capacity of $LiFePO_4/C$ cathode exhibited 165 mAh/g at 0.1 C. The initial discharge capacity decreased to 100 mAh/g at 10 C. At 30 C, it exhibited 60 mAh/g. Figure 5 shows the relation between cycle number and discharge capacity of $LiFePO_4/C$ cathode derived from citric acid at rate indicated. The cycling was carried out up to 500 cycles. It was clear that $LiFePO_4/C$ cathode had the excellent cycle stability. The discharge capacity of $LiFePO_4/C$ cathode maintained 84 % of initial discharge capacity after 600 cycles at rate of 1 C. The same tendency of cycle stability was also observed in the cycle data at rate of 5 C. The discharge capacity of $LiFePO_4/C$ cathode maintained 94 % of initial discharge capacity after 600 cycles at 20 C. Figure 6 shows the relation between cycle number and discharge capacity of $LiFePO_4/C$ cathode at 50 °C. The rechargeable test of coin cell was examined up to 100 cycles at rate of 1C. The coin cell was heated on the hot plate which was kept to 50 °C. The discharge capacity of $LiFePO_4/C$ cathode derived from citric acid exhibited 147 mAh/g and the cycle life of it was also stable. The discharge capacity of $LiFePO_4/C$ cathode maintained 96 % of initial discharge capacity after 100 cycles. It was found that $LiFePO_4/C$ cathode had high cycle stability at the elevated temperature.

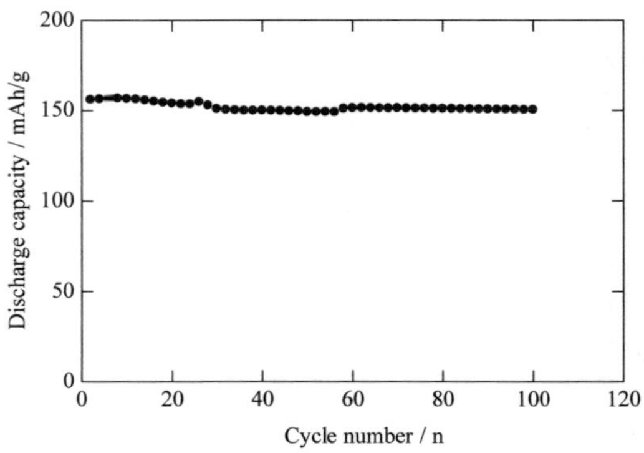

Fig. 6. Cycle performance of $LiFePO_4/C$ cathode at 50 °C

4. $Li_4Ti_5O_{12}/C$ anode materials

Various types carbons (Ohzuku, et.al, 1993, Ozawa et.al, 1994, Endo, et.al, 1996, Qiu, et.al, 1996, Buiel, et.al, 1999, Matsumura, et.al, 1995) have been always used as an anode material because they has better safety characteristic and long cycle life compared with lithium metal. It was well known that the carbon anode leads to the formation of dendrite at high rate charging. The solid electrolyte interface (SEI) layer on the carbon anode, which is usually formed at the potential below 0.8 V and accompanied over time with active lithium loss, an increase in impedance, a decrease in rechargeable capacity and fade in cycle life of lithium ion batteries. For an application of EVs and HVs, the oxide type anode is also expected as the candidate for anode materials because of better safety. Spinel type $Li_4Ti_5O_{12}$ has been demonstrated as an alternative anode material because it has a long plateau at 1.5 V and exhibited excellent cycle life due to the structure stability for the intercalation of Li ion. The disadvantage of $Li_4Ti_5O_{12}$ for anode was low electronic conductivity because $Li_4Ti_5O_{12}$ was ionic crystal with insulation. To improve the electric conductivity of it, the foreign metals with various valence numbers (Kubiak, et.al, 2003, Chen, et.al, 2001, Robertson, et.al, 1999, Mukai, et.al, 2005, Huang, et.al, 2006) or the carbon is added to $Li_4Ti_5O_{12}$ powders. Especially, many researchers have been reported that the addition of carbon is effective for the improvement of electrochemical properties (Gao, et.al, 2007, Huanga, et.al, 2006, Hao, et.al, 2007). So far, it was well known that the solution techniques such as spray drying, sol-gel enabled to homogeneously dope the carbon to $Li_4Ti_5O_{12}/C$ powders (Gao, et.al, 2006, Hao, et.al, 2006, Hao, et.al, 2005). Ju et al applied the spray pyrolysis to the preparation of $Li_4Ti_5O_{12}/C$ anode powders. They have been reported that $Li_4Ti_5O_{12}/C$ powders derived from spray pyrolysis exhibits higher rechargeable capacity and good cycle performance (Yang, et.al, 2006, Ju, et.al, 2009).

4.1 Preparation of $Li_4Ti_5O_{12}/C$ anode materials

Titanium tetraisoproxide ($Ti(iso\text{-}OC_3H_7)_4$, denoted as TTIP) and $LiNO_3$ were used as raw materials. They were dissolved in an atomic molar ratio of Li/Ti to prepare the starting aqueous solution. Organic compound as a carbon source was also added to starting solution. The concentration of starting solution ranged from 0.1 to 1 mol/dm^3. The concentration of organic compound was ranged from 0.1 to 0.4 mol/dm^3. Lactic acid, malic acid, citric acid and malonic acid were used as an organic compound. The mist of starting solution prepared was generated with an ultrasonic vibrator (1.6 MHz) and introduced into quartz tube (38 mmϕ × 2000 mm) in the electrical furnace with air carrier (6 dm^3/min). The mist was drying at 400 °C and then decomposed at 700 °C. The temperatures of electrical furnaces that were used to dry and pyrolysis were set to 400 °C and 700 °C, respectively. As-prepared $Li_4Ti_5O_{12}/C$ particles were continuously collected using the bag filter.

4.2 Preparation of $Li_4Ti_5O_{12}/C$ anode materials

Figure 7 shows typical SEM photograph and particle size distribution of as-prepared $Li_4Ti_5O_{12}/C$ powders formed by spray pyrolysis. SEM photograph reveal that as-prepared particles have spherical morphology with non-aggregation and that the microstructure is dense. No particles with irregular morphology or hollow microstructure were observed.

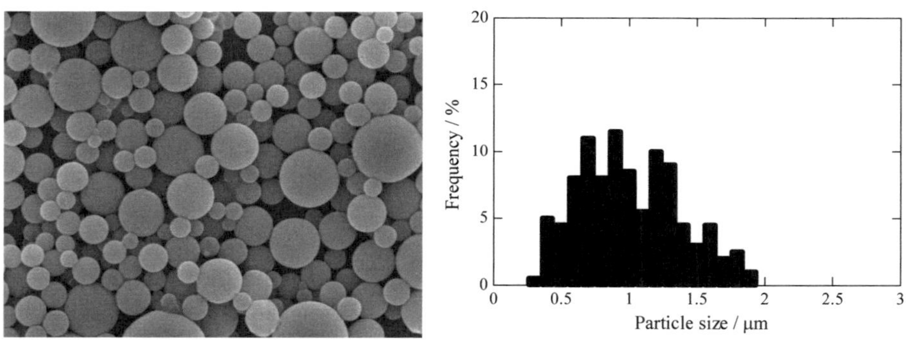

Fig. 7. SEM photograph and particle size distribution of $Li_4Ti_5O_{12}/C$ powders derived from spray pyrolysis

The particle size of as-prepared powders was about 1 μm. SEM photograph also indicate that the as-prepared particles had a broad size distribution. The particle size of all samples ranged from 300 nm to 2000 nm. The geometrical standard deviation (σ_g) of the as-prepared particles was 1.4. Table 1 summarizes the physical properties of $Li_4Ti_5O_{12}/C$ powders derived from various types of organic acids.

Type of acid	Particle size (nm)	Size distribution (P_g)	Atomic ratio Li/Ti ratio^a	SSA (m^2/g)^a	SSA (m^2/g)^b	Crystal phase^b	Lattice constant (nm)
Lactic acid	990	1.41	4:5	54.0	24.0	Spinel	0.8358
Citric acid	996	1.41	4:5	51.3	24.7	Spinel	0.8358
Malic acid	990	1.40	4:5	50.0	24.0	Spinel	0.8358

a : as-prepared
b : calcination at 700°C

Table 1. Physical properties of $Li_4Ti_5O_{12}/C$ powders derived from various types of organic acids

The average particle size and σ_g was independent on the organic acid used. The BET measurement results revealed that as-prepared powders had a high specific surface area that ranged from 50 to 60 m²/g. This result suggested that as-prepared powders had a porous microstructure that consisted of primary particles. After the calcination at 700 °C, their SSA decreased to about 20 m²/g and the primary particles were sintered. ICP analysis indicated that the Li/Ti ratio of as-prepared powders was in good agreement with that of the starting solution composition. This suggested that the Li^+ and Ti^{4+} ions were homogeneously blending in each mist; this acted as a microreactor. Figure 8 shows the typical differential thermal analysis-thermal gravimetry (DTA-TG) curves of as-prepared powders obtained from lactic acid. TG curves indicated that the weight loss of as-prepared powders was due to the volatility of carbon. The weight loss was approximately 13 wt%. The exothermic peak corresponding to the volatility of carbon was observed at 460 °C in the DTA curve. It was found that the $Li_4Ti_5O_{12}$ particles had high carbon content. In spray pyrolysis, the residence time of the particles in the electric furnace was less than 30 s. Therefore, because $Li_4Ti_5O_{12}$ particles were collected by a bag filter before the organic acid volatilized, it was considered that the carbon remained in the $Li_4Ti_5O_{12}$ particles.

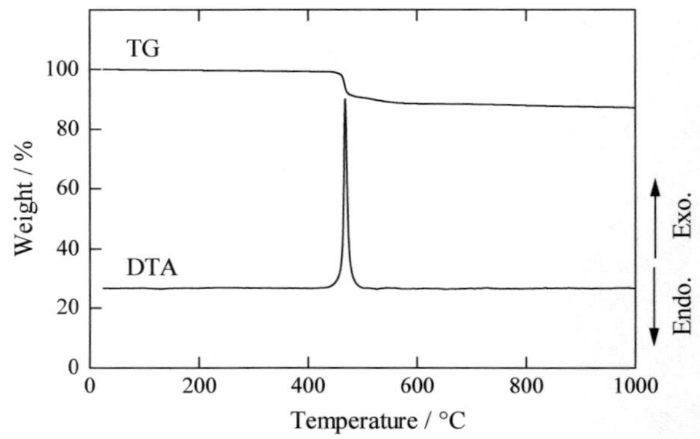

Fig. 8. DTG curve of $Li_4Ti_5O_{12}$/C powders

Table 2 summarizes the relationship between organic acid and carbon content. The carbon content of $Li_4Ti_5O_{12}$/C powders obtained from citric acid and malic acid was 11.8 wt% and 10.6 wt%, respectively.

Type of acid	Carbon content (wt%)
Lactic acid	12.8
Citric acid	11.8
Malic acid	10.6

Table 2. Relation between organic acid and carbon content

This suggests that the volatility of carbon from $Li_4Ti_5O_{12}$/C particles in the pyrolysis process has the following order: malic acid, citric acid, and lactic acid. It is known that carboxylic acid leads to the formation of the Ti^{4+} ion complex compound in the aqueous solution. Kakihana et al. already reported (Kakihana, et.al, 2004) chelating of Ti^{4+} ion by lactic acid in aqueous solution as shown in Fig.9. Therefore, we consider that the Ti^{4+} ion complex compound in our case was also formed by malic acid and citric acid, because a stable aqueous solution was obtained without the precipitation of titanium hydroxide. Volatilization of carbon is suppressed during the particle formation because of the chemical bonding of the Ti^{4+} ions with lactic acid. Similarly, the volatilization of carbon is suppressed by the chemical bonding of the Ti^{4+} ions with malic acid or citric acid, which is also carboxylic acid of the same type as lactic acid. Figure 10 shows the typical XRD patterns of $Li_4Ti_5O_{12}$/C powders obtained from lactic acid. The diffraction patterns of all samples were in good agreement with the spinel structure (space group: Fd3m), and other phases were not observed. As-prepared powders were already crystallized to $Li_4Ti_5O_{12}$. It was considered that Li_2O and TiO_2 were rapidly formed in the mist and their solid-state reaction occurred during the pyrolysis. As-prepared powders (a) were calcined in the range of 700 °C (b) to 800 °C (c); powders were well crystallized by the calcination under nitrogen. The lattice constant of calculated $Li_4Ti_5O_{12}$ was a = 0.8358 nm, which is in agreement with the values in the literature (Ohzuku, et.al, 1995).

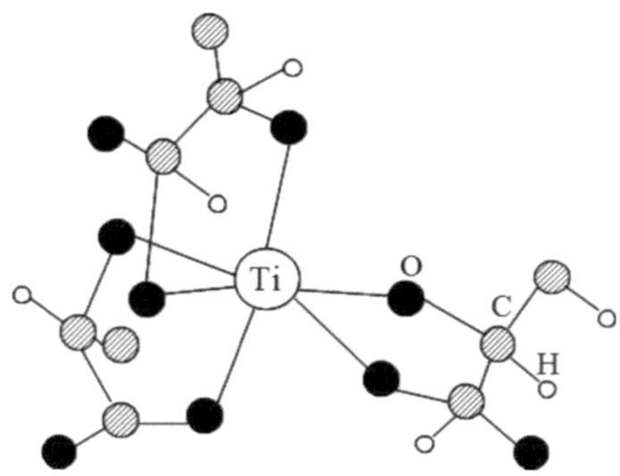

Fig. 9. Schematic diagram of Ti complex in the aqueous solution

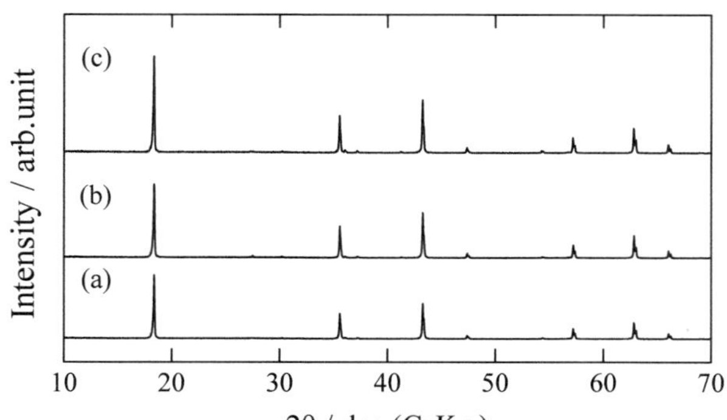

Fig. 10. XRD patterns of as-prepared $Li_4Ti_5O_{12}$/C powders a) and $Li_4Ti_5O_{12}$/C powders calcined at 700 °C b) and 800 °C c)

4.3 Electrochemical properties of $Li_4Ti_5O_{12}$/C anode materials

Figure 11 shows the rechargeable curves of $Li_4Ti_5O_{12}$/C anode at 1 C. The long plateaus were observed at 1.5 V in the rechargeable curves. When lactic acid was used as a carbon source, the charge and discharge capacity of the $Li_4Ti_5O_{12}$/C anode was 170 mAh/g and 165 mAh/g at 1 C, respectively. The efficiency of rechargeable capacity in this case was approximately 97%. These values were higher than those of carbon-coated $Li_4Ti_5O_{12}$ and $Li_4Ti_5O_{12}$/C prepared by spray pyrolysis. $Li_4Ti_5O_{12}$/C anode derived from citric acid exhibited a charge and discharge capacity of 157 mAh/g and 152 mAh/g at 1 C, respectively, and the efficiency of rechargeable capacity was approximately 97 %. $Li_4Ti_5O_{12}$/C anode derived from citric acid exhibited a charge and discharge capacity of 146 mAh/g and 140 mAh/g at 1 C, respectively, and the efficiency of rechargeable capacity was approximately 96 %.

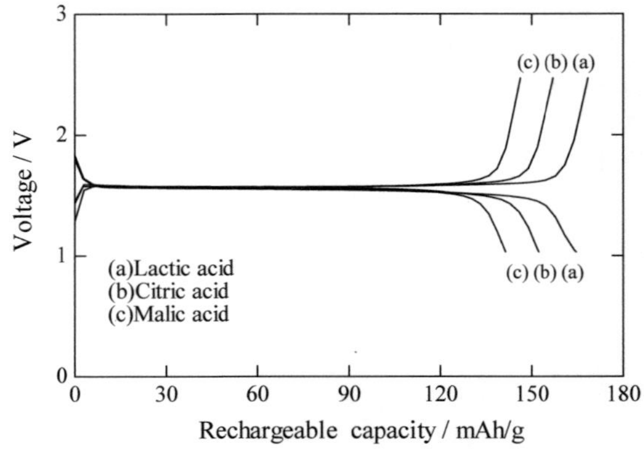

Fig. 11. Rechargeable curves of $Li_4Ti_5O_{12}$/C anode at 1C

$Li_4Ti_5O_{12}$/C anode derived from lactic acid exhibited the highest capacity and efficiency among all the organic acids used. It was thus confirmed that the rechargeable capacity was affected by the carbon content Figure 12 shows the change in the initial discharge capacity of the $Li_4Ti_5O_{12}$/C anode at the rechargeable rate indicated.

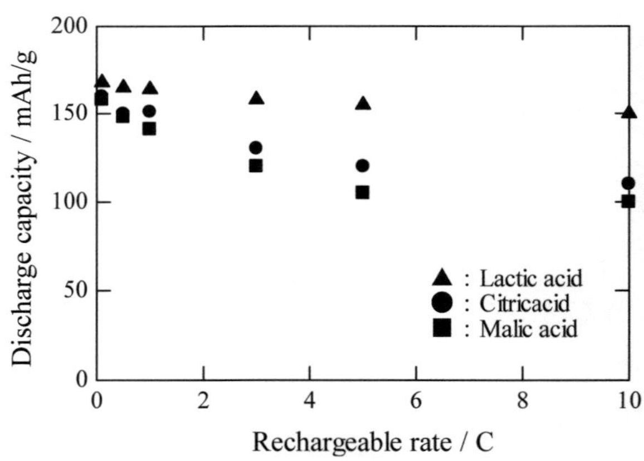

Fig. 12. Rate performance of $Li_4Ti_5O_{12}$/C anode

The initial discharge capacity of the $Li_4Ti_5O_{12}$/C anode gradually decreased with increasing rechargeable rate. The initial discharge capacity of the $Li_4Ti_5O_{12}$/C anode obtained from lactic acid decreased to 150 mAh/g at 10 C. The retention of the initial discharge capacity for 1 C was 91 %. It was found that the $Li_4Ti_5O_{12}$/C anode obtained from lactic acid had a relatively high discharge performance at a high rechargeable rate, which indicates superior rechargeable performance compared to that of $Li_4Ti_5O_{12}/C$ otained by spray pyrolysis and spray drying (Wen, et.al., 2005) and that of carbon-coated $Li_4Ti_5O_{12}/C$ (Wang, et.al., 2007). On the other hand, when citric acid and malic acid were used as the carbon source, the initial discharge capacity of the $Li_4Ti_5O_{12}/C$ anode decreased to 110 mAh/g and 100 mAh/g at 10 C, respectively. The efficiency in this case for 1 C was 68 % and 63 %, respectively. The rechargeable rate was influenced by the carbon content.

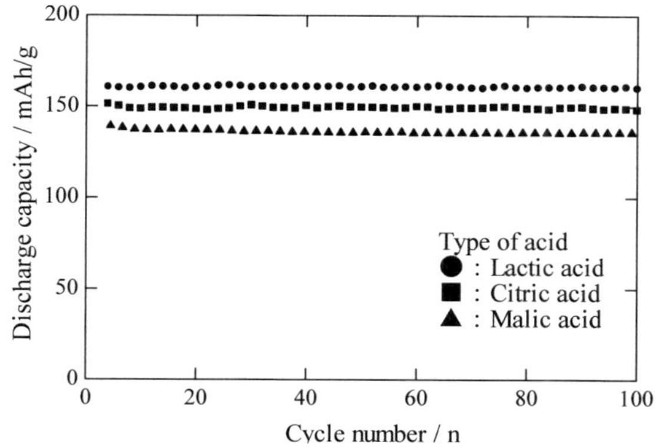

Fig. 13. Cycle performance of $Li_4Ti_5O_{12}/C$ anode at 1C

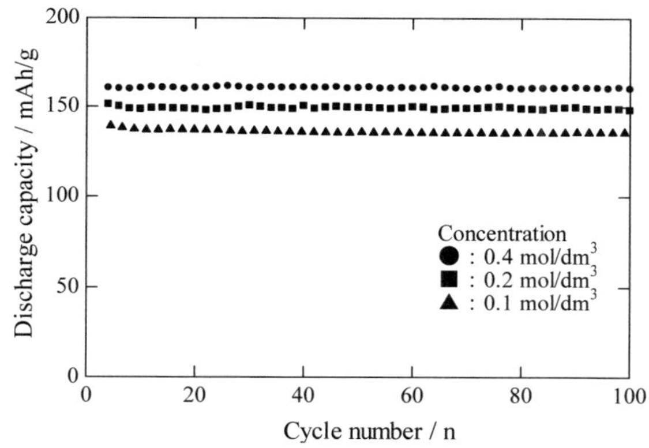

Fig. 14. Cycle performance of $Li_4Ti_5O_{12}/C$ anode with different carbon content

Figure 13 shows the relationship between the cycle number and the discharge capacity of the $Li_4Ti_5O_{12}/C$ anode at 1 C. The rechargeable test was conducted with up to 100 cycles at room temperature. It was clear that $Li_4Ti_5O_{12}/C$ anode had excellent cycle stability regardless of the organic acid type. The discharge capacity of $Li_4Ti_5O_{12}/C$ anode obtained from lactic acid maintained 98 % of the initial discharge capacity after 100 cycles at 1 C. When citric acid and malic acid were used, the rechargeable capacity of $Li_4Ti_5O_{12}/C$ anode reduced to 150 mAh/g and 138 mAh/g, respectively. The cycle performance showed high stability in the cycle data of both citric acid and malic acid. The retention ratio of the discharge capacity of $Li_4Ti_5O_{12}/C$ anode obtained from citric acid and malic acid was 94 % and 96 %, respectively. Figure 14 shows the relationship between the cycle number and the discharge capacity of the $Li_4Ti_5O_{12}/C$ anode prepared with different concentration of lactic acid.

The rechargeable rate was 1 C at 25 °C. When the concentration of lactic acid was 0.1 mol/dm^3, the carbon content in the $Li_4Ti_5O_{12}$/C anode was 9 wt% according to TG analysis and the initial discharge capacity of the anode was 141 mAh/g. When the concentration of lactic acid was 0.2 mol/dm^3, the carbon content in the $Li_4Ti_5O_{12}$/C anode was 11 wt% and the initial discharge capacity of the anode was 151 mAh/g. The initial discharge capacity of the $Li_4Ti_5O_{12}$/C anode increased to 162 mAh/g when the concentration of lactic acid was 0.4 mol/dm^3 (12.8 wt%). It was confirmed that the initial discharge capacity increased with increasing carbon content. The retention ratio of the discharge capacity after 100 cycles was more than 95 % for all $Li_4Ti_5O_{12}$/C anodes.

Figure 15 shows the relationship between the cycle number and discharge capacity of $Li_4Ti_5O_{12}$/C anode at 50 °C. The rechargeable test of the coin cell was examined at 1 C for up to 100 cycles while it was heated on the hot plate, which was kept at 50 °C. The discharge capacity of $Li_4Ti_5O_{12}$/C anode derived from lactic acid was 161 mAh/g and its cycle life was stable. The $Li_4Ti_5O_{12}$/C anode maintained 97 % of the initial discharge capacity after 100 cycles. It was found that $Li_4Ti_5O_{12}$/C anode had high cycle stability at an elevated temperature as well as at room temperature. It has been reported (Nakahara, et.al., 2003) that the rechargeable capacity and cycle stability of the $Li_4Ti_5O_{12}$/C anode at 50 °C are superior to those at 25 °C. This may result from the increase in the electric conductivity of $Li_4Ti_5O_{12}$/C at 50 °C.

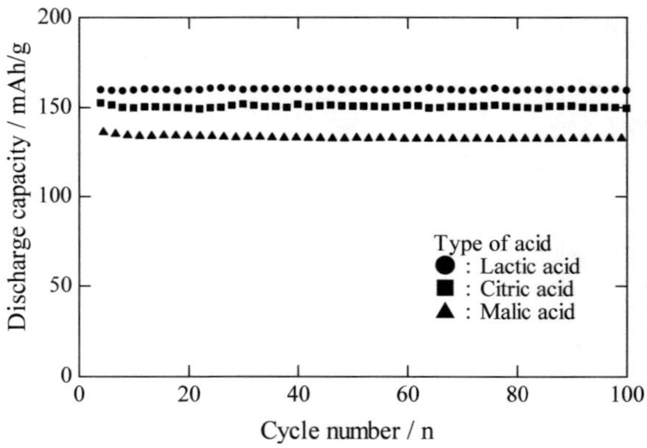

Fig. 15. Cycle performance of $Li_4Ti_5O_{12}$/C anode at 50 °C

5. Conclusions

$LiFePO_4$ cathode and $Li_4Ti_5O_{12}$ anode materials were successfully synthesized by spray pyrolysis using an aqueous solution with an organic acid. They had spherical morphology with a porous microstructure. The as-prepared powders had a high crystallinity with a homogeneous composition. The rechargeable properties of $LiFePO_4$/C cathode and $Li_4Ti_5O_{12}$/C anode were significantly improved by the addition of carbon. The rechargeable capacity of them was also dependent on the carbon content. The discharge capacity of $LiFePO_4$/C cathode and $Li_4Ti_5O_{12}$/C anode was 170 and 165 mAh/g at 1 C, respectively. They had also a high rechargeable capacity at high charging rate and a high retention ratio of rechargeable capacity. The high cycle stability of $LiFePO_4$ cathode and $Li_4Ti_5O_{12}$ anode was also maintained at the elevated temperature. It was concluded that the cathode and anode materials derived from spray pyrolysis were suitable as the electrode for lithium ion battery.

6. References

Aikiyo, H.; Nakane, K., Ogata, N. & Ogihara, T. (2001). *J. Ceram. Soc. Jpn.*, Vol.109, pp.197-200

Barker, J.; Saidi, M.Y. & Swoyer, J.L. (2003). *Electrochem. Solid State Lett.*, Vol.6, pp. A53-55

Bewlay, S.L.; Konstantinov, K., Wang, G.X., Dou, S.X. & Liu, H.K. (2004). *Mater. Lett.*, Vol.58, pp. 1788-1191

Buiel, E. & Dahn, J.R. (1999). *Electrochimi Acta*, Vol.45, pp. 121-130

Chen, C.H.; Vaughey, J.T. & Jansen, A.N. (2001). *J. Electrochem. Soc.*, Vol.148, pp. A102-A104

Dubois, B.; Ruffier, D. & Odier, P. (1989). *J. Am. Ceram. Soc.*, Vol.72, pp. 713-715

Endo, M.; Nishimura, Y., Takahashi, T., Takeuchi, K. & Dresselhaus, M.S. (1996). *J. Phys. Chem. Solids*, Vol.57, pp. 725-728

Gao, J.; Jiang, C., Ying, J. & Wan C. (2006). *J. Power Sources*, Vol.155, pp. 364-367

Gao, J.; Ying, J., Jiang, C. & Wan, C. (2007). *J. Power Sources*, Vol.166, pp. 255-259

Guyomard, D. & Trascon, J.M. (1994). *Solid State Ionics*, Vol.69, pp. 222-237

Hao, Y.; Lai, Q., Lu, J., Wang, H., Chen, Y. & Ji, X.Y. (2006). *J. Power Sources*, Vol.158, pp. 1358-1364

Hao, Y.; Lai, Q., Lua, J., Liu, D. & Ji, X. (2007). *J. Alloys Compd.*, Vol.439, pp. 330-336

Hao, Y.; Lai, Q., Xu, X., Liu, X. & Ji, X.Y. (2005). *Solid State Ionics*, Vol.176, pp. 1201-1206

Huang, S.; Wen, Z., Zhang, J., Gu, Z. & Xu, X. (2006). *Solid State Ionics*, Vol.177, pp. 851-855

Huanga, J. & Jiang, J. (2008). *Electrochimi. Acta*, Vol.53, pp. 7756-7759

Idemoto, Y.; Sekine, H., Ui, K. & Koura, N. (2004). *Electrochemistry*, Vol.70, pp. 564-568

Iida, N.; Nakayama, K., Lenggoro, I.W. & Okuyama, K. (2001). *J. Soc. Powder Technol. Jpn.*, Vol.38, pp. 542-547

Ishizawa, H.; Sakurai, O., Mizutani, N. & Kato, M. (1985). *Yogyo-Kyokai-shi*, Vol.93, pp. 382-386

Ju, S. & Y. Kang, (2009). *J. Phys. Chem. Solids*, Vol.70, pp. 40-44

Kakihana, M.; Tomita, K., Petrykin, V., Toda, M., Sasaki, S. & Nakamura, Y. (2004). *Inorg. Chem.*, Vol.43, pp. 4546-4548

Kubiak, P.; Garcia, A., Womes, M., Aldon, L., Fourcade, J.O., Lpens, P.E. & Jumas, J.C. (2003). *J. Power Sources*, Vol.119–121, pp. 626-630

Markovsky, B.; Talyossef, Y., Salitra, G., Aurbach, D., Kim, H.J. & Choi, S. (2004). *Electrochem.Commun.*, Vol.6, pp. 821-826

Matsumura, Y.; Wang, S., Shinohara, K. & Maeda, T. (1995). *Synthetic*, Vol.71, pp. 1757-1758

Messing, G.L.; Zhang, S.C. & Javanthi, G.V. (1993). *J. Am. Ceram. Soc.*, Vol.76, pp. 2707-2726

Mizushima, K.; Jones, P.C. Wiseman, P.J. & Goodenough, J.B. (1980). *Mat. Res. Bull.*, Vol.15, pp. 783-789

Mukai, K.; Ariyoshi, K. & Ohzuku, T. (2005). *J. Power Sources*, Vol.146, pp. 213-216

Nakahara, K.; Nakajima, R., Matsushima, T. & Majima, H. (2003). *J. Power Sources*, Vol.117, pp. 131-136

Ogihara, T.; Aikiyo, H., Ogata, N. & Mizutani, N. (1999). *Adv. Powder Technol.*, Vol.10, pp. 37-50

Ogihara, T.; Ogata, N., Yonezawa, S., Takashima, M. & Mizutani, N. (1998). *Denki Kagaku*, Vol.66, pp. 1202-1205

Ogihara, T.; Saito, Y., Yanagwa, T., Ogata, N., Yoshida, K., Takashima, M., Yonezawa, S., Mizuno, Y., Nagata, N. & Ogawa, K. (1993). *J. Ceram. Soc. Jpn.*, Vol.101, pp. 1159-1163

Ohzuku, T.; Iwakoshi, Y. & Sawai, K. (1993). *J. Electrochem. Soc.*, Vol.140, pp.2490-2498
Ohzuku, T.; Ueda, A. & Yamamoto, N. (1995). *J. Electrochem. Soc.*, Vol.142, pp. 1431-1435
Ozawa, K. (1994). *Solid State Ionics*, Vol.69, pp. 212-221
Padhi, A.K.; Nanjundaswamy, K.S. & Goodenough, J.B. (1997). *J. Electrochem. Soc.*, Vol.144, pp. 1188-1194
Park, S.H. & Sun, Y.K. (2004). *Electrochimi. Acta*, Vol.50, pp. 431-434
Park, S.H. & Sun, Y.K. (2004). *Electrochimi. Acta*, Vol.50, pp. 431-434
Park, S.H.; Yoon, C.S., Kang, S.G., Kim, H.S., Moon, S.I. & Sun, Y.K. (2004). *Electrochimi. Acta*, Vol.49, pp. 557–563
Pegeng, Z.; Huiqing, F., Yunfei, F., Zhuo, L. & Yongli, D. (2006). *Rare Metals*, Vol.25, pp. 100-104
Pluym, C.T.; Lyons, S.W., Powell, Q.H., Gurav, A.S. & Kodas, T. (1993). *Mater. Res. Bull.*, Vol.28, pp. 369-376
Qiu, W.; Zhou, R., Yang, L. & Liu, Q. (1996). *Solid State Ionics*, Vol.86, pp. 903–906
Robertson, A.D.; Trevino, L., Tukamoto, H. & Irvine, J.T. (1999). *J. Power Sources*, Vol.81-82, pp. 352-357
Roy, D.M.; Neurgaonkar, R.R., O'holleran, T.P. & Roy, R. (1977). *Am. Ceram. Soc. Bull.*, Vol.56, pp. 1023 -1024
Wang, D.; Li, H., Shi, S., Huang, X. & Chen, L. (2005). *Electrochimi. Acta*, Vol.50, pp. 2955-2958
Wang, G.; Gao, J., Fu, L., Zhao, N.H., Wu, Y.P. & Takamura, T. (2007). *J. Power Sources*, Vol.174, pp. 1109–1112
Wen, Z.; Gua, Z., Huanga, S., Yang, J., Lin, Z. & Yamamoto, O. (2005). *J. Power Source*, Vol.146, pp. 670-673
Yang, M.; Tean, T. & Wu, S. (2006). *J. Power Source*, Vol.159, pp. 307-311

3

Heterogeneous Nanostructured Electrode Materials for Lithium-Ion Batteries – Recent Trends and Developments

Xiangfeng Guan[1], Guangshe Li[1], Jing Zheng[2], Chuang Yu[2], Xiaomei Chen[1], Liping Li[2*] and Zhengwei Fu[1]
[1]State Key Laboratory of Structural Chemistry and
[2]Key Laboratory of Optoelectronic Material Chemistry and Physics, Fujian Institute of Research on the Structure of Matter, Chinese Academy of Sciences, Fuzhou, China

1. Introduction

Developing highly efficient, low cost, and environmentally benign energy storage devices and related materials is a key issue to meet the challenge of global warming, the finite nature of fossil fuels, and city pollution. Among the various available storage technologies, lithium-ion batteries (LIBs) currently represent the state-of-the-art technology in small rechargeable batteries because it can offer the largest energy density and output voltage of all known rechargeable battery technologies. A typical commercial lithium-ion battery consists of a negative electrode (anode, e.g., graphite) and a positive electrode (cathode, e.g., $LiCoO_2$), both of which are separated by a lithium-ion-conducting electrolyte. When the cell is charged, Li ions are extracted from the cathode, pass through the electrolyte, and are inserted into the anode. Discharge reverses the procedure. Although such batteries are commercially successful, these cells are still required to improve their performance to satisfy an increasing demand for portable and miniaturized electronics, uninterrupted power supplies for technological and rural needs, power tools, and even stationary storage batteries.

Since the recharging is completed with the lithium ion insertion/extraction process in both electrodes, the nature of two electrode materials is crucial to the performance improvement of batteries (Liang et al., 2009). At present, commercial batteries are mostly based on micrometer-size electrode materials, which are limited by their kinetics, lithium-ion intercalation capacities, and structural stability. The performance of currently available LIBs can only meet the requirements of applications to some degree (Guo et al., 2008). These have placed renewed demands on new and improved electrode materials for LIBs with a high energy density, high power density, longer cycle life, and improved safety.

Nanostructured approaches have been recently demonstrated to be highly effective in greatly improving the electrochemical performance of electrode materials. For instances, nanostructured materials decrease the diffusion length of lithium ion in the insertion/extraction process, which results in the higher capacities at high charge/discharge rates. Nanostructured materials can also increase the surface/interface storage due to the high contact between electrode material and electrolyte. Additionally, nanostructured materials can buffer the stresses caused by the volume variation occurring during the charge/discharge process, alleviating the problem of capacity fade and poor rate capability associated with the material breaking away into the electrolyte. However, nanostructured materials, especially single-phased nanostructured materials, are not an ultimate solution to meet the requirements of future LIBs. One of the primary reasons is that some of the intrinsic material properties of bulk phase, such as low conductivities, low energy densities at high charge/discharge rates, and weak mechanical stabilities, cannot be simply altered or improved by just transforming them into nanostructured materials (Liu et al., 2011). Moreover, without some surface protection, nanostructured materials may magnify the safety issues due to a high surface reactivity and may aggravate capacity fade due to aggregation.

Recently, heterogeneous nanostructured approaches have been introduced to overcome the above mentioned limitations when using single component nanostructured electrode materials. The heterogeneous nanostructured materials are composed of multi-nanocomponents, each of which is tailored to address a different demand (e.g. high lithium-ion/electrical conductivity, high capacity, and excellent structural stability) (Liu et al., 2011). There exists a synergistic effect by the interplay between particle shape, properties, and possible association of the individual components, which can be regulated to explore the full potential of the materials in terms of the performance (e.g. high energy density, high power density, longer cycle life, and improved safety). For instances, concentration-gradient core-shell nanostructured cathode material composed of $Li[Ni_{0.8}Co_{0.1}Mn_{0.1}]O_2$ and $Li[Ni_{0.46}Co_{0.23}Mn_{0.31}]O_2$ have been reported to exhibit a very high reversible capacity, excellent cycling, and safety characteristics, which are attributed to the synergic effects of core and shell components (Sun et al., 2009). Novel nanonet constructed by SnO_2 nanoparticles and ploy (ethylene glycol) chains has been prepared. The synergic properties and functionalities of both materials enable this composite material to exhibit unexpectedly high lithium storage over the theoretical capacity of SnO_2 (Xiong et al., 2011). Graphene anchored with Co_3O_4 nanoparticles have been synthesized and utilized as anode materials, which exhibits a large reversible capacity, excellent cyclic performance, high Coulombic efficiency, and good rate capability due to a strong synergic effect between Co_3O_4 nanoparticles and graphene nanosheet (Wu et al., 2010).

Therefore, developing the heterogeneous nanostructured electrode materials is considered to be the most promising avenue towards future LIBs with high energy density, high power density, longer cycle life, and improved safety. Even so, it is necessary first of all to know how heterogeneous nanostructured materials have impacts on the performance of the LIBs and what kinds of synergic effects these materials exhibit. In this chapter, we focus on the recent trends and developments of heterogeneous nanostructured electrode materials for uses in LIBs and provide an overview of the synthesis, synergic mechanism of heterogeneous nanostructured components, and a survey of promising candidates based on heterogeneous nanostructured materials for LIBs.

2. Heterogeneous nanostructured cathode materials

Heterogeneous nanostructured cathode materials are of great interest for LIBs because of the synergic properties arising from the intergrated multi-nanocomponents, each of which is tailored to address a different demand like high lithium-ion/electrical conductivity, high capacity, and excellent structural stability. In this section, we describe these heterogeneous cathode nanomaterials based on the type of void spaces available for lithium ion insertion: one-dimensional, two-dimensional, and three-dimensional transition-metal-oxide-based nanocomposites (Figure 1).

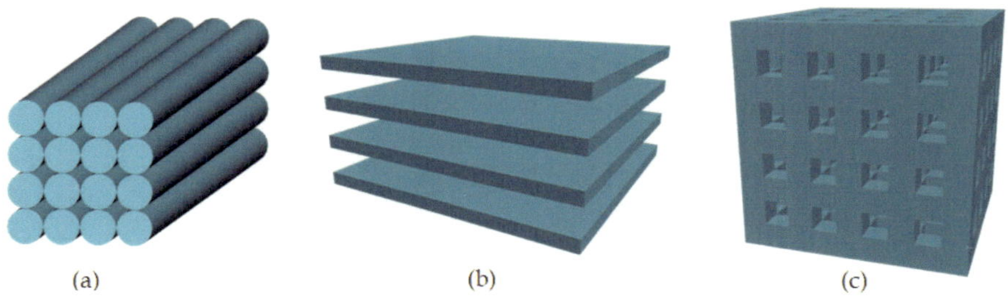

Fig. 1. Schematic representation of void spaces in a) one-dimensional, b) two-dimensional, and c) three-dimensional transition-metal-based oxides. (Redrawn from Ref.(Winter et al., 1998))

2.1 One-dimensional cathode nanocomposites

Since the first report by Padhi et al. (1997) on reversible electrochemical lithium insertion–extraction in $LiFePO_4$, olivine structured lithium transition metal phosphates, $LiMPO_4$ (M =Fe, Mn, Co, Ni), have attracted much attention as the promising candidate cathode materials for LIBs. As shown in Figure 2, the structure of $LiMPO_4$ consists of a distorted hexagonal close-packed (hcp) oxygen framework with 1/8 of the tetrahedral holes occupied by P, and 1/2 of the octahedral holes occupied by various metal atoms (Li and M).

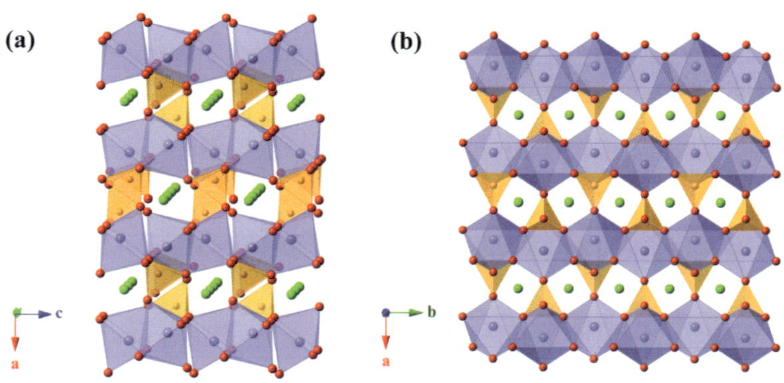

Fig. 2. Polyhedral representation of the structure of $LiMPO_4$ (space group Pnma) viewed a) along *b*-axis and b) along *c*-axis. MO_6 octahedra are shown in blue, phosphate tetrahedra in yellow, and lithium ions in green.

Crystalline $LiMPO_4$ has an orthorhombic unit cell (space group *Pmnb*). Layers of MO_6 octahedra are corner-shared in the bc plane and linear chains of LiO_6 octahedra are edge shared in a direction parallel to the b-axis. These chains are bridged by edge and corner shared phosphate tetrahedra, creating a stable crystal structure (Ellis et al., 2010). Because oxygen atoms are strongly bonded by both M and P atoms, the structure of $LiMPO_4$ is stable at high temperatures. The high lattice stability guarantees an excellent cyclic performance and operation safety for $LiMPO_4$. However, the strong covalent oxygen bonds lead to a low ionic diffusivity and poor electronic conductivity. Additionally, the Li diffusion in $LiMPO_4$ is widely believed to be one dimensional along the b-axis, as shown in Figure 3. These limit the high rate performance of olivine $LiMPO_4$. In this family compounds, lithium iron phosphate ($LiFePO_4$) and lithium manganese phosphate ($LiMnPO_4$) are more promising because they operate at 3.4–4.1 V versus Li/Li^+, which is not so high as to decompose the organic electrolyte but is not so low as to scarify the energy density.

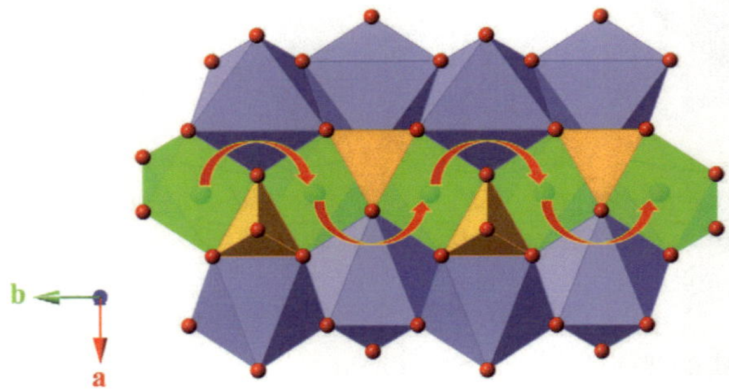

Fig. 3. Structure of $LiMPO_4$ that depicts the curved trajectory of Li ion transport along *b*-axis, shown with red arrows. The colour scheme is the same as that in Figure 2. (Redrawn from Ref. (Ellis et al., 2010))

2.1.1 $LiFePO_4$

$LiFePO_4$ is one of the most promising cathode materials because of its high operating voltage (~3.5 V vs. Li/Li^+) and large theoretical gravimetric capacity (~170 mAh/g), as well as its low cost, environmental friendliness, particularly high thermal stability, and strong overcharge tolerance as compared to other cathode materials. One impediment to the wider use of $LiFePO_4$, however, is its low lithium ion conductivity (lithium ion diffusion coefficient (DLi) is 10^{-14} cm^2s^{-1}) and electrical conductivity (10^{-9} Scm^{-1}). The slow lithium diffusion can be addressed by decreasing the particle dimensions to nanometer scale to reduce the diffusion paths. To overcome the low electronic conductivity, many conductive agents such as carbonaceous materials and polymers have been reported to extend the performance of $LiFePO_4$.

2.1.1.1 Carbonaceous materials

Carbonaceous materials such as amorphous carbon, carbon nanotube, and graphene have been used to form composites with $LiFePO_4$ to improve the electronic conductivity of electrodes. Amorphous carbon coating is one of the most important techniques used to improve the specific capacity, rate performance, and cycling life of $LiFePO_4$. The main role of amorphous carbon coating is to enhance the surface electronic conductivity of $LiFePO_4$ particles so that the active materials can be fully utilized at high current rates. The beneficial effects of carbon coating have been shown to be strongly related to the morphology. For example, Wang et al. (2008) coated $LiFePO_4$ nanoparticles with a carbon shell by in situ polymerization. As shown in Figure 4, these composites are composed of 20–40 nm $LiFePO_4$

nanoparticles covered with 1-2 nm thick carbon shells. The $LiFePO_4/C$ composite delivered a discharge capacity of 169 mAh/g at a 0.6 C rate, which is extremely closer to the theoretical value of 170 mAh/g for $LiFePO_4$. Even at a high discharge rate of 60 C, it still delivered a respectable specific capacity of 90 mAh/g. The composite also exhibits an excellent cycling performance, with less than 5% discharge capacity loss over 1100 cycles. These authors have proposed that an ideal core-shell network of $LiFePO_4/C$ composites should contain nano-size $LiFePO_4$ particles completely coated with a thin layer of porous conductive carbon, which serves as the electron pathway and porous nature that allow lithium ions to diffuse through.

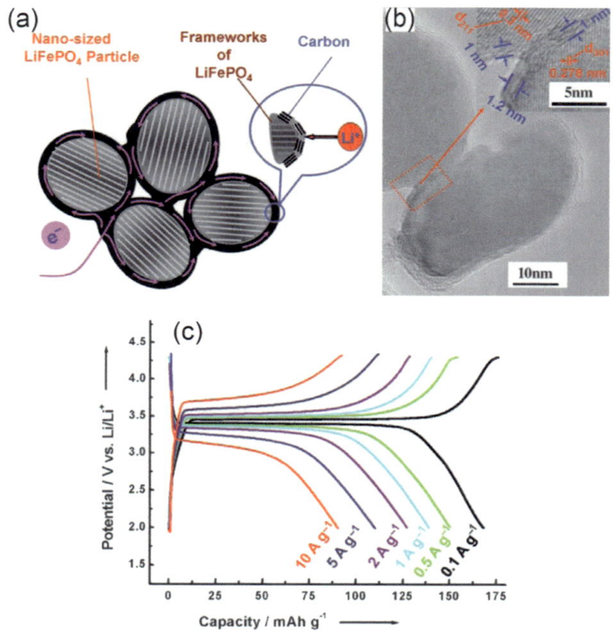

Fig. 4. a) Designed ideal structure for $LiFePO_4$ particles with nano-sizes and a complete carbon coating, b) a typical TEM image for some $LiFePO_4/C$ primary particles, c) charge/discharge tests for $LiFePO_4$/carbon composites at different current densities in a potential window of 2.0–4.3 V (vs. Li/Li^+) (Wang et al., 2008).

In-situ formed amorphous carbon coating layer can not only reduce the particle size of $LiFePO_4$ by inhibiting particle growth during the sintering (Huang et al., 2001), but also act as a reducing agent to suppress the oxidation from Fe^{2+} to Fe^{3+} during the sintering and thus simplify the atmosphere requirement in synthesis (Pent et al., 2011).

The beneficial effect of carbon coating has also been observed to depend on the uniformity, thickness, loading, and precursor of the coating. For instance, Konarova et al. (2008) prepared $LiFePO_4/C$ composites with different amounts of carbon by ultrasonic spray pyrolysis followed by a heat treatment. They have found that the discharge capacity is strongly affected by the amount of carbon residue and sintering temperature. After optimization, $LiFePO_4/C$ composites with 1.87 wt % of carbon exhibited a best electrochemical performance with capacities of 140 mAh/g at C/10 and 84 mAh/g at 5 C.

Vu et al. (2011) synthesized monolithic, three-dimensionally ordered macroporous and meso-/microporous (3DOM/m) $LiFePO_4$/C composite cathodes by a multiconstituent, dual templating method. In this composite, $LiFePO_4$ is dispersed in a carbon phase around an interconnected network of ordered macropores. The carbon phase enhances the electrical conductivity of the cathode and maintains $LiFePO_4$ as a highly dispersed phase during the synthesis and even the electrochemical cycling. The capacity of the composites is as high as 150 mAh/g at a rate of C/5, 123 mAh/g at 1 C, 78 mAh/g at 8 C, and 64 mAh/g at 16 C, showing no capacity fading over 100 cycles.

Carbon nanotubes (CNT) and graphene have advantages of high conductivity, nanotexture, and resiliency, which are very beneficial for LIBs, and therefore have been introduced to improve the electrochemical performances of $LiFePO_4$ cathodes. For example, recently, Zhou et al.(2010) synthesized hierarchically structured composites based on porous $LiFePO_4$ with CNT networks. Porous $LiFePO_4$-CNT composites were synthesized via a facile in-situ sol-gel method, where the carbon nanotubes were functionalized using a mixed acid method to ensure an uniform dispersion into the aqueous sol. As shown in Figure 5, $LiFePO_4$/CNT composites showed a significantly improved specific capacity and rate performance in comparison to the unmodified porous $LiFePO_4$. This performance enhancement is attributed to an improved electrochemical accessibility and a decrease in inert "dead" zones provided by the interpenetrating conductive CNT networks present in the composite structure.

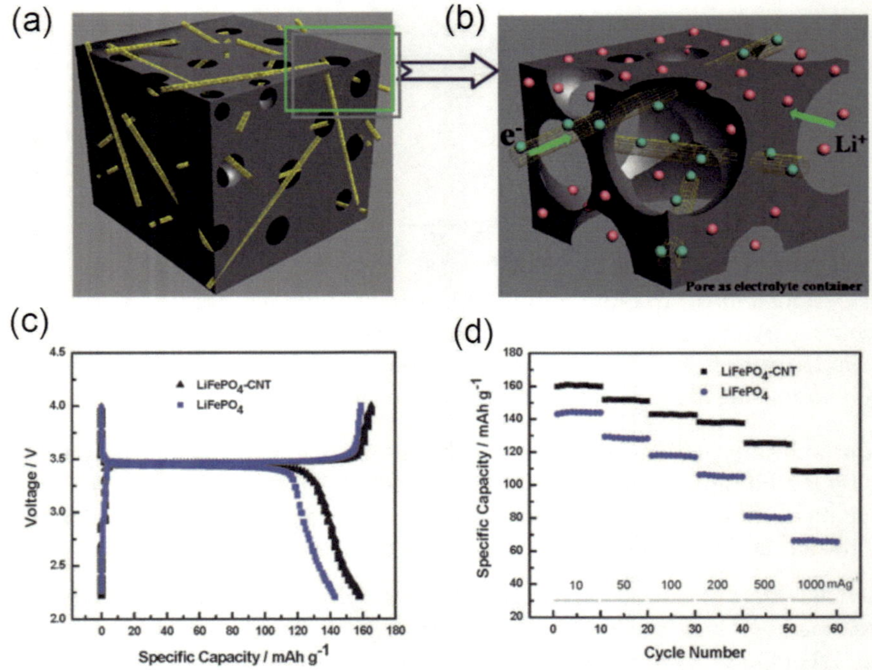

Fig. 5. a) Schematic illustrations of porous $LiFePO_4$-CNT composite, b) an enlarged zone from (a), showing the corresponding electron transport and ion diffusion mechanisms, c) charge/discharge profiles; and d) rate performance of porous $LiFePO_4$ (blue) and $LiFePO_4$-CNT composite (black). (Zhou et al., 2010)

More recently, Zhou et al. (2011) synthesized graphene-modified $LiFePO_4$ composites with $LiFePO_4$ nanoparticles and graphene oxide nanosheets by spray-drying and annealing processes. As indicated in Figure 6, the $LiFePO_4$ primary nanoparticles embedded in micro-sized spherical secondary particles were wrapped homogeneously and loosely with a graphene 3D network. Such a special nanostructure facilitates the electron migration throughout the secondary particles, while the presence of abundant voids between the $LiFePO_4$ nanoparticles and graphene sheets is beneficial for Li^+ diffusion. The composite cathode material could deliver a capacity of 70 mAh/g at 60 C discharge rate and showed a capacity decay rate of <15% when cycled under 10 C charging and 20 C discharging for 1000 times.

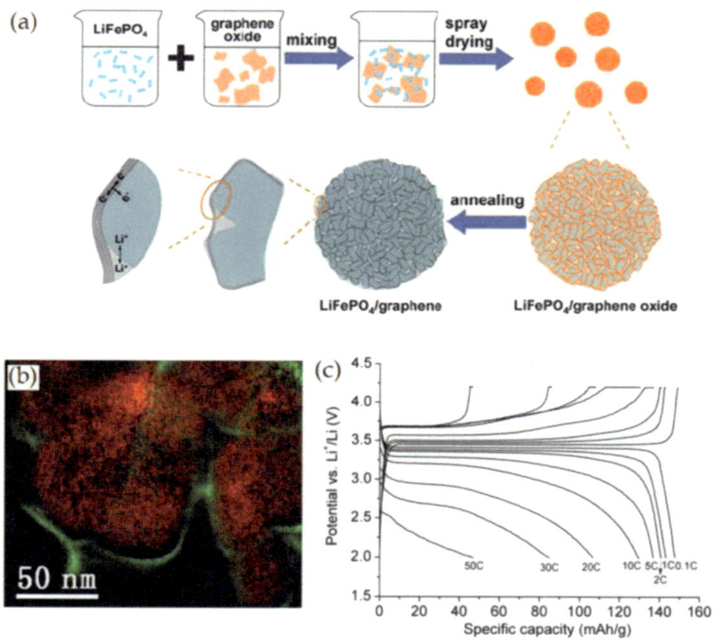

Fig. 6. a) Illustration of the preparation process and microscale structure of $LiFePO_4$/graphene composites; b) the corresponding elemental map of the same area showing graphene-sheets wrapped on $LiFePO_4$ nanoparticles, where red represents the $LiFePO_4$ nanoparticles, and the green represents graphene sheets; and c) rate discharge curves of $LiFePO_4$/graphene composite. (Zhou et al., 2011)

Though carbonaceous materials have showed some positive effects in improving the electrochemical performance of $LiFePO_4$ cathode, higher conductive carbon loading also needs more binder which generally reduces the energy density of the cells. Therefore, the influence of particle size, loading content, and mixing procedure of conductive carbonaceous materials on the performance of battery cells needs further attentions.

2.1.1.2 Polymers

Polymers have been proven to be an effective substance to improve the electrochemical properties of $LiFePO_4$ when forming composites with $LiFePO_4$. Huang and Goodenough

(2008) proposed a strategy of substituting the inactive carbon and Teflon (PTFE) binder with an electrochemically active polymer like polypyrrole (PPy) or polyaniline (PANI) to enhance the electrochemical performance of $LiFePO_4$. Significantly improved capacity and rate capability are achieved in such $LiFePO_4$/polymer composite cathodes, because polymers can increase the specific capacity and rate capability and lower the overpotential at high discharge rates. Recent work reported by Chen et al. (2011) also indicates that PANI can serve as a binder to make the carbon-coated $LiFePO_4$ (C-LFP) electrode surface smoother to endure a long cycling. In their work, they modified C-LFP with PANI: C-LFP was synthesized via a solid state reaction, whereas PANI was formed in situ by chemical oxidative polymerization of aniline with ammonium persulfate as an oxidizer to achieve the C-LFP/PANI composites. Specific capacities as high as 165 mAh/g at 0.2 C, 133 mAh/g at 7 C, and 123 mAh/g at 10 C are obtained in C-LFP/7 wt.% PANI composite.

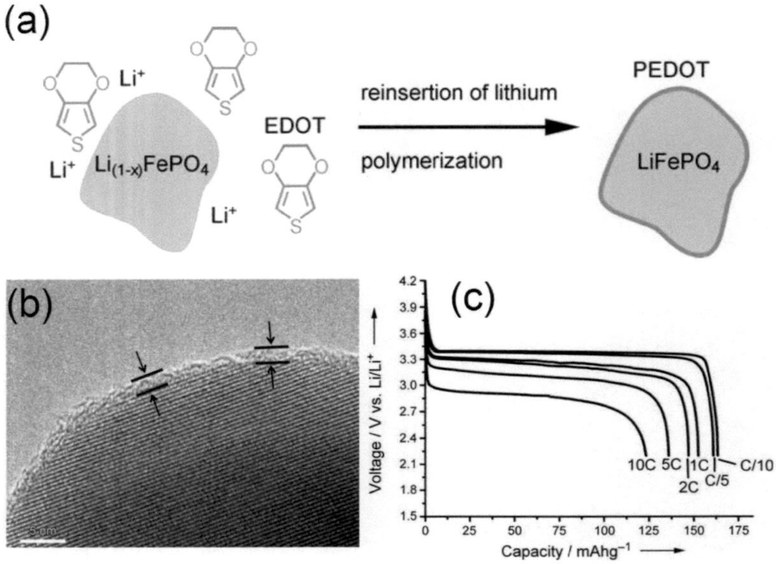

Fig. 7. a) Polymerization reaction. The reinsertion of lithium into $Li_{1-x}FePO_4$ leads to the oxidation of EDOT deposited on the solid surface as a conducting polymer PEDOT; b) TEM of PEDOT–$LiFePO_4$; and c) discharge curves of PEDOT/$LiFePO_4$/PVDF 5.9 : 86.6 : 7.5 in wt.%. (Charge conditions are 2.2–4.2 V, versus Li^+/Li). (Lepage et al., 2011)

Other means have also been used to synthesize polymer/$LiFePO_4$ composites, including electropolymerization from a suspension of $LiFePO_4$ particles (Boyano et al., 2010), polymerization using a chemical oxidant in the presence of the particle (Wang et al., 2005), or formation of a colloidal suspension of the polymer immediately before the introduction of $LiFePO_4$ particles (Murugan et al., 2008). However, coating $LiFePO_4$ with homogeneous polymer thin layer is still very difficult. This problem can possibly be solved by the soft chemistry approach reported by Lepage et al. (2011), which relies on the intrinsic oxidation power of $Li_{1-x}FePO_4$ rather than on an external oxidant as the driving force for the polymerization process, as shown in Figure 7a. The polymerization propagation requires the reinsertion of lithium into the partially delithiated $LiFePO_4$, as well as the transport of Li^+ ions and electrons through the deposited polymer coating like PEDOT. In the resultant

PEDOT/$LiFePO_4$, a very thin PEDOT layer is homogeneously covered on surface of $LiFePO_4$ nanoparticles (Figure 7b). PEDOT/$LiFePO_4$ shows a capacity of 163 mAh/g at C/10, which is similar to the theoretical capacity of 170 mAh/g for $LiFePO_4$. At higher rates of discharge, more specifically at 10 C (at constant current for a discharge in 6 min), the capacity is 123 mAh/g, approximately 70% of the theoretical capacity.

2.1.2 $LiMnPO_4$

Encouraged by the success of $LiFePO_4$, much research is now focused on the olivine $LiMPO_4$ (M=Mn, Co, and Ni) structures. Among them, $LiMnPO_4$ is of particular interest as it offers a higher potential of 4.1 V vs Li^+/Li compared to 3.4 V vs Li^+/Li of $LiFePO_4$, the expected safety features, and abundant resources. Hence, $LiMnPO_4$ can be an ideal cathode material. Unfortunately, besides a low ionic conductivity (D_{Li} <10^{-14} cm^2s^{-1}), $LiMnPO_4$ also shows an electronic conductivity of <10^{-10} Scm^{-1} much lower than that of $LiFePO_4$ (i.e., 1.8 × 10^{-9} Scm^{-1} at 25 °C), rendering it difficult to obtain decent electrochemical activity. Similar to the $LiFePO_4$ electrode, a leading approach to improve the performance of $LiMnPO_4$ cathode materials is to use nanosized particles and a carbon coating. For instances, Drezen et al. (2007) synthesized $LiMnPO_4$ (140~200 nm) using a sol-gel method followed by carbon coating through a dry ball milling. The $LiMnPO_4$/C nanocomposites could deliver reversible capacities of 156 and 134 mAh/g at C/100 and C/10, respectively. At faster charging rates, the electrochemical performance was further improved when smaller $LiMnPO_4$ particles were used. Martha et al. (2009) synthesized $LiMnPO_4$/C nanocomposites with a polyol method followed by ball-milling. The nanocomposites were composed of 25-30 nm platelet-like $LiMnPO_4$ particles covered by a carbon film about 15 nm thick. The $LiMnPO_4$/C cathode showed a good rate capability and delivered a practical capacity of 140 mAh/g and 120 mAh/g at 0.1 C and 0.5 C, respectively.

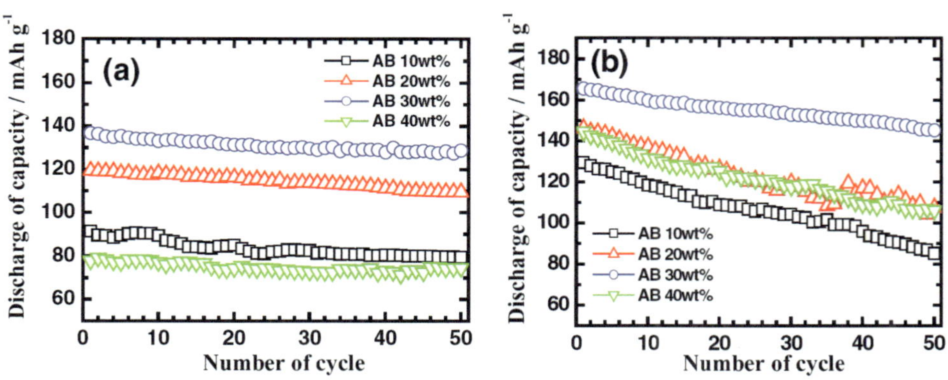

Fig. 8. Cycling stability of Li/C-$LiMnPO_4$ cells with different amounts of acetylene black at 0.5 C rate: a) cycled at 25 °C and b) cycled at 55 °C. The cell was charged at a constant current rate of C/20 to 4.5 V and kept at 4.5 V until C/100. (Oh et al., 2010)

Recently, Baknev et al. (2010) prepared a $LiMnPO_4$/C composite cathode by a combination of spray pyrolysis and a wet ball milling. The composite cathode delivered discharge capacities of 153 mAh/g at 0.05 C and 149 mAh/g at 0.1 C at room temperature and exhibited initial discharge capacities of 132 and 80 mAh/g at 1 and 5 C at 50 °C. These authors also demonstrated that particle size and homogeneity of the carbon distribution are critical in determining the reversible capacity and rate capability of $LiMnPO_4$. More recently, Oh et al. (2010) prepared nanostructured C-$LiMnPO_4$ powders by ultrasonic spray pyrolysis followed by a ball milling and found that the content of acetylene black (AB) carbon has a great influence on the electrochemical properties of the C-$LiMnPO_4$ nanocomposites. As shown in Figure 8, when AB carbon content is 30 wt%, the C-$LiMnPO_4$ composites exhibited a best electrochemical performance, delivering discharge capacities of 158 mAh/g and 107 mAh/g at rates of 1/20 C and 2 C, respectively. Additionally, the capacity retention of the 30 wt% AB electrode after 50 cycles was 94.2% at 25 ° C and 87.7% at 55 ° C, with its initial capacity at 0.5 C rate being 137 mAh/g and 166 mAh/g, respectively. While, an excessive amount of carbon causes the carbon particles to segregate and decreases the electrochemical properties of the C-$LiMnPO_4$.

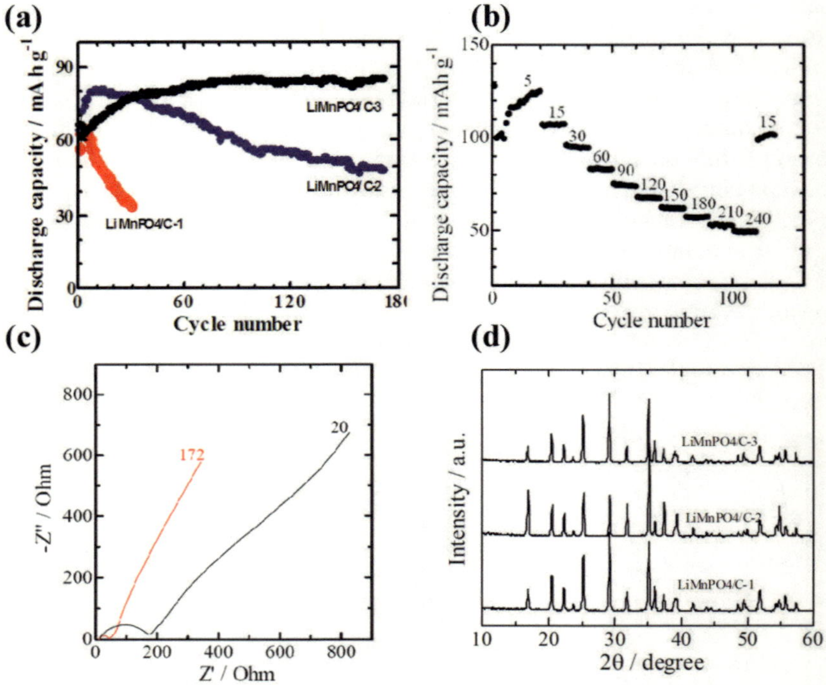

Fig. 9. a) Cycle performances of the samples; b) discharge capacity of $LiMnPO_4$/C-3 at different current densities; c) impedance results of $LiMnPO_4$/C-3 after 20 and 172 cycles; and d) XRD patterns of the samples.

In our recent experiments, we have found that the procedures of mixing and annealing treatments significantly affect the electrochemical performance of $LiMnPO_4$/C cathode. We prepared $LiMnPO_4$ nanoparticles by a hydrothermal method, the same as the method described by our previous work (Fang et al., 2007) except that LiOH was used as pH modifier instead of $NH_3 \cdot H_2O$. Then, the $LiMnPO_4$/C composites were synthesized by three ways: (1) hydrothermally synthesized $LiMnPO_4$ was milled with 20 wt% of carbon black for 4 h to obtain carbon-$LiMnPO_4$ composites, named as $LiMnPO_4$/C-1; (2) hydrothermally synthesized $LiMnPO_4$ was milled with 20 wt% of carbon black for 4h and followed by annealing at 700 °C for 2 h in N_2. The obtained composites were ball-milled for 4 h again to obtain carbon-$LiMnPO_4$ composites, named as $LiMnPO_4$/C-2; (3) hydrothermally synthesized $LiMnPO_4$ was milled with 20 wt% of carbon black for 4h and followed by annealing at 700 °C for 2 h in N_2 to obtain $LiMnPO_4$/C-3. As indicated by Figure 9a, among three samples, $LiMnPO_4$/C-3 exhibited the best electrochemical properties, which delivered a discharge capacity of 80 mAh/g after 180 cycles at a current density of 30 mA/g. $LiMnPO_4$/C-3 also showed an excellent capacity retention at different charge/discharge current density (Figure 9b). The excellent electrochemical performance of $LiMnPO_4$/C-3 may result from the decreased internal impedance during the charge/discharge cyclings, as shown in Figure 9c. In order to study the influence of different synthetic procedures on crystallinity of $LiMnPO_4$/C, the samples were characterized by X-ray diffraction (XRD). As shown in Figure 9d, $LiMnPO_4$/C-1 and $LiMnPO_4$/C-2 showed a broad noncrystalline peaks in the diffraction range of 15~40°, which is not observed in $LiMnPO_4$/C-3. It indicates that $LiMnPO_4$/C-3 shows a highest crystallinity. Therefore, annealing in N_2 could effectively reduce the surface defects as produced by ball-milling and improve the crystallinity, which may result in a higher discharge capacity and enhance the electrode stability through suppressing the internal impedance of the electrode from increasing. Similar phenomenon has also been observed in other cathodes (Dimesso et al., 2011). Therefore, the influence of particle size, loading content, mixing procedure of conductive carbons on the performance of battery cells needs further detailed investigations.

2.2 Two-dimensional cathode nanocomposites

Layered-type transition metal oxides ($LiMO_2$, M = V, Mn, Fe, Co, and Ni) is an important two-dimensional insertion compound for cathode materials. It adopts the α-$NaFeO_2$-type structure, which can be regarded as a distorted rock salt superstructure, as shown in Figure 10. In a cubic close-packed oxygen array, lithium and transition-metal atoms are distributed in the octahedral interstitial sites. MO_2 layers are formed consisting of edge-sharing [MO_6] octahedra. In between these MO_2 layers, lithium resides in octahedral [LiO_6] coordination, leading to the alternating planes (111) of the cubic rock-salt structure. (Winter et al, 1998).

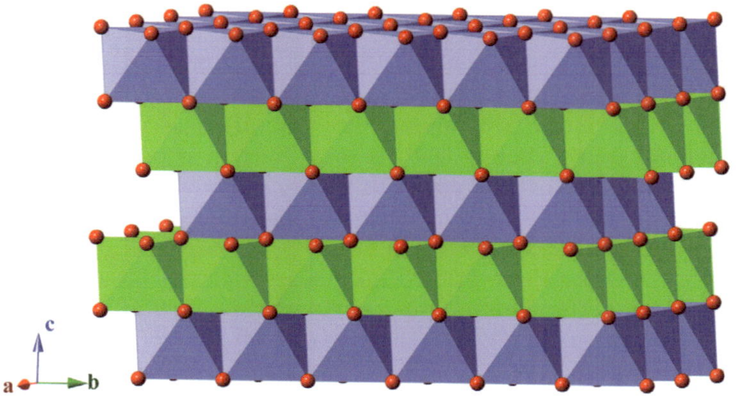

Fig. 10. Two-dimensional crystal structure of $LiMO_2$ (M = Ni, Co, V) of α-$NaFeO_2$ -type. MO_6 octahedra are shown in blue, LiO_6 octahedra in green, and oxygen ions in red.

Among the above-mentioned isostructural dioxides, particularly lithium cobalt oxide ($LiCoO_2$) and mixed manganese-nickel-cobalt dioxide ($Li(Mn, Ni, Co)O_2$) have gained an industrial importance as the electrode materials.

2.2.1 $LiCoO_2$

$LiCoO_2$ is the most widely used cathode materials in commercial LIBs owing to its favourable electrochemical attributes, including a good capacity retention, favourable rate capability, and high structural reversibility below 4.2 V vs. Li/Li^+. Layered Li_xCoO_2 materials are typically charged up to 4.2 V ($0.5 < x < 1$) because further increase in the charge cut-off voltage results in a phase transition from monoclinic to hexagonal. The phase transition leads to an anisotropic volume change of the host lattice, which causes a structural degradation and large capacity fade during the repeated cyclings. In addition, the dissolution of Co^{4+} when $x>0.5$ is also attributed to the degradation of $LiCoO_2$. Therefore, the specific capacity of this type of material is limited to the range of 137 to 140 mAh/g, although the theoretical capacity of $LiCoO_2$ is 273 mAh/g (Wang et al., 2008). In this regard, great efforts have been made to increase the specific capacity of $LiCoO_2$ and improve the capacity retention. The construction of heterogeneous nanostructure by coating the chemically stable compounds on the surface of $LiCoO_2$ is the most effective strategy to improve the electrochemical performance of $LiCoO_2$ at high cut-off voltage. There has been extensively research by coating $LiCoO_2$ material with various metal oxides, metal phosphates, and other chemically stable compounds.

2.2.1.1 Metal oxide

Various metal oxide coatings such as Al_2O_3 (Oh et al., 2010), CeO_2 (Ha et al., 2006), ZrO_2 (Liu et al., 2010), $Li_4Ti_5O_{12}$ (Yi et al, 2011), etc., have been studied. These metal oxides behave as an effective physical protection barrier to prohibit the chemical reactions between the electrolyte and $LiCoO_2$ cathode. They also prevent the structural degradation of $LiCoO_2$ during the cyclings. Therefore, both the cycle life and rate capability of $LiCoO_2$ cathode were improved by coating these oxides. Interestingly, some of them such as Al_2O_3 and ZnO can act as HF scavenger to reduce the acidity of non-aqueous electrolyte, suppressing metal dissolution from the $LiCoO_2$ materials. It is well known that $LiPF_6$, the dominant lithium salt for LIBs, is sensitive to trace of moisture. HF is formed when $LiPF_6$ decomposes in the presence of moisture, which can result in the dissolution of the transition metal and surface corrosion of cathode materials (Zhou et al. 2005). For example, Al_2O_3 and ZnO can react with a trace amount of HF in the electrolyte to reduce the active concentration of HF by the formation of AlF_3 and ZnF_2, suppressing Co dissolution and F- concentration in the electrolyte. In general, excess coating leads to a loss of rate or power capability. Therefore, control over the thickness of coating materials is very important. Recently, Scott et al (2011) modified $LiCoO_2$ with coating ultrathin Al_2O_3 film as a surface protective layer by atomic layer deposition (ALD) technique. The conformal ~1-2 nm thick film which has clearly different lattice fringes from those in the bulk region is clearly observed for Al_2O_3 coated $LiCoO_2$ (see Figure 11a). The coated nano-$LiCoO_2$ electrodes with 2 ALD cycles deliver a discharge capacity of 133 mAh/g with a currents of 1400 mA/g (7.8 C), corresponding to a 250% improvement in reversible capacity compared to the uncoated $LiCoO_2$, as shown in Figure 11b. The simple ALD process is broadly applicable and provides new opportunities for the battery industry to design other novel nanostructured electrodes that are highly durable even while cycling at high rate.

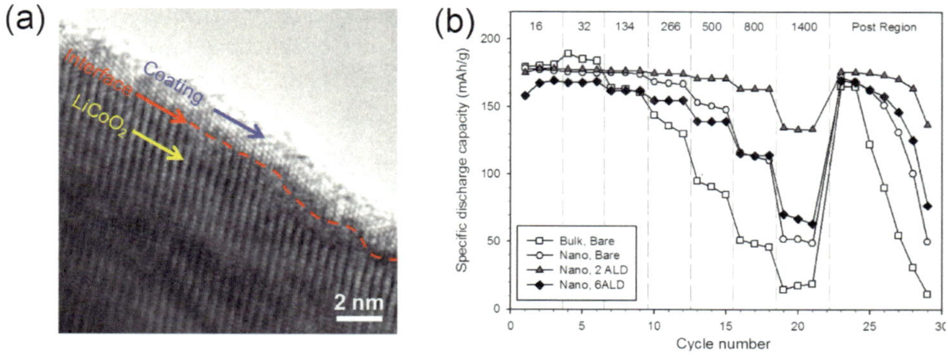

Fig. 11. a) HR-TEM images of the Al_2O_3-coated $LiCoO_2$ nanoparticles by 6 ALD cycles on the bare powders; and b) variations in discharge capacities versus charge/discharge cycle number for different $LiCoO_2$ electrodes cycled at different rates between 3.3 and 4.5 V (vs Li/Li^+) at room temperature. Current densities (mA/g) are indicated at the top. (Scott et al.,2011)

2.2.1.2 Metal phosphates

$AlPO_4$ and $LiFePO_4$ have been reported as an effective coating material for improving the electrochemical properties of $LiCoO_2$. $AlPO_4$ can be coated on $LiCoO_2$ by direct dispersing $LiCoO_2$ in $AlPO_4$ precursors and followed by annealing at elevated temperatures. $AlPO_4$ coating can behave as a protective layer to prohibit the chemical reactions between the electrolyte and $LiCoO_2$, suppressing Co dissolution and the lattice distortion during cycling with a charge cut-off voltage of 4.6V. The crystallinity of the $AlPO_4$-coating layer and the inter-diffusion at the interface can affect the electrochemical properties of the $AlPO_4$-coated $LiCoO_2$ (Kim et al. 2006). The same group also reported that different phases of $AlPO_4$ have great influences on the electrochemical properties of $LiCoO_2$. They coated $LiCoO_2$ with three types of $AlPO_4$ phase, i.e., amorphous, tridymite, and cristobalite phases, by spin coating method. The $LiCoO_2$ thin film coated with amorphous nanoparticles showed the best cycle-life performance by effectively suppressing the degradation of Li^+-diffusion kinetics (Kim et al., 2007). Wang et al. (2007) modified $LiCoO_2$ using $LiFePO_4$ coating with a thickness ranging 10–100 nm by impregnation method. The coating of $LiFePO_4$ serves as both the protecting layer and the active cathode material with good conductivity. The $LiFePO_4$-coated $LiCoO_2$ exhibited an improved electrochemical performance during the charge/discharge cycles, especially at high potentials and high temperature. For example, at 60 °C and at a rate of 1 C, 5.0 wt.% $LiFePO_4$-coated $LiCoO_2$ cathodes showed a better capacity retention (132 mAh/g) with a 4.2 V charge-cutoff after 250 cycles, while the uncoated $LiCoO_2$ cathode showed 5% capacity retention under the same conditions after only 150 cycles.

2.2.1.3 Other chemically stable compounds

Highly ionic conductive solid electrolytes have been used as the protective layer of $LiCoO_2$. Recently, Choi et al. (2010) modified the surfaces of $LiCoO_2$ with a lithium phosphorus oxynitride (LiPON) glass-electrolyte thin film. Homogeneous LiPON thin films with 0.5~0.7 μm in thickness were deposited on the surfaces of $LiCoO_2$ by a radio frequency (RF) magnetron sputtering method. They found that the LiPON coating improved the rate capability and the thermal stability of the charged $LiCoO_2$ cathode. The ion conducting

LiPON film enhances the lithium-ion migration through the interface between the surface of $LiCoO_2$ particles and the electrolyte, suppressing the surface reaction between the electrode surface and electrolyte during cyclings.

2.2.2 $Li(Mn,Ni,Co)O_2$

Metals Mn, Ni, and Co can all be accommodated in the layered metal oxide structure, giving a range of compositions $Li[Mn_xNi_yCo_z]O_2$ ($x + y + z = 1$). A solution of $Li[Mn_xNi_yCo_z]O_2$ may possess improved performances, such as thermal stability due to the synergetic effect of three ions. Recently, intense efforts have been directed towards the development of $Li[Mn_xNi_yCo_z]O_2$ as a possible replacement for $LiCoO_2$. However, layered $Li[Mn_xNi_yCo_z]O_2$ structures tend to become unstable at high levels of delithiation, which is attributed largely to the highly oxidizing nature of tetravalent Co and Ni that results in the loss of oxygen and the migration of the transition metal ions to the lithium-depleted layer. In spite of numerous valuable works, the cycle life and thermal stability of $Li[Mn_xNi_yCo_z]O_2$ still need further improvement. Several strategies have been used to improve the electrochemical performances of $Li[Mn_xNi_yCo_z]O_2$ cathode, which can be summarized as follows:

2.2.2.1 Surface modifications

Coating of the surfaces of $Li[Mn_xNi_yCo_z]O_2$ with Al_2O_3, $LiAlO_2$, $AlPO_4$, and ZrO_2 has been widely attempted, leading to significantly improved electrochemical properties. For example, Hu et al. (2009) coated the surface of $LiNi_{1/3}Co_{1/3}Mn_{1/3}O_2$ with a uniform nano-sized layer of ZrO_2. ZrO_2-coated $LiNi_{1/3}Co_{1/3}Mn_{1/3}O_2$ exhibits an improved rate capability and cycling stability under a high cut-off voltage of 4.5 V, owing to the ability of ZrO_2 layer in preventing the direct contact of the active material with the electrolyte. In this case, ZrO_2 play a similar role as it does in ZrO_2 coated $LiCoO_2$ (Chen et al., 2002).

2.2.2.2 Mixing with other layered components

Thackeray et al. (2007) proposed a strategy to stabilize a layered $Li[Mn_xNi_yCo_z]O_2$ electrode by integrating a structurally compatible component that was electrochemically inactive between 4 and 3 V, aiming to increase the stability of layered electrodes over a wider compositional range without compromising the power or cycle life of the cells. Li_2MnO_3 is the best stabilized component for $Li[Mn_xNi_yCo_z]O_2$ electrode because it is electrochemically inactive over a potential window of 2.0~4.4 V in which $Li[Mn_xNi_yCo_z]O_2$ component is electrochemically active and operates as a true insertion electrode. In Li_2MnO_3, all manganese ions are tetravalent and cannot be oxidized further. Lithium insertion into Li_2MnO_3, with a concomitant reduction of the manganese ions, is also prohibited because there are no energetically favourable interstitial sites for the guest ions. Under such conditions, Li_2MnO_3 component acts as a stabilizing unit in the electrode structure. (Thackeray et al., 2007) When charged to higher potentials, typically 4.5 V, it is possible to activate the Li_2MnO_3 component during the initial charge by removing Li^+ with a concomitant loss of oxygen. The removal of two Li^+ from the Li_2MnO_3 component and the reinsertion of only one Li^+ into the resulting MnO_2 component, exceptionally high reversible capacities (230–250 mAh/g) can be obtained. For example, Li et al. (2011) synthesized a high-voltage layered $Li[Li_{0.2}Mn_{0.56}Ni_{0.16}Co_{0.08}]O_2$ cathode material, a solid solution between Li_2MnO_3 and $LiMn_{0.4}Ni_{0.4}Co_{0.2}O_2$, by co-precipitation method followed by a high-temperature annealing. After the initial decay, no obvious capacity fading was observed when cycling the material at different rates. Steady-state reversible capacities of 220 mAh/g at 0.2 C, 190 mAh/g at 1 C, 155 mAh/g at 5 C, and 110 mAh/g at 20 C were achieved in long-term cycle tests within the voltage cutoff limits of 2.5 and 4.8 V at 20 C.

2.2.2.3 Heterogeneous core-shell structure

Recently, Sun and co-workers proposed a novel core/shell concept that construct layered structured $LiNi_xCo_yMn_zO_2$ cathode with a core enriched in Ni for high capacity and a shell enriched in Mn for high stability and cycling performance. They reported that core–shell cathode material $Li[(Ni_{0.8}Co_{0.1}Mn_{0.1})_{0.8}(Ni_{0.5}Mn_{0.5})_{0.2}]O_2$ exhibited an excellent cyclability and thermal stability (Sun et al., 2005). As shown in Figure 12, the core-shell structure consists of a Ni-rich $LiNi_{0.8}Co_{0.1}Mn_{0.1}O_2$ core that delivered a high capacity and a Mn-rich $Li[Ni_{0.5}Mn_{0.5}]O_2$ shell that provides a structural and thermal stability in highly delithiated states. After that, Sun et al. (2009) found a structural mismatch between the core and the shell: Voids of tens of nanometres between the core and shell appear in the core-shell powders after cycling. A shell with a gradient in chemical composition was proposed to grow onto the surfaces of a core material, forming a constant chemical composition. The concentration-gradient cathode material ($Li(Ni_{0.64}Co_{0.18}Mn_{0.18})O_2$) thus achieved shows a high capacity of 209 mAh/g, which retains 96% after 50 charge/discharge cycles under an aggressive test profile (55 °C between 3.0 and 4.4 V), as shown in Figure 13. In a latest work by Koenig and Sun (Koenig et al., 2011), particles with internal gradients in transition metal composition was synthesized using a co-precipitation reaction with a control over the process conditions. By this, compositions and structures can be rationally integrated into the synthesis process, and further gradient core-shell structure $LiNi_xCo_yMn_zO_2$ materials will become a future choice for cathode in LIBs.

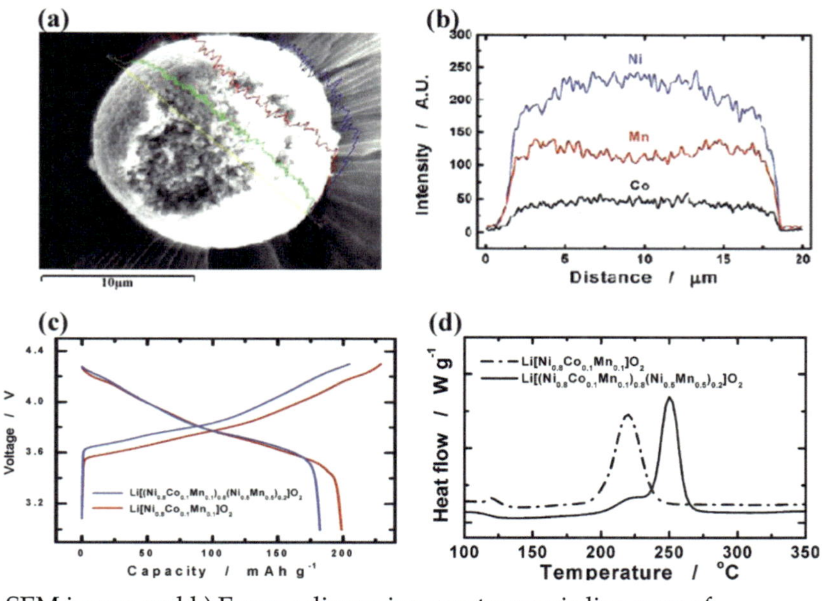

Fig. 12. a) SEM image and b) Energy dispersive spectroscopic line scan of $Li[(Ni_{0.8}Co_{0.1}Mn_{0.1})_{0.8}(Ni_{0.5}Mn_{0.5})_{0.2}]O_2$ particles; c) initial charge/discharge curves of $Li/Li[Ni_{0.8}Co_{0.1}Mn_{0.1}]O_2$ and $Li/Li[(Ni_{0.8}Co_{0.1}Mn_{0.1})_{0.8}(Ni_{0.5}Mn_{0.5})_{0.2}]O_2$ cell in a voltage range of 3.0-4.3 V; and d) differential scanning calorimetry traces of $Li[(Ni_{0.8}Co_{0.1}Mn_{0.1})_{0.8}(Ni_{0.5}Mn_{0.5})_{0.2}]O_2$ and $Li[Ni_{0.8}Co_{0.1}Mn_{0.1}]O_2$ at a charged state to 4.3 V. (Sun et al., 2005)

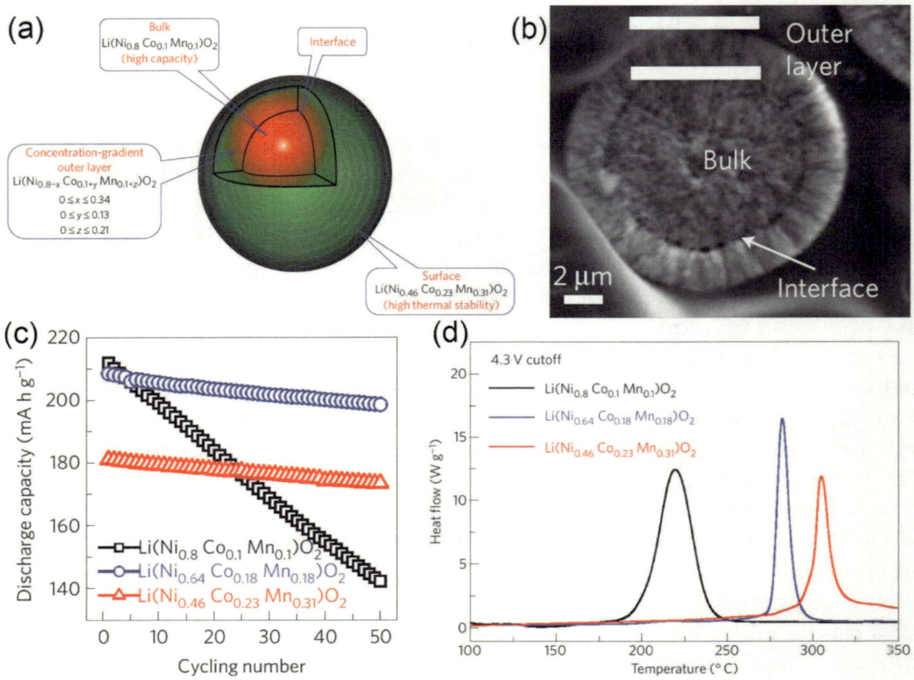

Fig. 13. a) Schematic diagram of cathode particle with Ni-rich core surrounded by concentration-gradient outer layer; b) a scanning electron micrograph of a typical particle; c) cycling performance of half cells based on $Li(Ni_{0.8}Co_{0.1}Mn_{0.1})O_2$, $Li(Ni_{0.46}Co_{0.23}Mn_{0.31})O_2$ and concentration-gradient material cycled between 3.0 and 4.4 V at 55 °C by applying a constant current rate of 0.5 C (95 mA/g); and d) differential scanning calorimetry traces showing a heat flow from the reaction between the electrolyte and different electrode materials when charged to 4.3 V. (Sun et al., 2009)

2.3 Three-dimensional cathode nanocomposites

Three-dimensional framework structures have many crosslinked channels which must be sufficiently large to accommodate the lithium ion. The advantages of three-dimensional frameworks are represented by i) a possibility of avoiding, for steric reasons, the co-insertion of bulky species such as solvent molecules; and ii) the smaller degree of expansion/contraction of the framework structure upon lithium insertion/extraction (Winter et al., 1998).

2.3.1 $LiMn_2O_4$

$LiMn_2O_4$ is regarded as one of the most promising cathodes because of its high output voltage, high natural abundance, low toxicity, and low cost. $LiMn_2O_4$ crystallizes in a spinel structure with the space group of Fd3m. In the $LiMn_2O_4$ spinel structure, a cubic close-packed (ccp) array of oxygen ions occupy the 32e position, Mn ions are located in the 16d site, and Li in the 8a site. Mn ions have an octahedral coordination to oxygen, and the MnO_6 octahedra share edges in a three-dimensional host for Li guest ions. The 8a tetrahedral site is situated furthest away from the 16d site of all the interstitial tetrahedral (8a, 8b and 48f) and octahedra (16c). Each of the 8a-tetrahedron faces is shared with an adjacent, vacant 16c site. This combination of structural features in the stoichiometric spinel compound constitutes a very stable structure, as shown in Figure 14. $LiMn_2O_4$ has a three-dimensional lithium diffusion path, in which every plane is suitable to exchange lithium ion from active materials to the electrolyte.

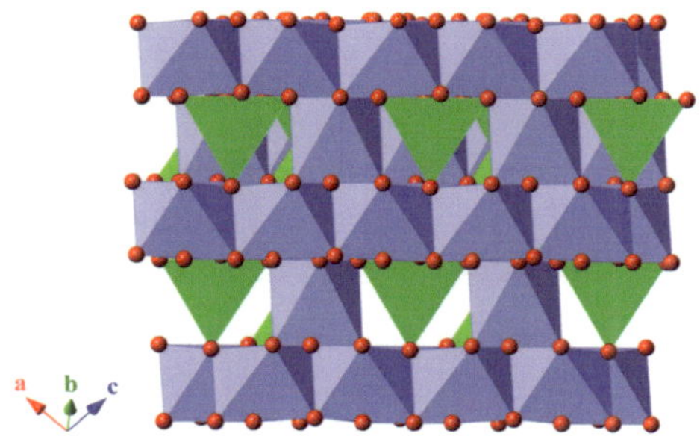

Fig. 14. Three-dimensional crystal structure of $LiMn_2O_4$. MnO_6 octahedra are shown in blue, LiO_4 tetrahedra in green, and oxygen ions in red.

$LiMn_2O_4$ has a theoretical capacity of 148 mAh/g and a voltage plateau of about 4 V. A typical cyclic voltammogram of $LiMn_2O_4$ is shown in Figure 15. Two couples of reversible redox peaks are observed during the charge/discharge processes. The split of the redox peaks into two couples shows that the electrochemical reaction of the extraction and insertion of lithium ions occurs in two stages. The first oxidation peak is ascribed to the removal of Li^+ from half of the tetrahedral sites in which Li–Li interactions exist, whereas the second oxidation peak is attributed to the removal of Li^+ from the remaining tetrahedral sites where no Li–Li interactions exist. The whole process can be described as follows:

$$0.5Li^+ + 0.5e + 2\lambda MnO_2 \Leftrightarrow Li_{0.5}Mn_2O_4 \tag{1}$$

$$0.5Li^+ + 0.5e + Li_{0.5}Mn_2O_4 \Leftrightarrow LiMn_2O_4 \tag{2}$$

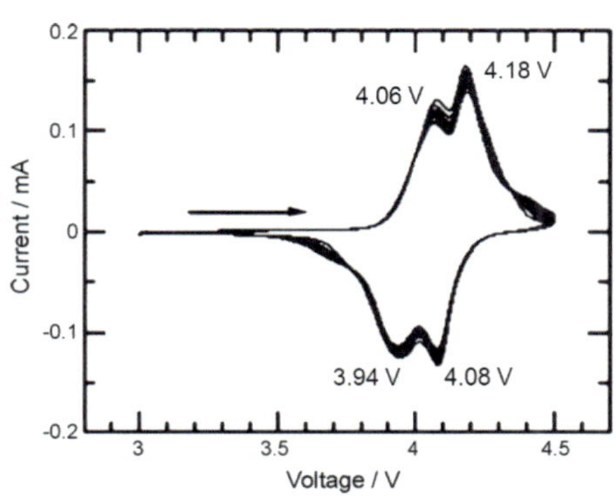

Fig. 15. Cyclic voltammograms of $LiMn_2O_4$ within the voltage range from 3.0 to 4.5 V versus Li/Li^+. (Fang et al., 2008)

Since the application of $LiMn_2O_4$ in high power systems requires the development of fast kinetic electrodes which appears nowadays possible thanks to the use of nanostructured morphologies, various synthesis methods have been widely investigated to prepare nanostructured $LiMn_2O_4$ and these nanostructured methods have been demonstrated to be highly effective in greatly enhancing the electrochemical performance of $LiMn_2O_4$ electrode materials. It is found that the electrochemical performance of $LiMn_2O_4$ electrodes strongly depends on the morphology and the crystalline phase. For example, in our previous work (Fang et al., 2008), we verified that the electrochemical performance of $LiMn_2O_4$ could be improved by altering the starting materials, as shown in Figure 16. In particular, α-MnO_2 nanorods are proved to be a quite promising starting material for the preparation of highly crystallized and high performance $LiMn_2O_4$. After that, Ding et al. (2011) and Lee et al. (2010) prepared single-crystalline $LiMn_2O_4$ nanotubes and nanowires from β-MnO_2 and α-MnO_2 precursors, respectively, which also leads to an improved electrochemical performance.

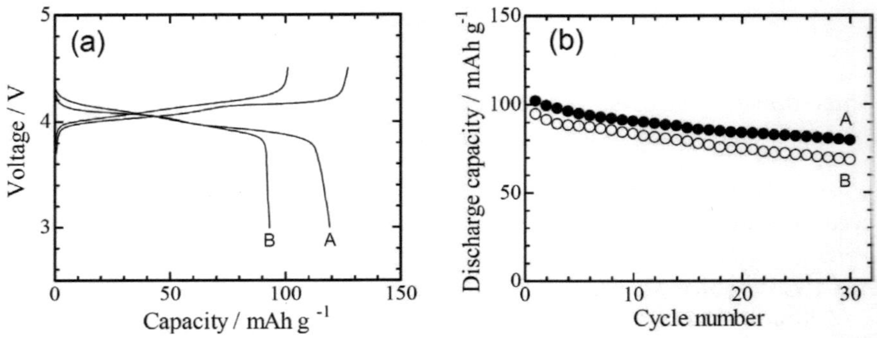

Fig. 16. a) Charge/discharge curves and b) cycling performance of $LiMn_2O_4$ prepared from α-MnO_2 nanorods (A) and commercial MnO_2 (B) in the voltage range from 3.0 to 4.5 V versus Li/Li^+ at a current density of 20 mA/g. (Fang et al., 2008)

Unfortunately, $LiMn_2O_4$ shows a fast capacity fading during the cyclings, which cannot be ultimately solved by nanostructured approaches only. The capacity fading of $LiMn_2O_4$ has been ascribed to the dissolution of Mn, Jahn–Teller distortion of Mn^{3+}, and decomposition of electrolyte solution on electrode. Of the above factors, dissolution of surface Mn^{3+} with the electrolyte is now regarded as the most important reason for the capacity fading. Surface modification of the cathode electrode has been a successful strategy to solve the capacity fading problem, which could decrease the surface area to retard the side reactions between the electrode and electrolyte and to further diminish the Mn dissolution during cycling. In recent years, various types of coating materials, including oxides, metals, fluorides, phosphate, polymers, carbon materials, and other electrode materials have been explored to improve the electrochemical stability of $LiMn_2O_4$ cathodes. For example, Sahan et al. (2008) reported that Li_2O–$2B_2O_3$(LBO)-coated $LiMn_2O_4$ electrode prepared by a solution method shows an excellent cycling behaviour without any capacity loss even after 30 cycles at room temperature and a 1-C rate, as plotted in Figure 17. The improved cycling stability is also achieved by coating with [Li, La]TiO_3 (Jung et al., 2011) and $Li_4Ti_5O_{12}$ (Liu et al., 2007). Recent developments in the surface modification have been reviewed by Yi et al. (2009). Despite the numerous studies in this area, a functional understanding of coated $LiMn_2O_4$ electrodes remains incomplete. In the case of metal oxides, it appears likely that some of the metal oxides such as MgO, Al_2O_3, ZrO_2

may act as a physical protection barrier to depress Mn^{2+} dissolution by forming epitaxial layers on the underlying coated spinel, improving the structural stability of the cathode. It has also been suggested that ZnO coatings may act as HF scavenger to be capable of neutralizing the acid species. However, it is not yet clear which of these various processes is most important to the stabilization of a $LiMn_2O_4$ electrode surface. Recently, Ouyang et al., (2010) reported the results of ab initio studies of the atomic and electronic structure of clean and Al_2O_3 covered $Li_xMn_2O_4$ (001) surfaces (x=0 or 1). They gave a clear picture that only Mn^{3+} ions occur at the clean $LiMn_2O_4$ (001) surfaces due to a lower coordination with O atoms, while Mn atoms in bulk $LiMn_2O_4$ exhibit a mixed Mn^{3+}/Mn^{4+} oxidation states. An Al_2O_3 ad-layer inhibits the formation of Mn^{3+} at the $Li_xMn_2O_4$ (001) surfaces and thus has a more complex role than just separating the cathode from the electrolyte as is commonly assumed. Key variables that influence the coating efficacy remain to be identified and optimized for most of the coating materials that have been employed to date.

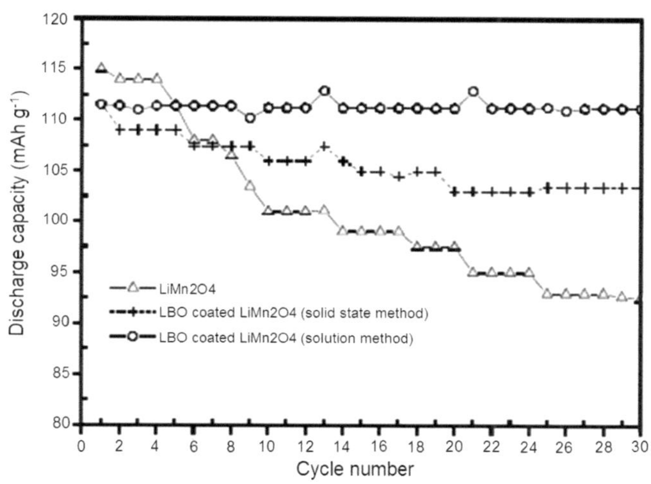

Fig. 17. Cycling performance of different materials at 1C discharge rate at room temperature. (Sahan et al., 2008)

2.3.2 $LiNi_{0.5}Mn_{1.5}O_4$

Ni-doped spinel oxide, $LiNi_{0.5}Mn_{1.5}O_4$, is another promising three-dimensional cathode material for future high power battery applications due to its large reversible capacity (147 mAh/g) at a high operating voltage around 4.7 V. $LiNi_{0.5}Mn_{1.5}O_4$ spinel is fundamentally different from pure spinels as all redox activity takes place on Ni and Mn remains in +4 state.

$LiNi_{0.5}Mn_{1.5}O_4$ has two different crystal structures of the space groups of *Fd-3m* or $P4_332$ depending on Ni ordering in the lattice. As shown in Figure 18, *Fd-3m* type $LiNi_{0.5}Mn_{1.5}O_4$ is a disordering distribution, the Ni and Mn, Li and O atoms are occupied in the 16d octahedral sites, 8a tetrahedral sites and 32e sites, respectively. In this case, Ni and Mn atoms are randomly distributed in the 16d sites. The $P4_332$ type $LiNi_{0.5}Mn_{1.5}O_4$ is has an ordered distribution, in which Ni, Mn, and Li atoms occupythe 4a, 12d, and 4c sites, respectively. O ions occupy the 8c and 24e sites. In this case, Ni and Mn atoms are ordered regularly (Santhanam et al., 2010). It has been reported that $LiNi_{0.5}Mn_{1.5}O_4$ with *Fd-3m* space group has a better electrochemical performance than that of spinel with $P4_332$ space group due to a 2.5 orders of magnitude faster electronic conductivity.

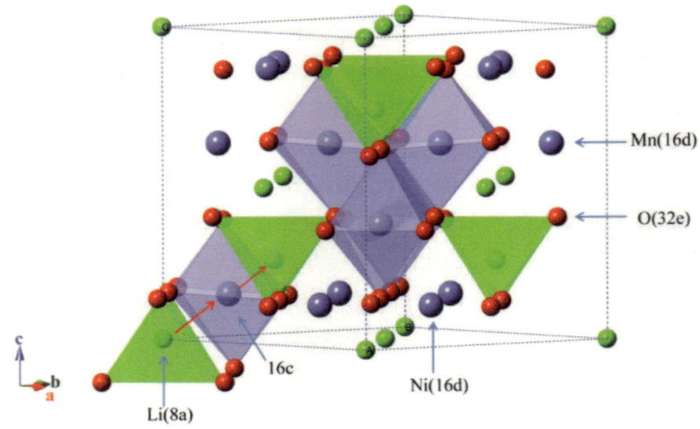

Fig. 18. Spinel structure of $LiNi_{0.5}Mn_{1.5}O_4$ (*Fd-3m*) showing the diffusion path of Li^+. (Redrawn from Ref. (Xia et al., 2007))

A typical cyclic voltammogram of *Fd-3m* type $LiNi_{0.5}Mn_{1.5}O_4$ is shown in Figure 19. Three well-defined reversible peaks are observed during the charge/discharge processes. The appearance of 4 V peak is due to Mn^{3+} which is formed by oxygen loss during the high-temperature calcinations. Two major peaks appearing between 4.5 and 5 V during charge/discharge cyclings are due to the redox couples Ni^{2+}/Ni^{3+} and Ni^{3+}/Ni^{4+} or ordering of lithium and vacancies at x = 0.5. For the ordered spinel, the 4V peaks are absent because oxidation states of Ni and Mn were +2 and +4, respectively (Santhanam, et al., 2010).

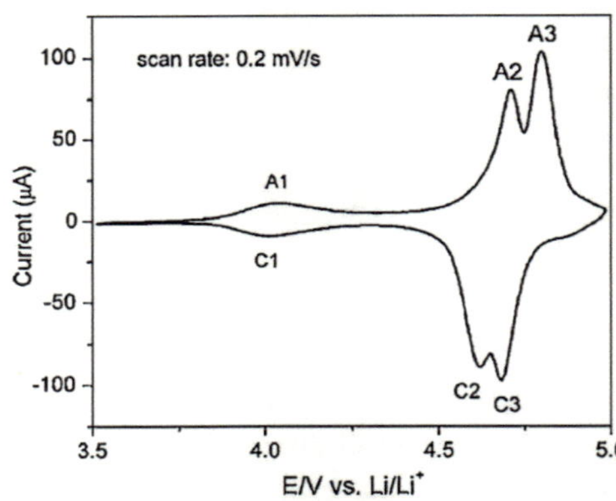

Fig. 19. Cyclic voltammogram of $LiNi_{0.5}Mn_{1.5}O_4$ electrode cycled between 3.5 and 5 V vs. Li/Li^+ at 0.2 mVs^{-1} scan rate (Xia et al., 2007)

There are two main problems with this $LiNi_{0.5}Mn_{1.5}O_4$ spinel material: One is that it is difficult to prepare pure spinel $LiNi_{0.5}Mn_{1.5}O_4$ due to the formation of $Li_xNi_{1-x}O$ as a second phase at high calcination temperatures. Developing low-temperature synthesis method of nanostructure $LiNi_{0.5}Mn_{1.5}O_4$ is an effective strategy to solve this problem. In our previous work, we discovered a low-temperature solid-state route for single-phase $LiNi_{0.5}Mn_{1.5}O_4$ (Fang et al., 2007). As shown in Figure 20, $LiNi_{0.5}Mn_{1.5}O_4$ shows a dimension of about 200 nm and has a cubic spinel structure, which can be indexed in a space group of *Fd-3m*. The discharge capacity of $LiNi_{0.5}Mn_{1.5}O_4$ reaches up to 134 mAh/g at a current density of 20 mA/g, while it retains as high as 110 mAh/g even at a high current density of 900 mA/g.

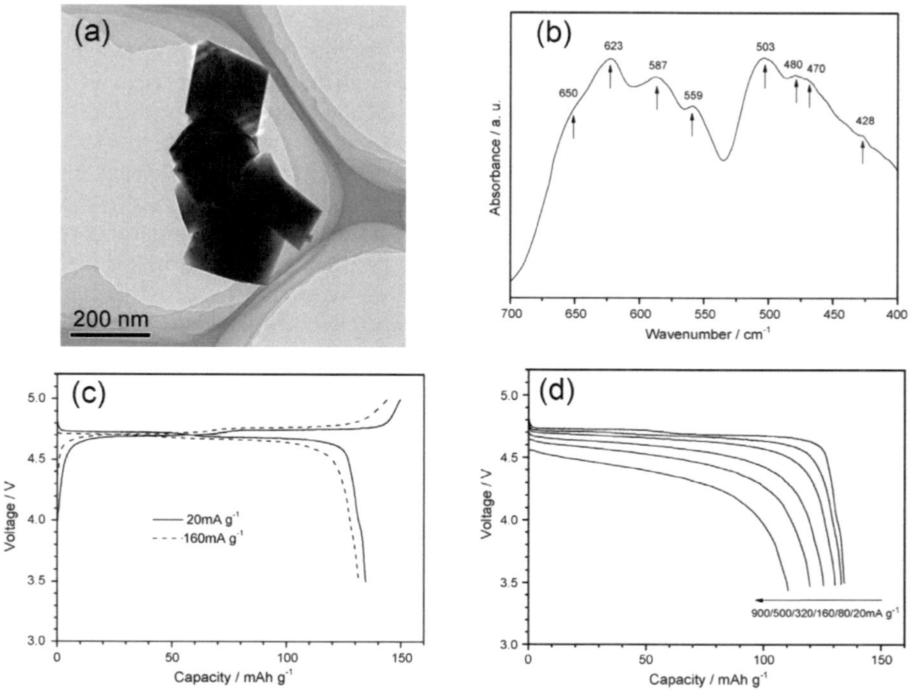

Fig. 20. a) Transmission electron micrograph and b) IR spectrum of $LiNi_{0.5}Mn_{1.5}O_4$; c) Charge/discharge curves of $LiNi_{0.5}Mn_{1.5}O_4$ cycled at current densities of 20 mA/g (solid line) and 160 mA/g (dash line); and d) Discharge curves of the as-prepared $LiNi_{0.5}Mn_{1.5}O_4$ at various current densities. (Fang et al., 2007)

The another one problem is that $LiNi_{0.5}Mn_{1.5}O_4$ still undergoes decomposition of electrolyte and dissolution of Mn and Ni, particularly under high voltage and elevated temperature conditions. The construction of heterogeneous nanostructure by coating some chemically stable compounds on $LiNi_{0.5}Mn_{1.5}O_4$ is a most widely used strategy to enhance the electrochemical properties of $LiNi_{0.5}Mn_{1.5}O_4$. This strategy has the advantages of the delivery of the original capacity due to no reduction in the amount of electrochemically active elements in the parent oxide and the suppression of the corrosion reactions between the high voltage spinel cathode and electrolyte. In this regard, various materials such as metal oxides, metal, fluoride, and carbon have been used to modify the surfaces of $LiNi_{0.5}Mn_{1.5}O_4$.

2.3.2.1 Coating by metal oxides

Surface modifications of $LiNi_{0.5}Mn_{1.5}O_4$ spinel material by metal oxides such as ZnO (Arrebola et al., 2010), ZrO_2 (Wu et al., 2010), Bi_2O_3 (Liu et al., 2009), and SiO_2 (Fan et al., 2007) have been investigated. Metal oxides coating act as a protective layer against the attack by HF that generated from the decomposition of $LiPF_6$ salt in the electrolyte, which effectively suppresses the corrosion reaction. For example, Sun et al. (2002) reported that the surface modification of $LiNi_{0.5}Mn_{1.5}O_4$ by ZnO significantly improves the capability at elevated temperature. ZnO-coated electrode delivers a capacity of 137 mAh/g without any capacity loss even after 50 cycles at 55 °C. ZnO coating layer scavenged fluorine anions by transforming HF to ZnF_2, which keeps the original particle morphology and reduces an increase in interfacial impedance between cathode and electrolyte. Arrebola et al (2010) have also confirmed the role of scavenging HF of ZnO coating layer, in which ZnO coating is beneficial for enhancing the rate capability of $LiNi_{0.5}Mn_{1.5}O_4$. In the case of SiO_2 coated $LiNi_{0.5}Mn_{1.5}O_4$ reported by Fan et al. (2007), $LiNi_{0.5}Mn_{1.5}O_4$ spinel is modified by a porous, nanostructured, amorphous SiO_2 surface layer, which is a good HF consumer that can neutralize HF and yields Si-F species. Nanocomposites displayed an obviously improved capacity retention rate at 55 °C, which is beneficial from the lower HF content in electrolyte resulting from the reaction between SiO_2 and HF and the mechanical separation effect of the SiO_2 coating result in relatively lower content of LiF in the surfaces of the coated $LiNi_{0.5}Mn_{1.5}O_4$ cathode materials. Very recently, Liu et al. (2009) modified the surface of the 5 V cathode material by Bi_2O_3 coating, which rendered both excellent rate capability and improved cycling performance. Further investigation indicates that Bi_2O_3 is reduced on the cathode surfaces during the electrochemical cycling to metallic Bi, an electronic conductor. Thus, "Bi_2O_3" modified layer acts as a protection shell and a fast electron transfer channel as well, which renders an excellent rate capability and good cycling performance.

2.3.2.2 Coating by metals

Metals such as Zn (Alcantara et al., 2004), Au (Arrebola et al., 2007) and Ag (Arrebola et al., 2005) have also been coated onto the surfaces of spinel $LiNi_{0.5}Mn_{1.5}O_4$ which significantly improves the electrochemical performance. For instance, Arrebola et al. (2007) found that Au coating on $LiNi_{0.5}Mn_{1.5}O_4$ has a beneficial effect in increasing the capacity by hindering the unwanted reactions and simultaneously preventing the active material particles from reacting with the decomposition products such as HF. These authors studied the adverse effect of Ag treatment on the electrochemical performance of $LiNi_{0.5}Mn_{1.5}O_4$. It is found that Ag treatment has only limited benefit at low current densities, while at moderate and high current densities, the capacity delivered is relatively low, even lower than that of the untreated spinel.

2.3.2.3 Coating by fluoride

Coating of spinel $LiNi_{0.5}Mn_{1.5}O_4$ with BiOF can also improve the electrochemical performances (Kang et al., 2010). For instance, BiOF-coated $LiNi_{0.5}Mn_{1.5}O_4$ shows a significantly improved capacity retention of 84.5% for 70 cycles at 55 °C, while the uncoated one exhibits a capacity retention of only 31.3%. The rate capability of the BiOF-coated $LiNi_{0.5}Mn_{1.5}O_4$ is also significantly enhanced. Such improvements in electrochemical performance are attributed to the scavenging HF by BiOF layer from the electrolyte.

2.3.2.4 Coating by carbon

Yang et al. (2011) coated the spinel $LiNi_{0.5}Mn_{1.5}O_4$ with a small amount of conductive carbon by the carbonization of sucrose. It is found that carbon coating greatly enhances the discharge capacity, rate capability, and cycling stability of the $LiNi_{0.5}Mn_{1.5}O_4$ without degrading the spinel structure. $LiNi_{0.5}Mn_{1.5}O_4$ modified with optimal 1 wt% sucrose can deliver a large capacity of 130 mAh/g at 1 C discharge rate with a high retention of 92% after 100 cycles and a high 114 mAh/g at 5 C discharge rate. The improved electrochemical performances upon carbon-coating are the consequence of the suppression of SEI layer as well as the enhancements in the electronic conductivity and Li^+ diffusion.

2.4 Vanadium pentoxide (V_2O_5)

V_2O_5 is a typical intercalation compound with a layered structure, and therefore is a very promising electrode material for LIBs as it offers the essential advantages of low-cost, abundant sources, high energy density, and better safety. As seen from Figure 21, V_2O_5 has an orthorhombic crystal structure which can be described as layers of VO_5 square pyramids that share edges and corners. The sixth V-O bond in the *c*-direction consists of the weak electrostatic interactions, which facilitates the insertion of various ions and molecules between the layers. Li-ion insertion and electronic transport occur more easily along the *a-b* plane rather than through the layers of the *c*-axis (Chan et al., 2007). Lithium intercalates along the *a-b* plane into the interlayer space follows the equation:

$$V_2O_5 + xLi^+ + xe^- \rightarrow Li_2V_2O_5 \quad (3)$$

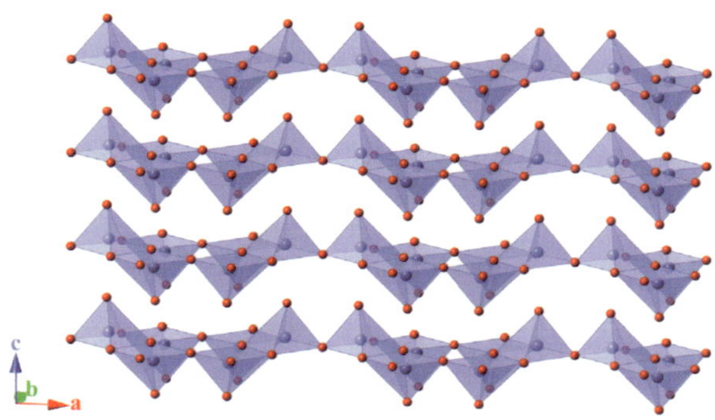

Fig. 21. Layered crystal structure of bulk V_2O_5 with boxed region indicating the unit cell. O atoms are in red and VO_5 units in blue.

The theoretical capacity of V_2O_5 with two lithium insertions/extractions is about 294 mAh/g, much higher than those of the commonly used cathode materials like $LiFePO_4$ (170 mAh/g), making it a promising cathode material for high power batteries. Since the nanostructured approaches have been recently demonstrated to be highly effective in enhancing the electrochemical performance of electrode materials, various nanostructured V_2O_5 has been studied to show the improved electrochemical properties as featured by high capacity and remarkable reversibility. For example, Sides et al. (2005) reported the preparation of monodisperse V_2O_5 fibers with an improved electrochemical performance because of the reduced lithium ion diffusion path and increased electrolyte-accessible surface area. Cao et al. (2005) have synthesized V_2O_5 with highly ordered superstructures, in which nanoparticles interconnect to form nanorods, and these rods circle around to form hollow microspheres. The specific capacity of these V_2O_5 hollow microspheres is 286.4 mAh/g at 0.2 C charge/discharge rate, which is very close to the theoretical capacity of V_2O_5. However, the Li-ion diffusion coefficient (10^{-12} cm^2/s) and the electronic conductivity (10^{-2}–10^{-3} S/cm) in crystalline V_2O_5 are inherently too low to sustain a large specific capacity at high charge/discharge rates, which cannot be simply altered by the nanostructured strategies mentioned above. In order to improve the electrochemical performance, many heterogeneous nanostructured forms of V_2O_5 composites have been reported in the literature, and the usefulness of these designs has been demonstrated.

2.4.1 Mixed with conducting secondary phase

To enhance the high rate capability of V_2O_5 cathode materials, the creation of heterogeneous nanostructures from nanoscale V_2O_5 and conducting secondary phase is one of the most widely used strategies. There are three kinds of mainly adopted conducting materials, viz. conducting conjugated polymers (CCPS), carbon materials, and various metal oxides. CCPS possess a high inherent conductivity and is able to undergo practically fully reversible redox processes. The heterogeneous nanostructures from V_2O_5 and CCPS combine a high capacity and ability to intercalate lithium ions in V_2O_5, and high cyclability, conductivity, and other properties of CCPS. For example, Posudievsky et al., (2011) prepared two-component guest-host nanocomposites which are composed of conducting polymers (polyaniline, polypyrrole and polythiophene) and V_2O_5. The nanocomposites are capable of reversible cycling as the positive electrode in a lithium ion cell, and retain their capacity over one hundred full charge/discharge cycles. The higher discharge capacity, stability in prolonged charge/discharge cycling, and improved diffusion of lithium ions in the nanocomposites, are the consequences of pillaring by the conducting polymer macromolecules in the layers of the inorganic matrix.

Carbon coating has been applied to V_2O_5 cathode to give good electronic conductivity and stability (Odani et al., 2006; koltypin et al., 2007). For example, koltypin et al. (2007) have prepared carbon-coated V_2O_5 nanoparticles. The carbon coated V_2O_5 electrodes demonstrate a higher capacity, much better rate capability, and very good stability upon cycling and aging at elevated temperatures when compared to the electrodes that comprise the microparticles of V_2O_5.

Other than CCPS and carbon, inorganic oxides such as TiO_2 (Takahashi et al, 2006) and WO_3 (Cai et al, 2011) have been introduced with an aim to improve the electrochemical properties by modifying the crystallinity of V_2O_5. For example, Cai et al (2011) found that V_2O_5-WO_3 composites films show the enhanced Li-ion intercalation properties compared to pure V_2O_5 or WO_3 films. At a high current density of 1.33 A/g, V_2O_5-WO_3 film with a V_2O_5/WO_3 molar ratio of 10/1 exhibits the highest capacities of 200 mAh/g at the first cycle and 132 mAh/g after 50 cycles, while pure V_2O_5 film delivers discharge capacities of 108 mAh/g at the first cycle and 122 mAh/g after 50 cycles. The enhanced Li-ion intercalation properties of the composite films are ascribed to the reduced crystallinity, increased porosity, and the enhanced surface area.

2.4.2 Thin layer loaded on a continuous conductive matrix

To improve the high rate electrochemical property of V_2O_5, rapid and continuous electronic and ionic transportation path are required. Reducing the dimension of V_2O_5 is a most effective method to shorten the Li ion diffusion path and fasten the Li ion transportation, since Li-ion diffusion coefficient of V_2O_5 (10^{-12} cm^2/s) cannot be altered simply by nanostructure strategy. Also, the disadvantage of the low electronic conductivity (10^{-2}–10^{-3} S/cm) of V_2O_5 can be effectively overcome by the introduction of conductive secondary phase. So, if uniformly ultra thin layer of V_2O_5 is formed on a continuous conductive matrix, rapid and continuous electronic and ionic transportation path can be achieved. For example, Yamada et al. (2007) coated thin V_2O_5 layer on porous carbon, as shown in Figure 22a. In their work, the porous carbons were prepared using SiO_2 colloidal crystals as the templates with three dimensionally interconnected pores, and then V_2O_5/carbon composites were achieved by immersing the porous carbons in V_2O_5 sol under a reduced pressure. Nanoporous V_2O_5/carbon composites thus prepared exhibit a large capacity of more than 100 mAh/g and good rate capability of 80% at 5.0 A/g (Figure 22b). The good performance is explained by electric double layer capacitance of large surface area and high rate lithium insertion to V_2O_5 gel.

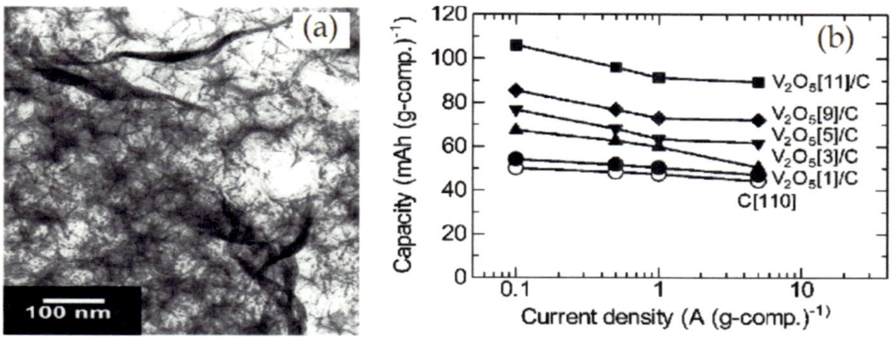

Fig. 22. a) TEM image of V_2O_5/C composites; and b) rate capability of V_2O_5[n]/C composites. (Yamada et al., 2007)

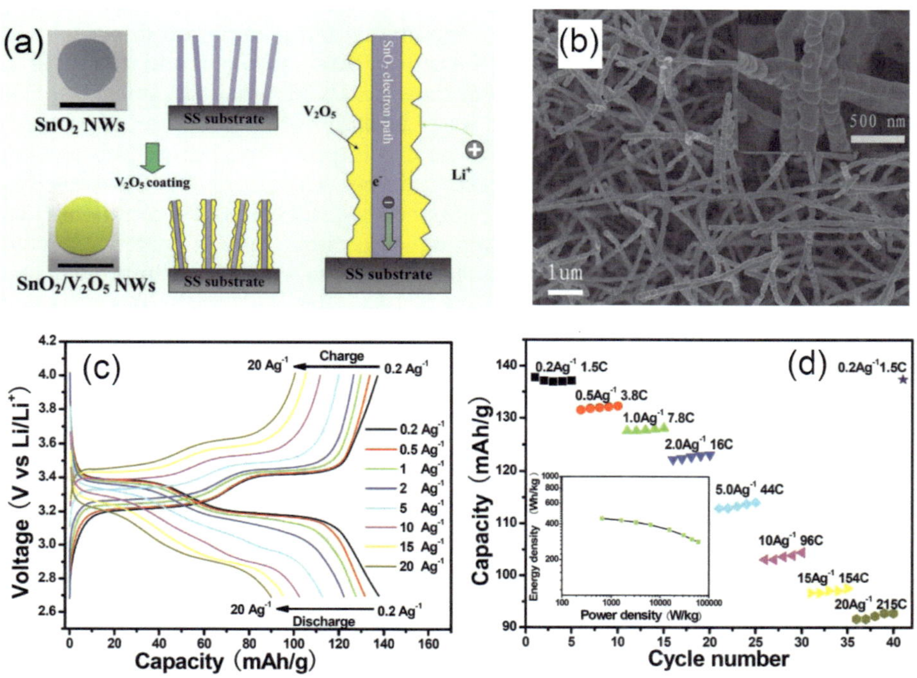

Fig. 23. a) Schematic diagram showing the strategy for coating V_2O_5 on SnO_2 nanowires (NWs) grown on a stainless-steel (SS) substrate; b) FE-SEM image of the typical V_2O_5 loaded SnO_2 nanowires; the inset indicates the diameter is about 100 nm; c) the galvanostatic charge/discharge curves of the SVNs electrode at different current densities; and d) rate performance at different current densities, and the inset showing the Ragone plots. (Yan et al., 2011)

Recently, Yan et al. (2011) have adopted a simple gas-phase-based method to synthesize V_2O_5 loaded SnO_2 nanowires on the stainless-steel substrate (Figures 23a and b). V_2O_5 loaded SnO_2 nanowires exhibit a high rate capability and very good cycling stability (Figure 23c). SnO_2/V_2O_5 core/shell-nanowires deliver a high power density of about 60 kW/kg while the energy density remains at 282 wh/kg (Figure 23d). The improved electrochemical performance can be ascribed to the synergic effect of V_2O_5 thin layer and SnO_2 nanowires. V_2O_5 layer coated on SnO_2 is very thin, which could shorten the ion-diffusion time and enable a high utilization of V_2O_5. The free-standing SnO_2 nanowires create porous channels, providing effective electrolyte transport and a large contact interface between the electrolyte and active materials. These findings open up new opportunities in the development of high-performance energy-storage devices used for micro-systems, thin-film devices, and other functional devices.

3. Heterogeneous nanostructured anode materials

Due to the low theoretical capacity of 372 mAh/g for graphitic carbon, intensive research has been conducted to search for the alternative anode materials with higher capacities and rate capabilities. Many other anode materials, such as Sn, Sb, Si, Ge, SnO_2, and Co_3O_4 have been reported to have larger lithium storage capacities than graphitic carbon via the formation of alloys with lithium or through the reversible reactions with lithium ions. However, these anode materials suffer from the huge volume variation during the lithium insertion/extraction process which leads to the pulverization problem and moreover a rapid deterioration in capacity. Reducing the size of these anode materials to nanometer scale can partially suppress the volume variation, and designing the structure of these anode materials as hollow spheres or porous particles can improve the lithium storage capacity and their initial Coulombic efficiency to a certain degree. Unfortunately, the capacity decay of the electrode along with the charge/discharge cycling still cannot be completely avoided (Liang et al., 2009). Recently, heterogeneous nanostructured anode materials, which are composed of multi-nanocomponents have been shown to exhibit an improved rate and cyclying performance due to the synergic effect between the functional nanocomponents. In this section, we introduced the recent trends and developments of these heterogeneous nanostructured anode materials, including carbonaceous nanocomposites, polymer/inorganic oxides nanocomposites, and binder-free thin-film nanocomposites.

3.1 Carbonaceous composites anode materials

Carbonaceous composites materials have been widely used for anodes. Though there are many different types of carbon materials, such as amorphous carbon, carbon nanotubes, and graphene, etc., all of them can not only significantly enhance the electronic conductivity of electrode materials, but also lead to the stabilized SEI films on anodes, which result in the improved rate and cycling performance. However, the synergic effect between carbon materials and other lithium intercalation materials is very complicated. Even when the same carbon material is adopted, the synergic effect could be significantly different due to the varied combination states of nanocomponents.

3.1.1 Core-shell nanocomposites

Core–shell nanocomposites for carbon-based anodes refer to a core energy dense nanoparticle coated with a thin amorphous carbon shell, which enhances the electrical conductivity, prevents aggregation, improves chemical stability, and buffers the stress of the inner nanoscale active material. Zhang et al. (2008) reported tin nanoparticles encapsulated elastic hollow carbon spheres (TNHCs) with an uniform size. As shown in Figure 24, in the TNHCs composites, multiple tin nanoparticles with a diameter of less than 100 nm are encapsulated in one thin hollow carbon sphere with a thickness of only 20 nm. This special structure results in both the content of Sn up to 70% by weight and the void volume in carbon shell as high as 70–80% by volume. This void volume and elasticity of thin carbon spherical shell efficiently accommodate the volume change of tin nanoparticles due to the Li-Sn alloying-dealloying reactions, and thus prevent the pulverization of electrode. As a result, this composite shows a decent cycle ability with a high initial capacity of more than 800 mAh/g, and after 100 cycles it shows only a moderate decrease in its specific capacity to 550 mAh/g. Coaxial SnO_2@carbon hollow nanospheres have also been reported by Lou et al. (2009). In this new type of nanoarchitecture, the carbon shell is tightly attached to the SnO_2 shell, beneficial for the mechanical reinforcement and the enhanced electronic conduction. The composites deliver a discharge capacity of around 460 mAh/g after 100 cycles with little capacity fade.

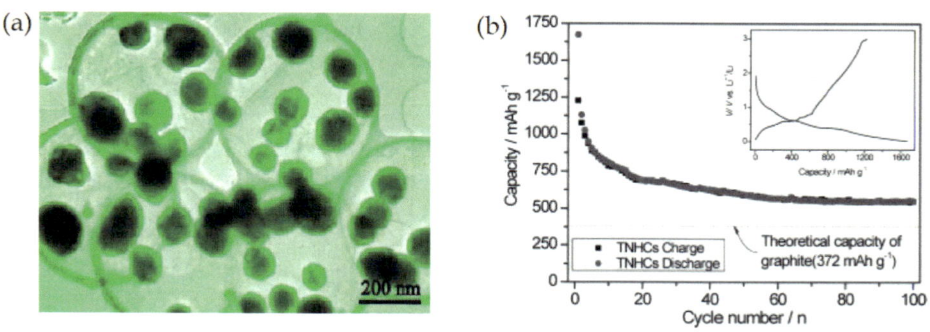

Fig. 24. a) TEM image of TNHCs; b) discharge/charge capacity profiles of TNHCs in 5 mV-3 V (vs.Li^+/Li) voltage window and C/5. Inset shows the first discharge/charge profiles of TNHCs cycled at a rate of C/5. (Zhang et al., 2008)

Recently, Yang et al. (2011) reported TiO_2(B) @carbon composite nanowires by a two-step hydrothermal process with subsequent heat treatment in argon. As shown in Figure 25, the nanostructures exhibit a unique feature of having TiO_2(B) encapsulated inside and an amorphous carbon layer coating outside. The carbon shell accelerates the diffusion of both electrons and lithium ions, ensuring a high electrode–electrolyte contact area and eventually making a greater contribution to the capacity. TiO_2(B) core also acts as an effective mechanical support to alleviate the stress produced during lithium intercalation/ deintercalation and preserve the one dimensional core/shell structure even after 100 charge/discharge cycles. The composite nanowires exhibit a high reversible capacity of 560 mAh/g after 100 cycles at a current density of 30 mA/g, and excellent cycling stability and rate capability.

Above reports have fully demonstrated the benefits of heterogeneous core-shell nanostructures. In this core-shell strategy, the void volume in the core-shell configuration should be optimized according to the actual volume variation of electrode materials during the lithium insertion and extraction, in order to avoid decreasing the volume energy density of LIBs in practical uses.

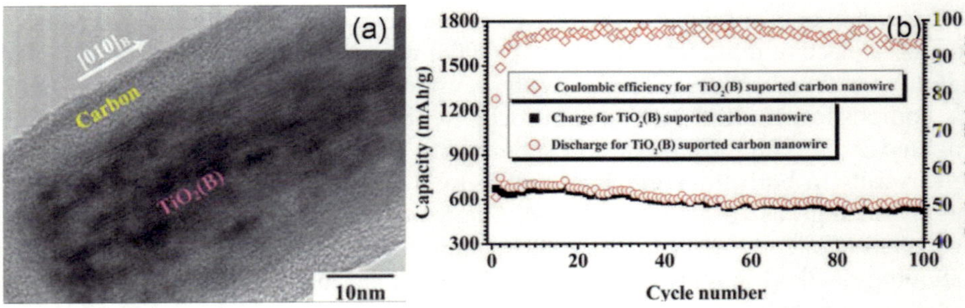

Fig. 25. a) HRTEM image of a section of a TiO_2(B)@carbon composite nanowire; b) capacity vs. cycle number curves from the first cycle to the 100th cycle for the TiO_2(B)@carbon composite nanowires at a current density of 30 mA/g. (Yang et al., 2011)

3.1.2 Carbon-nanotube-based nanocomposites

Unique properties of carbon nanotube (CNT) such as high conductivity, nanotexture, resiliency, are very helpful for LIBs, where pure CNT and/or their composites play the role of electrode materials. CNT provides an ideal conducting path to ensure a good charge propagation and can serve as a perfect resilient skeleton with stretchable properties that gives a significant improvement in electrochemical cyclability especially when volumetric changes take place during the charge/discharge processes. In this regard, CNTs have been widely used as the conductive additives or conductive matrix for the anode materials. There are two main strategies to synthesize CNT heterogeneous nanostructures: One is interconnecting the electrode materials with the exterior surface of CNT by various mixing procedures. For example, Zhang et al. (2009) prepared CNT@SnO_2 hybrid nanostructures with excellent lithium storage performance by uniformly loading SnO_2 nanoparticles onto a cross-stacked CNT network. Xia et al. (2010) prepared caterpillar-like nanoflaky MnO_2/CNT nanocomposites, in which highly porous interconnected MnO_2 nanoflakes are uniformly coated on the surface of CNT. This nanoflaky MnO_2/CNT nanocomposite electrode exhibits a large reversible capacity of 801 mAh/g for the first cycle without capacity fading for the first 20 cycles (Figure 26).

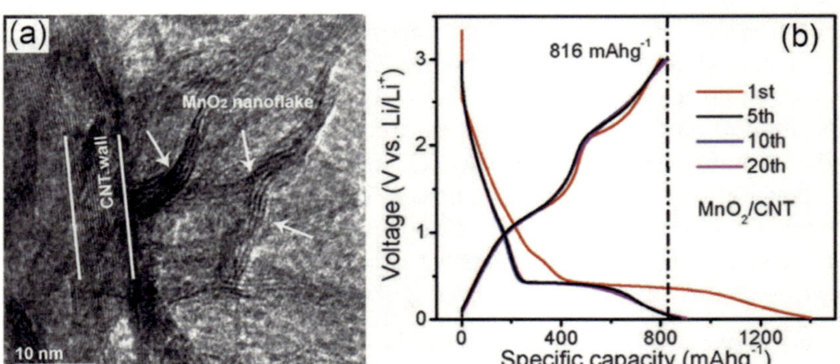

Fig. 26. a) TEM images of one single CNT coated with porous MnO_2 layer; and b) voltage profiles of a MnO_2/CNT nanocomposite electrode cycled between 0.01 and 3 V at a current density of 200 mA/g. (Xia et al., 2010)

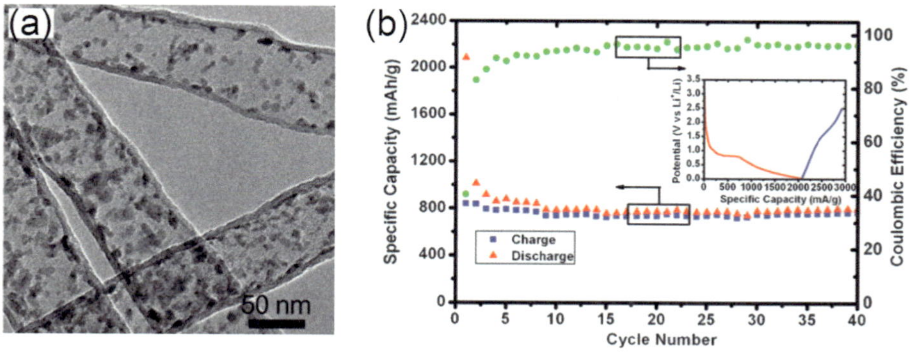

Fig. 27. a) TEM images of the Fe_2O_3-filled CNTs; and b) cycling performance of the Fe_2O_3-filled CNT measured at a rate of 35 mA/g between 0.02 and 2.5 V (vs. Li^+/Li). (Yu et al., 2010)

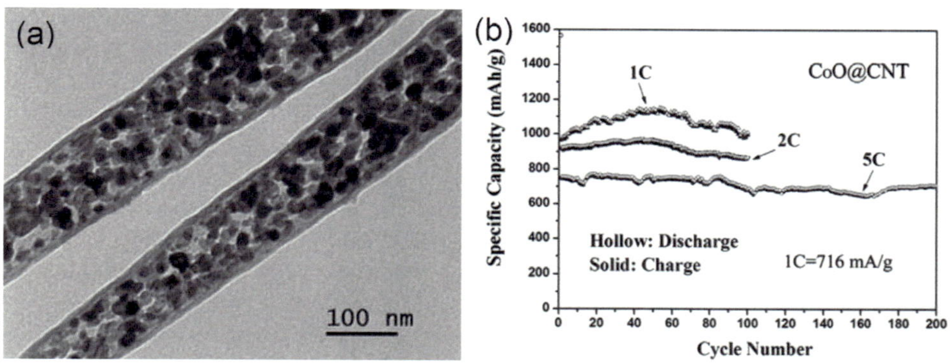

Fig. 28. a) TEM images of CoO@CNT; and b) cycling performance of CoO@CNT at high-rates. (Wu et al., 2011)

The other strategy is encapsulating the electrode materials in the interior space of CNT (Wang et al., 2009; Wu et al., 2010; Yu et al., 2010; Wu et al., 2011). For example, Yu et al. (2010) synthesized Fe_2O_3/CNT nanocomposites by AAO template method. As shown in Figure 27, in this nanocomposite, Fe_2O_3 nanoparticles with a mean diameter of ~9 nm are filled homogeneously into the hollow core of high aspect ratio CNT. The Fe_2O_3-filled CNTs exhibit a high initial discharge capacity of 2081 mAh/g, and a reversible capacity of 768 mAh/g is obtained after 40 cycles, while the Coulombic efficiency steadily keeps at higher than 95%. Fe_2O_3-filled CNTs also show a superior cycling performance and rate capability, which can be attributed to the small size of Fe_2O_3 particles and the confinement effect as well as the good electrical conductivity of CNTs. Very recently, Wu et al. (2011) reported a self-assembled echinus-like nanostructure that consists of mesoporous CoO nanorod@carbon nanotube core–shell materials, as shown in Figure 28a. This core–shell nanostructure shows a high capacity (703–746 mAh/g in 200 cycles) and a long cycle life (0.029% capacity loss per cycle) at a high current rate of 3580 mA/g, as shown in Figure 28b. The excellent electrochemical energy storage should be attributed to the increased electrical conductivity, mechanical stability and electrochemical activity of porous CoO materials in the presence of a carbon nanotube overlayer. This core-shell nanostructure not only increases the electronic and ionic conductivity due to the presence of highly conductive CNT, but also improves the stability during the repeated ionic intercalation due to the flexible CNT, which act as a buffer to alleviate the volume expansion of metal oxides.

3.1.3 Graphene-based nanocomposites

In the family of carbon nanostructures, graphene is the youngest member but has attracted enormous recent interest. Graphene is a two-dimensional macromolecular sheet of carbon atoms with a honey comb structure. It is an excellent substrate to host active nanomaterials for LIBs applications due to its high conductivity, large surface area, flexibility, and chemical stability. In this regard, chemically modified graphene materials have been used to form the heterogeneous nanostructured materials with Si (Chou et al., 2010), SnO_2 (Paek et al., 2009), Co_3O_4 (Yang et al., 2010), TiO_2 (Li et al., 2011), Fe_2O_3 (Zhu et al., 2011), Mn_3O_4 (Wang et al., 2010), and $SnSe_2$ (Choi et al., 2011). Improved electrochemical performances are thus achieved, owing to the synergic effect between graphene and active nanoparticles. In these nanocomposites, the ultrathin flexible graphene layers can provide a support for anchoring well-dispersed active nanoparticles and work as a highly conductive matrix for enabling good contact between them, which can also effectively prevent the volume expansion/contraction and aggregation of the active nanoparticles during Li charge/discharge processes. Meanwhile, the anchoring of active nanoparticles on graphene effectively reduces the degree of restacking of graphene sheets and consequently keeps a highly active surface area and to some extent, increases the lithium storage capacity and cyclic performance (Wu et al., 2010). For example, Paek et al. (2009) reported SnO_2/graphene nanocomposites, in which graphene nanosheets are homogeneously distributed between the loosely packed SnO_2 nanoparticles. This nanocomposite exhibits a charge capacity of 570 mAh/g after 30 cycles with about 70 % retention of the reversible capacity. Recently, Zhang et al. (2010) synthesized Fe_3O_4/graphene composites by depositing Fe^{3+} in the interspaces of graphene sheets. In the composites, Fe_3O_4 nanoparticles are dispersed on graphene sheets, which show a high reversible capacity, as well as a significantly enhanced cycling performance (about 650 mAh/ g after 50 cycles) and high rate capabilities (350 mAh/g at 5 C).

Though graphene sheets can effectively buffer the strain from the volume change of metals or metal oxides during the charge/discharge processes and preserve the high electrical conductivity of the overall electrode, the metal and metal oxide nanoparticles are still prone to strong aggregation during the cycle processes because of non-intimate contact between graphene layers and active nanoparticles, leading to a slow capacity fading. An alternative strategy for solving the aggregation problem of metal and metal oxides is to confine them within individual graphene shells. Yang et al. (2010) reported graphene-encapsulated Co_3O_4 nanoparticles prepared by co-assembly between negatively charged graphene oxide and positively charged oxide nanoparticles. This assembly enables a good encapsulation of electrochemically active metal oxide nanoparticles by graphene sheets, thus leading to a remarkable lithium-storage performance such as highly reversible capacity and excellent cycle performance. A very high and stable reversible capacity of about 1100 mAh/g in the initial 10 cycles, and 1000 mAh/g after 130 cycles is achieved for the graphene-encapsulated Co_3O_4 electrode, which is superior over those of Co_3O_4/graphene composite or bare Co_3O_4 electrodes (Figure 29).

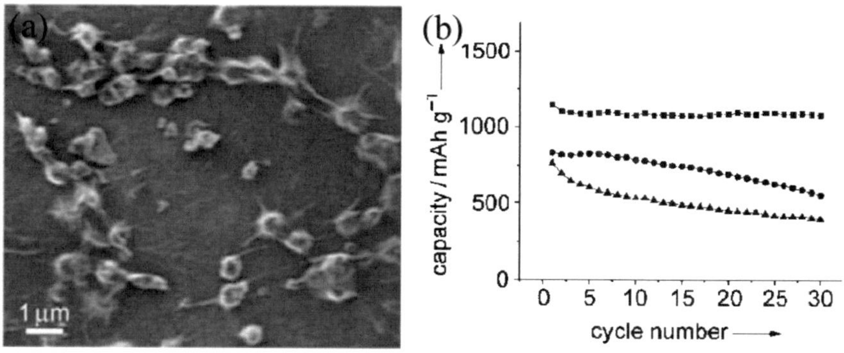

Fig. 29. a) SEM images of graphene-encapsulated Co_3O_4 and b) comparison of the cycle performance of graphene-encapsulated Co_3O_4 (■), mixed Co_3O_4/graphene composite (•), and bare Co_3O_4 electrodes (▲) over 30 cycles (current density=74 mA/g). (Yang et al. 2010)

3.2 Polymer/inorganic oxides nanocomposites

Heterogeneous nanostructures that consist of the composites of inorganic oxides and organic conducting polymers have been proven to be an effective approach to improve the high-rate electrochemical properties of LIBs. Organic conducting polymers provide a conducting backbone for the nanoscale inorganic oxides and accelerate the charge transfer between them, limit formation of a surface electrochemical interface (SEI) layer to improve the chemical stability, and buffer the stress of the inner nanoscale inorganic oxides to improve the structural stability.

3.2.1 Polypyrrole (PPy)

PPy is a chemical compound formed from a number of connected pyrrole ring structures and it is the first polyacetylene-derivative to show a high conductivity. Chew et al. (2007) reported a novel nano-silicon/PPy composite by chemical ploymerization. The cycling stability of Si/PPy electrodes is significantly improved when compared to that of the bare Si anodes. PPy plays a key role in improving the electrochemical performance of the composites: (i) PPy in the composites is a conducting polymer, which increases the conductivity of the samples. (ii) PPy can act as a conductive binder, increasing the contact among particles, therefore the particle-to-particle resistance will be decreased, thus reducing the irreversible reactions with the electrolyte. (iii) PPy is an effective component that buffers the great volume changes during the cycling process, thus improving the cyclability of the Si electrode. Similarly, Cui et al. (2011) have synthesized SnO_2@ PPy nanocomposites by an one-pot oxidative chemical polymerization method. The nanocomposites exhibit high discharge/charge capacities and favourable cycling. For SnO_2@PPy nanocomposites with 79 wt% SnO_2, the electrode reaction kinetics is controlled by the diffusion of Li^+ in the nanocomposite. The calculated diffusion coefficiency of lithium ions in SnO_2@PPy nanocomposite with 79 wt% SnO_2 is 6.7×10^{-8} cm^2 s^{-1}, while the lithium-alloying activation energy at 0.5 V is 47.3 kJ mol^{-1}, which is obviously lower than that for the bare SnO_2. The enhanced electrode performance with the SnO_2@PPy nanocomposite is originated from the advantageous nanostructures that allow a better structural flexibility, shorter diffusion length, and easier interaction with lithium.

3.2.2 Polyaniline (PANI)

PANI is a conducting polymer of the semi-flexible rod polymer family. It has the advantages of ease of fabrication, high conductivity, and good flexibility, so it can be used to prepare the composite anode of varied microstructures. Nickel foam-supported NiO/PANI composites with a porous net-like morphology have been prepared by a chemical bath deposition technique (Huang et al., 2008). These composites are constructed by NiO nanoflakes coated by PANI layer. As an anode for lithium ion batteries, NiO/PANI composites exhibit a weaker polarization as compared to NiO film. The specific capacity after 50 cycles for NiO/PANI film is 520 mAh/g at 1 C, higher than that of 440 mAh/g for NiO film. These improvements are attributed to the fact that PANI enhances the electrical conduction of the electrode and keeps the NiO film stable during cyclings. Similarly, He et al. (2008) prepared SnO_2-PANI composite by microemulsion polymerization method. The composite is composed of amorphous PANI and SnO_2 nanoparticles, in which SnO_2 is coated with PANI. SnO_2-PANI composite to show a reversible capacity of 657.6 mAh/g with a capacity loss per cycle at only 0.092% after 80 cycles. Recently, Cai et al. (2010) reported Si/PANI nanocomposite synthesized by chemical polymerization of aniline and nano-silicon. Si/PANI composite maintains a capacity of 1870 mAh/g up to 25 cycles with a slight loss per cycle, showing an excellent rate performance. The high stable cyclability and improved rate performance are attributed to the fact that the nano-silicon particles are electrically connected by the nest-like PANI so that dispersity of the nano-silicon will be greatly enhanced and the conductivity could be improved. Moreover, Si/PANI composite accommodates a large volume expansion because of the flexible nest-like PANI matrix.

3.2.3 Poly(ethylene glycol) (PEG) /polyethylene oxide (PEO)

PEG or PEO refers to an oligomer or polymer of ethylene oxide. Two names are chemically synonymous, but historically PEG has tended to refer to oligomers and polymers with a molecular mass below 20,000 g/mol, while PEO to polymers with a molecular mass above 20,000 g/mol. PEG or PEO has a chain structure of $HO\text{-}CH_2\text{-}(CH_2\text{-}O\text{-}CH_2\text{-})_n\text{-}CH_2\text{-}OH$. PEG and PEO have been widely investigated as the electrolytes in lithium batteries. The ionic transport mechanism is that Li^+ ions move from one coordination site to another revolved by the polymer chains. On one hand, the reaction between lithium and PEG is not reversible at ambient conditions. On the other, PEO is stable even in contact with lithium at 60 °C. Hence, until very recently, Xiao et al. (2010) and Xiong et al. (2011) reported the improved lithium storage by introduction of PEO and PEG, respectively, there have been no literature reports that ethylene oxide units in polymers are used for reversible lithium storage. In the case of MoS_2/PEO nanocomposite, PEO is inserted into the interlay spacing of exfoliated MoS_2 to form $Li_{0.12}(PEO)_yMoS_2$. A small amount of PEO ($Li_{0.12}(PEO)_{0.05}MoS_2$) not only doubles the initial capacity of the pristine MoS_2 but significantly improves the capacity retention. This nanocomposite delivers a charge/discharge capacity of above 900 mAh/g after 50 cycles, as shown in Figure 30, during which PEO stabilizes the disordered structure of the exfoliated MoS_2 throughout the cycling regime to accommodate more Li^+ ions and accelerates the Li^+ transportation between MoS_2 sheets and electrolyte. While higher amounts of PEO may decrease the lithium accessibility and electronic conductivity, resulting in a higher internal resistance and charge transfer. Xiong et al. reported a novel nanonet that is composed of SnO_2 nanoparticles and PEG chains. The synergic properties and functionalities of these two components enable this composite to exhibit an unexpectedly high lithium storage over the theoretical capacity of SnO_2. Lithium is stored in the PEG-SnO_2 nanocomposites, where the crosslinked SnO_2 particles pave the electron paths and the revolved EO segments provide the coordination centers. When the composite electrode is charged, those Li^+ ions coordinated with EO segments are reduced and deposited, and more Li^+ ions from electrolytes will be reduced at this site to form the lithium clusters. When the composite electrode is discharged, lithium clusters transform into Li^+ ions to release the extra capacity.

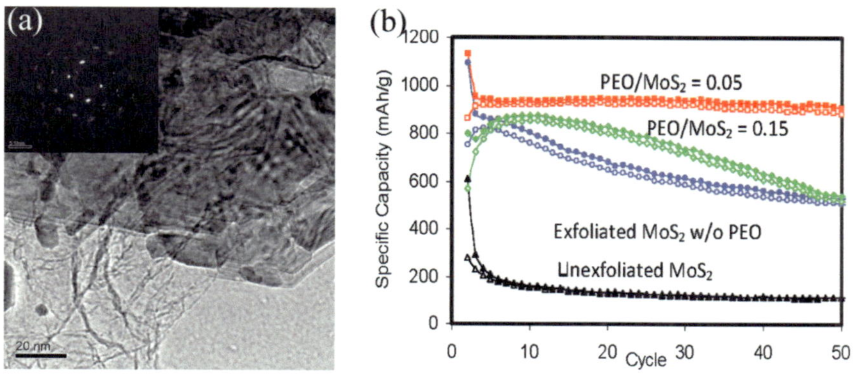

Fig. 30. a) TEM images for exfoliated PEO/MoS_2 = 0.05 composite; and b) comparison of cycling stability for unexfoliated MoS_2 and exfoliated MoS_2 with different amounts of PEO. All cells were cycled between 0.01 and 3.0 V at 50 mA/g. Solid symbols, Li insertion; open symbols, Li extraction. (Xiao et al., 2010)

3.3 Heterogeneous nanoarchitectured thin-film electrodes

Researchers have developed a new structural strategy for high performance electrodes, namely, "nanoarchitectured thin-film electrode," which is composed of a direct growth of highly conductive one-dimensional (1D) metal nanorods onto a metallic current collector followed by attachment of active materials on the surface of 1D metal nanorods. Due to the merit of this unique structure, LIBs can achieve high specific capacities and superior rate capabilities because of the formation of 3D conductive networks by direct connections among active materials, highly conductive 1D pathways, and a metallic current collector. Additionally, the large free space secured by 1D nanostructures effectively alleviates the huge volume changes that occur during Li^+ insertion/extraction, leading to a suppressed electrode degradation (Park et al., 2011). This strategy obviates the use of binders, electron conducting additives, and solvents traditionally used for electrode fabrication. Following this strategy, many heterogeneous nanoarchitectured electrodes have been synthesized, such as Sn coated TiO_2 nanotube array (Kim et al., 2010), NiO coated silicon nanowire array (Qiu et al., 2010), carbon-coated ZnO nanorod array (Liu, et al., 2009), MnO_2/carbon nanotube array (Reddy et al., 2009), Cu_6Sn_5-coated TiO_2 nanotube arrays (Xue et al., 2011), etc. For example, carbon-coated SnO_2 nanorod array directly grown on Fe-Co-Ni alloy substrate (Ji et al. 2010) shows a good capacity retention at 585 mAh/g after 50 cycles at 500 mA/g and cyclability (stable 320 mAh/g at 3,000 mA/g). Ortiz et al. (2010) fabricated nanoarchitectured crystalline SnO nanowires /TiO_2 nanotubes composites electrodes. As shown in Figure 31, the composites consist of tin and SnO nanowires grown onto TiO_2

nanotubes by anodization of titanium and tin electrodeposition. The 2 μm length tin-based nanowires supported on self-organized TiO_2 provide a real capacities of 95 and 140 μA h cm^{-2} (~675 mAh/g) under high rates of 4 C and 2 C, keeping 70 and 85% initial capacity over 50 cycles, respectively.

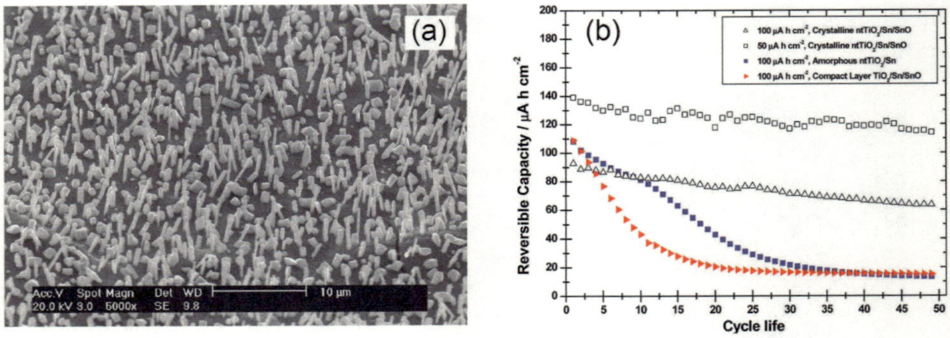

Fig. 31. a) SEM image of tin nanowires supported onto TiO_2 nanotube arrays; and b) cycle life performance under galvanostatic regime at a current density of 100 μAcm^{-2}. (Oritz et al., 2010)

These heterogeneous nanoarchitectured thin-film electrodes have demonstrated to display a stable capacity over many cycles, whereas an increase in film thickness is urgently required to provide the sufficiently active material because stress-induced cracking would lead to a poor cycling performance and rate capability owing to an increase in film thickness. An alternative strategy for tackling this problem is to design the functional structures that buffer the stress during the charge/discharge processes. Zhang et al. (2010) synthesized a Ni nanocone-array (NCA) supported Si architecture composed of many cylinders with regular domes on top to serve as the anode material for LIBs. In this configuration, Ni NCAs facilitate the charge collection and transport, supporting the electrode structure and acting as the inactive confining buffers. These nanostructured Si electrodes show an impressive electrochemical performance with a high capacity of around 2400 mAh/g at 0.2 C rate over 100 cycles with superior capacity retention of 99.8% per cycle. These Ni NCAs also exhibit an excellent lithium storage capability at high charging and discharging rates of 1 C or 2 C (Figure 32).

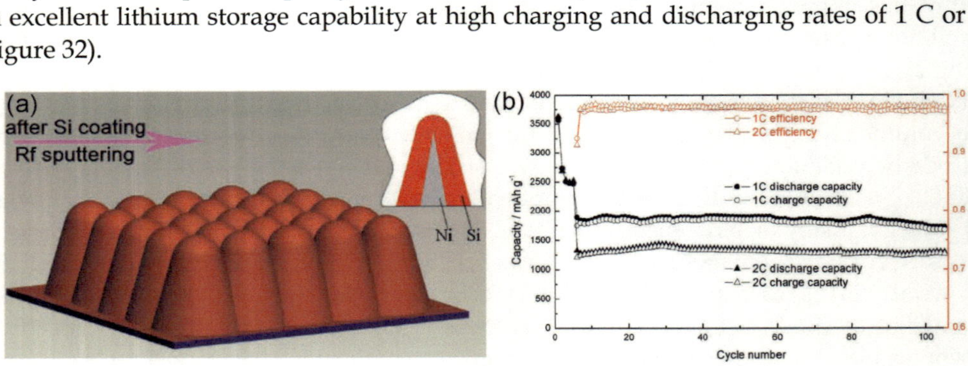

Fig. 32. a) Schematic diagram that illustrates the fabrication of a nickel nanocone-array supported silicon anode architecture; and b) discharge/charge capacity and Coulombic efficiency versus cycle number for a nanostructured Si electrode at 1 C and 2 C rates after 0.2 C for 5 cycles with a cut-off voltage of 10 mV–1.6 V. (Zhang et al., 2010)

Recently, Krishnan et al. (2011) synthesized a functionally strain-graded C-Al-Si anode composite using DC magnetron sputtering. As shown in Figure 33, the composites consist of an array of nanostructures each comprising an amorphous carbon nanorod (~170 nm long) with an intermediate layer of aluminium (~13 nm thick) that is finally capped by a silicon nanoscoop (~40 nm thick) on the very top. The gradation in strain arises from the graded levels of volumetric expansion in these three components on alloying with lithium. The introduction of aluminium as an intermediate layer enables the gradual transition of strain from carbon to silicon, thereby minimizing the mismatch at interfaces between differentially strained materials and enabling the stable operation of electrode under high-rate charge/discharge conditions. C-Al-Si composites can provide an average capacity of about 412 mAh/g with a power output of 100 kW/kg continuously over 100 cycles. Even when the power output is as high as 250 kW/kg, the average capacity over 100 cycles is still at the level of 90 mAh/g.

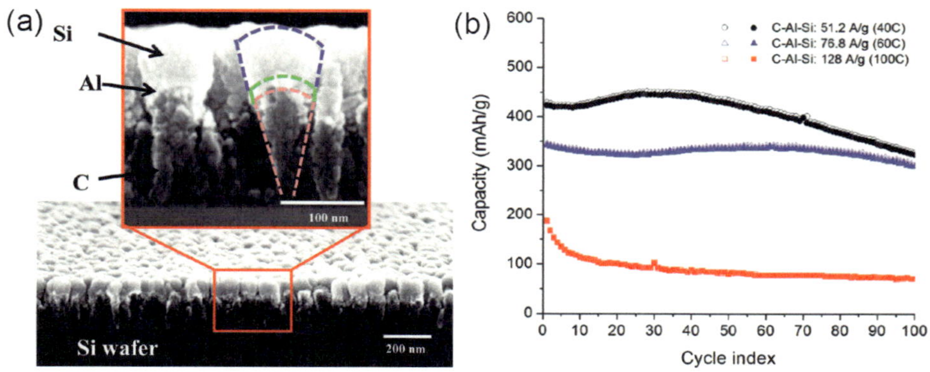

Fig. 33. a) Cross-sectional SEM view of C-Al-Si nanoscoop structures deposited on a Si wafer. A magnified cross-section image is also shown with the C, Al, and Si regions demarcated; and b) Capacity as a function of cycle index shown for C-Al-Si electrodes at 51.2, 76.8, and 128 A/g over 100 cycles. The empty symbols represent the discharge capacity while the filled symbols represent the charge capacity in each case. (Krishnan et al., 2011)

4. Summary and outlook

In this chapter, we have reviewed the recent trends and developments of using heterogeneous nanostructures as the electrode materials for LIBs. It has been illustrated that heterogeneous nanostructured electrodes materials have advantages over the conventional single component nanostructured electrode materials for LIBs, though one has to face some significant challenges before these heterogeneous nanostructured can be large-scale utilized in practical LIBs. There are many novel heterogeneous nanostructured cathode and anode materials that have been reported to show a high rate capability. But the dream is still there as to design and/or develop novel electrodes with much higher performance at high charge/discharge rates.

For example, one has optimized the performance of plenty of LIBs by extensively studying the influences of the morphology, particle size, loading content, and mixing procedure of different components, while the synergic mechanism of heterogeneous nanostructured components remains to be further clarified and key variables that determine the interface interaction between different components have to be identified. Over the last five years, tremendous progress has been made in addressing these challenges. Through a survey of those studies from the viewpoints of the synergic mechanism of heterogeneous nanostructured components, one can find that maximizing both ion and electron transportations in the cathodes and anodes while maintaining the good cycling stability can be fundamentally important in designing and developing novel electrodes for HEVs, PHEVs or EVs. Despite the variations of the reported performance, a consistent trend has been emerging that high power density of 60 kW/kg and high energy density of 282 Wh/kg for SnO_2/V_2O_5 core/shell composite thin-film cathodes and average capacities of ~412 mAh/g and a power output of~100 kW/kg for C-Al-Si composites thin-film anodes with rather impressive cycling lifetimes up to 100 cycles at practical charge/discharge rate can be reproducibly achieved. High rate performance and long cycling lifetimes have most frequently been observed when four critical factors are properly addressed: i) use of nanoscale-dimension electroactive materials that allow a rapid lithium ion diffusion; ii) efficient electronic and ionic conduction from electro-active materials to the current collectors that can be maintained during the cycling; iii) incorporation of void/free space or softer inactive material matrices or additives to accommodate the volume variation; iv) good adhesion between different heterogeneous components by constructing special gradient structure can be formed, which can effectively tolerate the different expansion and contraction efficiency of heterogeneous components during the continual lithium insertion/extraction.

Certainly, it is realized that all current heterogeneous nanostructures have their own drawbacks. For example, heterogeneous nanoarchitectured thin-film electrodes can become mechanically unstable when grown excessively long in the axial direction. This will limit their uses as on-chip power sources for MEMS devices and other applications where the areal footprint is at a premium and a high areal energy density is needed. Here, we tend to propose a future heterogeneous nanoarchitectured thin-film electrode as shown in Figure 34. This proposed heterogeneous nanoarchitectured thin-film electrode is composed of a well ordered branched array with a conformal polymer coating layer grown on a conductive substrate. The backbone is an active electrode material, while the branched is active electrode materials with the role of optimizing the lithium ion/electron transportation or inactive electrode materials to depress the side interaction between electrode and electrolyte. The conformal polymer coating layer acts as a conductive layer and binders. There are a few advantages of this proposed structure: (i) the interconnectivity of the electrode material provides a 3-D continuous lithium ion and electron pathway and can increase the mechanical strength of branched arrays; (ii) the interconnected pores accommodate the volume variations of the electrode material and allow the electrolyte to rapidly penetrate and uniformly contact the electrode material; (iii) 3-D interconnected coating of polymer facilitates a fast electron transfer into the 3-D continuous framework, allowing the electrons to directly transport to the electrode materials due to its conformality, effectively alleviating the stress due to its resiliency that ensures mechanically stability for longer arrays, and acting as a binder to strengthen the adhesion between the branched and backbone materials; and (iv) merits of the material being directly contacted with a conductive substrate, in which electrons can easily transfer from the conductive substrate to the electroactive backbone and branched material as well as to the coating conductive polymer shell.

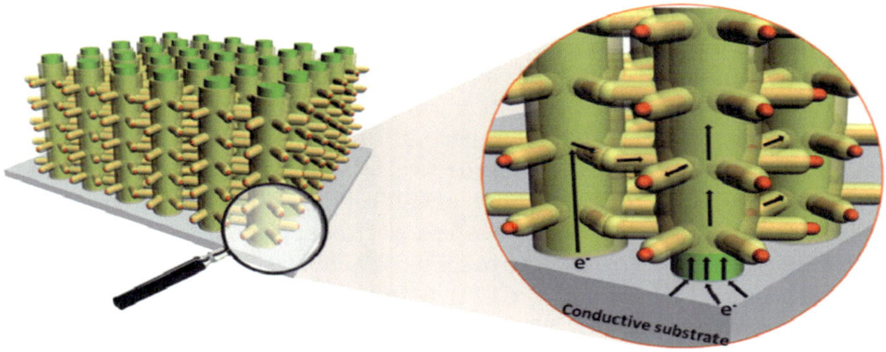

Fig. 34. Future heterogeneous nanoarchitectured thin-film electrodes. The backbone material is in green, the branch material in red, and conductive polymer in yellow.

The progress described here suggests that future research should explore more structurally and compositionally complex hierarchical composite nanostructures with internal void/pore space. Considering the potential large-scale energy applications of LIBs, such as in hybrid and plug-in electric vehicles or electrical power storage, mass production possibility, manufacturability, safety factors, and cost of the heterogeneous nanostructrued materials will also be important to the eventual success in their practical applications.

It is imperative that we as scientists and engineers put all our efforts into exploiting more technologically advanced heterogeneous nanostructures to fulfil the future requirements of LIBs.

5. Acknowledgment

This work was financially supported by NSFC (No. 91022018, 21025104), National Basic Research Program of China (No. 2011CBA00501, 2011CB935904), Fujian Program (No. 2009HZ0004-1), and FJIRSM fund (No. SZD09003-1, 2010KL002).

6. References

Alcantara, R.; Jaraba, M.; Lavela, P. & Tirado, J.-L. (2004). X-ray diffraction and electrochemical impedance spectroscopy study of zinc coated $LiNi_{0.5}Mn_{1.5}O_4$ electrodes. *Journal of Electroanalytical Chemistry,* Vol.566, No.1, (May 2004), pp.187–192, ISSN 0022-0728

Arrebola, J.; Caballero, A.; Hernan, L. & Morales, J. (2010). Re-examining the effect of ZnO on nanosized 5V $LiNi_{0.5}Mn_{1.5}O_4$ spinel: An effective procedure for enhancing its rate capability at room and high temperatures. *Journal of Power Sources,* Vol.195, No.13, (July 2010), pp. 4278–4284, ISSN 0378-7753

Arrebola, J.; Caballero, A.; Hernan, L.; Morales, J.; Castellon, E.R. & Barrado, J.-R.-R. (2007). Effects of coating with gold on the performance of nanosized $LiNi_{0.5}Mn_{1.5}O_4$ for lithium batteries. *Journal of the Electrochemical Society,* Vol.154, No.3, (January 2007), pp.A178–A184, ISSN 0013-4651

Arrebola, J.; Caballero, A.; Hernan, L.; Morales, J. & Castellon, E.-R. (2005). Adverse effect of Ag treatment on the electrochemical performance of the 5 V nanometric spinel

$LiNi_{0.5}Mn_{1.5}O_4$ in lithium cells. *Electrochemical and Solid State Letters,* Vol.8, No.6, (April 2005), pp.A303–A307, ISSN 1099-0062

Bakenov, Z. & Taniguchi, I. (2010). Electrochemical performance of nanocomposite $LiMnPO_4$/C cathode materials for lithium batteries. *Electrochemistry Communications,* Vol.12, No.1, (January 2010), pp. 75–78, ISSN 1388-2481

Boyano, I.; Blazquez, J.; de Meatza, I.; Bengoechea, M.; Miguel, O.; Grande, H.; Huang, Y. & Goodenough, J. (2010). Preparation of C-$LiFePO_4$/polypyrrole lithium rechargeable cathode by consecutive potential steps electrodeposition. *Journal of Power Sources,* Vol.195, No.16, (August 2010), pp. 5351–5359, ISSN 0378-7753

Cai, C.; Pol, V.; Guan, D. & Wang, Y. (2011). Solution processing of V_2O_5–WO_3 composite films for enhanced Li-ion intercalation properties. *Journal of Alloys and Compounds,* Vol.509, No.3, (January 2011), pp. 909–915, ISSN 0925-8388

Cai, J.-J.; Zuo, P.-J.; Cheng, X.-Q.; Xu, Y.-H. & Yin, G.-P. (2010). Nano-silicon/polyaniline composite for lithium storage. *Electrochemistry Communications,* Vol.12, No.11, (November 2010), pp. 1572–1575, ISSN 1388-2481

Cao, A.-M.; Hu, J.-S.; Liang, H.-P. & Wan, L.-J. (2005). Self-assembled vanadium pentoxide (V_2O_5) hollow microspheres from nanorods and their application in lithium-ion batteries. *Angewandte Chemie-International Edition,* Vol.44, No.28, (June 2005), pp.4391–4395, ISSN 1433-7851

Chan, C.; Peng, H.; Twesten, R.; Jarausch, K.; Zhang, X. & Cui, Y. (2007). Fast, completely reversible Li insertion in vanadium pentoxide nanoribbons. *Nano Letters,* Vol.7, No.2, (February 2007), pp. 490–495, ISSN 1530-6984

Chew, S.-Y.; Guo, Z.-P.; Wang, J.-Z.; Chen, J.; Munroe, P.; Ng, S.-H.; Zhao, L. & Liu, H.-K. (2007). Novel nano-silicon/polypyrrole composites for lithium storage. *Electrochemistry Communcations,* Vol.9, No.5, (May 2007), pp. 941–946, ISSN 1388-2481

Chen, W.; Qie, L.; Yuan, L.; Xia, S.; Hu, X.; Zhang, W. & Huang, Y. (2011). Insight into the improvement of rate capability and cyclability in $LiFePO_4$/polyaniline composite cathode. *Electrochimica Acta,* Vol.56, No.6, (February 2011), pp. 2689–2695, ISSN 0013-4686

Chen, Z.-H. & Dahn, J.-R. (2002). Effect of a ZrO_2 coating on the structure and electrochemistry of Li_xCoO_2 when cycled to 4.5 V. *Electrochemcial and Solid State Letters,* Vol.5, No.10, (October 2002), pp.A213–A216, ISSN 1099-0062

Choi, J.; Jin, J.; Jung, I.-G.; Kim, J.-M.; Kim, H.-J. & Son, S.-U. (2011). $SnSe_2$ nanoplate-graphene composites as anode materials for lithium ion batteries. *Chemical Communications,* Vol.47, No.18, (March 2011), pp.5241–5243, ISSN 1359-7345

Choi, K.; Jeon, J.; Park, H. & Lee, S. (2010). Electrochemical performance and thermal stability of $LiCoO_2$ cathodes surface-modified with a sputtered thin film of lithium phosphorus oxynitride. *Journal of Power Sources,* Vol.195, No.24, (December 2010), pp. 8317–8321, ISSN 0378-7753

Chou, S.-L.; Wang, J.-Z.; Choucair, M.; Liu, H.-K.; Stride, J.-A. & Dou, S.-X. (2010). Enhanced reversible lithium storage in a nanosize silicon/graphene composite. *Electrochemistry Communications,* Vol.12, No.2, (February 2010), pp.303–306, ISSN 1388-2481

Cui, L.; Shen, J.; Cheng, F.; Tao, Z. & Chen, J. (2011). SnO_2 nanoparticles@polypyrrole nanowires composite as anode materials for rechargeable lithium-ion batteries.

Journal of Power Sources, Vol.196, No.4, (February 2011), pp. 2195–2201, ISSN 0378-7753

Dimesso, L.; Jacke, S.; Spanheimer, C. & Jaegermann, W. (2011). Investigation on 3-dimensional carbon foams/$LiFePO_4$ composites as function of the annealing time under inert atmosphere. *Journal of Alloys and Compounds,* Vol.509, No.9, (March 2011), pp. 3777–3782, ISSN 0925-8388

Ding, Y.-L.; Xie, J.; Cao, G.-S.; Zhu, T.-J.; Yu, H.-M. & Zhao, X.-B. (2011). Single-crystalline $LiMn_2O_4$ nanotubes synthesized via template-engaged reaction as cathodes for high-power lithium ion batteries. *Advanced Functional Materials,* Vol.21, No.2, (January 2011), pp. 384–355, ISSN 1616-301X

Drezen, T.; Kwon, N.; Bowen, P.; Teerlinck, I.; Isono, M. & Exnar, I. (2007). Effect of particle size on $LiMnPO_4$ cathodes. *Journal of Power Sources,* Vol.174, No.2, (December 2007), pp. 949–953, ISSN 0378-7753

Ellis, B.; Lee, K. & Nazar, L. (2010). Positive Electrode Materials for Li-Ion and Li-Batteries. *Chemistry of Materials,* Vol.22, No.3, (February 2010), pp. 691–714, ISSN 0897-4756

Fan, Y.; Wang, J.; Tang, Z.; He, W. & Zhang, J. (2007). Effects of the nanostructured SiO_2 coating on the performance of $LiNi_{0.5}Mn_{1.5}O_4$ cathode materials for high-voltage Li-ion batteries. *Electrochimica Acta,* Vol.52, No.11, (March 2007), pp. 3870–3875, ISSN 0013-4686

Fang, H.; Li, L. & Li, G. (2007). A low-temperature reaction route to high rate and high capacity $LiNi_{0.5}Mn_{1.5}O_4$. *Journal of Power Sources,* Vol.167, No.1, (May 2007), pp. 223–227, ISSN 0378-7753

Fang, H.; Li, L. & Li, G. (2007). Hydrothermal synthesis of electrochemically active $LiMnPO_4$. *Chemistry Letters,* Vol.36, No.3, (March 2007), pp. 436–437, ISSN 0366-7022

Fang, H.; Li, L.; Yang., Y; Yan, G. & Li, G. (2008). Low-temperature synthesis of highly crystallized $LiMn_2O_4$ from alpha manganese dioxide nanorods. *Journal of Power Sources,* Vol.184, No.2, (October 2008), pp. 494–497, ISSN 0378-7753

Guo, Y.; Hu, J. & Wan L. (2008). Nanostructured Materials for Electrochemical Energy Conversion and Storage Devices. *Advanced Materials,* Vol.20, No.23, (December 2008), pp. 2878–2887, ISSN 0935-9648

Ha, H.-W.; Yun, N.-J.; Kim, M.-H.; Woo, M.-H. & Kim, K. (2006). Enhanced electrochemical and thermal stability of surface-modified $LiCoO_2$ cathode by CeO_2 coating. *Electrochimica Acta,* Vol.51, No.16, (April 2006), pp.3297–3302, ISSN 0013-4686

He, Z.-Q.; Xiong, L.-Z.; Liu, W.-P.; Wu, X.-M.; Chen, S. & Huang, K.-L. (2008). Synthesis and electrochemical properites of SnO_2-polyaniline composite. *Journal of Central South University Technology,* Vol.15, No.2, (April 2008), pp. 214–217, ISSN 1005-9784

Huang, H.; Yin, S. & Nazar, L. (2001). Approaching theoretical capacity of $LiFePO_4$ at room temperature at high rates. *Electrochemical and Solid State Letters,* Vol.4, No.10, (October 2001), pp. A170–A172, ISSN 1099-0062

Huang, X.-H.; Tu, J.-P.; Xia, X.-H.; Wang, X.-L. & Xiang, J.-Y. (2008). Nickel foam-supported porous NiO/polyaniline film as anode for lithium ion batteries. *Electrochemistry Communications,* Vol.10, No.9, (September 2008), pp. 1288–1290, ISSN 1388-2481

Huang, Y. & Goodenough, J. (2008). High-rate $LiFePO_4$ lithium rechargeable battery promoted by electrochemically active polymers. *Chemistry of Materials,* Vol.20, No.23, (December 2008), pp. 7237–7241, ISSN 0897-4756

Hu, S.-K.; Cheng, G.-H.; Cheng, M.-Y.; Hwang, B.J. & Santhanam, R. (2009). Cycle life improvement of ZrO_2-coated spherical $LiNi_{1/3}Co_{1/3}Mn_{1/3}O_2$ cathode material for lithium ion batteries. *Journal of Power Sources,* Vol.188, No.2, (March 2009), pp.564–569, ISSN 0378-7753

Ji, X.; Huang, X.; Liu, J.; Jiang, J.; Lin, X.; Ding, R.; Hu, Y.; Wu, F. & Li, Q. (2010). Carbon-coated SnO_2 nanorod array for lithium-ion battery anode material. *Nanoscale Research Letters,* Vol.5, No.3, (March 2010), pp.649–653, ISSN 1931-7573

Jung, K.-H.; Kim, H.-G. & Park, Y.-J. (2011). Effects of protecting layer [Li,La]TiO_3 on electrochemical properties of $LiMn_2O_4$ for lithium batteries. *Journal of Alloys and Compounds,* Vol.509, No.12, (March 2011), pp. 4426–4432, ISSN 0925-8388

Kang, H.-B.; Myung, S.-T.; Amine, K.; Lee, S.-M. & Sun, Y.-K. (2010). Improved electrochemical properties of BiOF-coated 5V spinel $Li[Ni_{0.5}Mn_{1.5}]O_4$ for rechargeable lithium batteries. *Journal of Power Sources,* Vol.195, No.7, (April 2010), pp. 2023–2028, ISSN 0378-7753

Kim, B.; Kim, C.; Ahn, D.; Moon, T.; Ahn, J.; Park, Y. & Park, B. (2007). Nanostructural effect of $AlPO_4$-nanoparticle coating on the cycle-life performance in $LiCoO_2$ thin films. *Electrochemical and Solid-State Letters,* Vol.10, No.2, (December 2006), pp. A32–A35, ISSN 1099-0062

Kim, B.; Kim, C.; Kim, T.-G.; Ahn, D. & Park, B. (2006). The effect of $AlPO_4$-coating layer on the electrochemical properties in $LiCoO_2$ thin films. *Journal of the Electrochemical Society,* Vol.153, No.9, (July 2006), pp.A1773–A1777, ISSN 0013-4651

Kim, H.-S.; Kang, S.-H.; Chung, Y.-H. & Sung, Y.-E. (2010). Conformal Sn coated TiO_2 nanotube arrays and its electrochemical performance for high rate lithium-ion batteries. *Electrochemical and Solid-State Letters,* Vol.13, No.2, (November 2009), pp.A15–A18, ISSN 1099-0062

Koenig, G.-M.; Belharouak, I.; Deng, H.; Sun, Y. & Amine, K. (2011). Composition-tailored synthesis of gradient transition metal precursor particles for lithium-ion battery cathode materials. *Chemistry of Materials,* Vol.23, No.7, (April 2011), pp. 1954–1963, ISSN 0897-4756

Koltypin, M.; Pol, V.; Gedanken, A. & Aurbach, D. (2007). The study of carbon-coated V_2O_5 nanoparticles as a potential cathodic material for Li rechargeable batteries. *Journal of the Electrochemical Society,* Vol.154, No.7, (May 2007), pp. A605–A613, ISSN 0013-4651

Konarova, M. & Taniguchi, I. (2008). Preparation of $LiFePO_4$/C composite powders by ultrasonic spray pyrolysis followed by heat treatment and their electrochemical properties. *Materials Research Bulletin,* Vol.43, No.12, (December 2008), pp. 3305–3317, ISSN 0025-5408

Krishnan, R.; Lu, T.-M. & Koratkar, N. (2011). Functionally strain-graded nanoscoops for high power Li-ion battery anodes. *Nano Letters,* Vol.11, No.2, (December 2010), pp. 377–384, ISSN 1530-6984

Lee, H.-W.; Muralidharan, P.; Ruffo, R.; Mari, C.; Cui, Y. & Kim, D. (2010). Ultrathin spinel $LiMn_2O_4$ nanowires as high power cathode materials for Li-ion batteries. *Nano Letters,* Vol.10, No.10, (October 2010), pp. 3852–3856, ISSN 1530-6984

Lepage, D.; Michot, C.; Liang, G.; Gauthier, M. & Schougaard, S. (2011). A Soft Chemistry Approach to Coating of $LiFePO_4$ with a Conducting Polymer. *Angewandte Chemie-International Edition,* DOI: 10.1002/anie.201101661, ISSN 1433-7851

Liang, M. & Zhi L. (2009). Graphene-based electrode materials for rechargeable lithium batteries. *Journal of Materials Chemistry,* Vol.19, No.33, (May 2009), pp. 5871–5878, ISSN 0959-9428

Li, N.; Liu, G.; Zhen, C.; Li, F.; Zhang, L.-L. & Cheng, H.-M. (2011). Battery performance and photocatalytic activity of mesoporous anatase TiO_2 nanospheres/graphene composites by template-free self-assembly. *Advanced Functional Materials,* Vol.21, No.9, (May 2011), pp.1717–1722, ISSN 1616-301X

Li, J.; Klopsch, R.; Stan, M.-C.; Nowak, S.; Kunze, M.; Winter, M. & Passerini, S. (2011). Synthesis and electrochemical performance of the high voltage cathode material $Li[Li_{0.2}Mn_{0.56}Ni_{0.16}Co_{0.08}]O_2$ with improved rate capability. *Journal of Power Sources,* Vol.196, No.10, (May 2011), pp.4821–4825, ISSN 0378-7753

Liu, D.-Q.; Liu, X.-Q. & He, Z.-Z. (2007). The elevated temperature performance of $LiMn_2O_4$ coated with $Li_4Ti_5O_{12}$ for lithium ion battery. *Materials Chemistry and Physics,* Vol.105, No.2-3, (October 2007), pp. 362–366, ISSN 0254-0584

Liu, G.-Q.; Kuo, H.-T.; Liu, R.-S.; Shen, C.-H.; Shy, D.-S.; Xing, X.-K. & Chen, J.-M. (2010). Study of electrochemical properties of coating ZrO_2 on $LiCoO_2$. *Journal of Alloys and Compounds,* Vol.496, No.1-2, (April 2010), pp.512–516, ISSN 0925-8388

Liu, J.; Li, Y.; Ding, R.; Jiang, J.; Hu, Y.; Ji, X.; Chi, Q.; Zhu, Z. & Huang, X. (2009). Carbon/ZnO nanorod array electrode with significantly improved lithium storage capability. *Journal of Physical Chemistry C,* Vol.113, No.13, (April 2009), pp.5336–5339, ISSN 1932-7447

Liu, J. & Manthiram, A. (2009). Understanding the improvement in the electrochemical properties of surface modified 5 V $LiMn_{1.42}Ni_{0.42}Co_{0.16}O_4$ spinel cathodes in lithium-ion cells. *Chemistry of Materials,* Vol.21, No.8, (Apirl 2009), pp.1695–1707, ISSN 0897-4756

Liu, R.; Duay, J. & Lee, S. (2011). Heterogeneous nanostructured electrode materials for electrochemical energy storage. *Chemical Communications,* Vol.47, No.5, (November 2010), pp. 1384–1404, ISSN 1359-7345

Lou, X.; Li, C. & Archer, L. (2009). Designed synthesis of coaxial SnO_2@carbon hollow nanospheres for highly reversible lithium storage. *Advanced Materials,* Vol.21, No.24, (June 2009), pp. 2536–2539, ISSN 0935-9648

Martha, S.-K.; Markovsky, B.; Grinblat, J.; Gofer, Y.; Haik, O.; Zinigrad, E.; Aurbach, D.; Drezen, T.; Wang, D.; Deghenghi, G. & Exnar, I. (2009). $LiMnPO_4$ as an advanced cathode material for rechargeable lithium batteries. *Journal of the Electrochemical Society,* Vol.156, No.7, (May 2009), pp. A541–A552, ISSN 0013-4651

Murugan, A.; Muraliganth, T. & Manthiram, A. (2008). Rapid microwave-solvothermal synthesis of phospho-olivine nanorods and their coating with a mixed conducting polymer for lithium ion batteries. *Electrochemistry Communications,* Vol.10, No.6, (January 2008), pp. 903–906, ISSN 1388-2481

Odani, A.; Pol, V.; Plo, S.; Koltypin, M.; Gedanken, A. & Aurbach, D. (2006). Testing carbon-coated VO_x prepared via reaction under autogenic pressure at elevated temperature as Li-insertion materials. *Advanced Materials,* Vol.18, No.11, (June 2006), pp. 1431–1436, ISSN 0935-9648

Oh, S.-M.; Oh, S.-W.; Yoon, C.-S.; Scrosati, B.; Amine, K. & Sun, Y.-K. (2010). High-performance carbon-$LiMnPO_4$ nanocomposite cathode for lithium batteries.

Advanced Functional Materials, Vol.20, No.19, (October 2010), pp. 3260–3265, ISSN 1616-301X

Oh, Y.; Ahn, D.; Nam, S. & Park, B. (2010). The effect of Al_2O_3-coating coverage on the electrochemical properties in $LiCoO_2$ thin films. *Journal of Solid State Electrochemistry,* Vol.14, No.7, (July 2010), pp. 1235–1240, ISSN 1432-8488

Ortiz, G.; Hanzu, I.; Lavela, P.; Knauth, P.; Tirado, J. & Djenizian, T. (2010). Nanoarchitectured TiO_2/SnO: a future negative electrode for high power density Li-ion microbatteries? *Chemistry of Materials,* Vol.22, No.5, (March 2010), pp.1926–1932, ISSN 0897-4756

Ouyang, C.; Zeng, X.; Sijivancanin, Z. & Baldereschi, A. (2010). Oxidation states of Mn atoms at clean and Al_2O_3-covered $LiMn_2O_4$(001) surfaces. *Journal of Physical Chemistry C,* Vol.114, No.10, (March 2010), pp. 4756–4759, ISSN 1932-7447

Paek, S.-M.; Yoo, E. & Honma, I. (2009). Enhanced cyclic performance and lithium storage capacity of SnO_2/graphene nanoporous electrodes with three-dimensionally delaminated flexible structure. *Nano Letters,* Vol.9, No.1, (January 2009), pp. 72–75, ISSN 1530-6984

Padhi, A.; Nanjundaswamy, K. & Goodenough, J. (1997). Phospho-olivines as positive-electrode materials for rechargeable lithium batteries. *Journal of the Electrochemical Society,* Vol.144, No.4, (April 1997), pp. 1188–1194, ISSN 0013-4651

Park, K.-S.; Kang, J.-G.; Choi, Y.-J.; Lee, S.; Kim, D.-W. & Park, J.-G. (2011). Long-term, high-rate lithium storage capabilities of TiO_2 nanostructured electrodes using 3D self-supported indium tin oxide conducting nanowire arrays. *Energy & Enviromental Science,* Vol.4, No.5, (May 2011), pp.1796–1801, ISSN 1754-5692

Peng, W.; Jiao, L.; Gao, H.; Qi, Z.; Wang, Q.; Du, H.; Si, Y.; Wang, Y. & Yuan, H. (2011). A novel sol–gel method based on $FePO_4 \cdot 2H_2O$ to synthesize submicrometer structured $LiFePO_4$/C cathode material. *Journal of Power Sources,* Vol.196, No.5, (March 2011), pp. 2841–2847, ISSN 0378-7753

Posudievsky, O.; Kozarenko, O.; Dyadyun, V.; Jorgensen, S.; Spearot, J.; Koshechko, V. & Pokhodenko, V. (2011). Characteristics of mechanochemically prepared host–guest hybrid nanocomposites of vanadium oxide and conducting polymers. *Journal of Power Sources,* Vol.196, No.6, (March 2011), pp. 3331–3341, ISSN 0378-7753

Qiu, M.-C.; Yang, L.-W.; Qi, X.; Li, J. & Zhong, J.-X. (2010). Fabrication of ordered NiO coated Si nanowire array films as electrodes for a high performance lithium ion battery. *Applied Materials & Interfaces,* Vol.2, No.12, (December 2010), pp.3614–3618, ISSN 1944-8244

Reddy, A.-L.-M.; Shaijumon, M.-M.; Gowda, S.-R. & Ajayan, P.-M. (2009). Coaxial MnO_2/carbon nanotube array electrodes for high-performance lithium batteries. *Nano Letters,* Vol.9, No.3, (March 2009), pp.1002–1006, ISSN 1530-6984

Sahan, H.; Goktepe, H.; Patat, S. & Ulgen, A. (2008). The effect of LBO coating method on electrochemical performance of $LiMn_2O_4$ cathode material. *Solid State Ionics,* Vol.178, No.35-36, (February 2008), pp. 1837–1842, ISSN 0167-2738

Santhanam, R. & Rambabu, B. (2010). Research progress in high voltage spinel $LiNi_{0.5}Mn_{1.5}O_4$ materials. *Journal of Power Sources,* Vol.195, No.17, (May 2010), pp.5442–5451, ISSN 0378-7753

Scott, I.; Jung, Y.; Cavanagh, A.; Yan, Y.; Dillon, A.; George, S. & Lee, S. (2011). Ultrathin coatings on nano-$LiCoO_2$ for Li-ion vehicular applications. *Nano Letters,* Vol.11, No.2, (February 2011), pp. 414–418, ISSN 1530-6984

Sides, C.-R. & Martin, C.-R. (2005). Nanostructured electrodes and the low-temperature performance of Li-ion batteries. *Advanced Materials,* Vol.17, No.1, (January 2005), pp. 125–128, ISSN 0935-9648

Sun, Y.-K.; Lee, Y.-S.; Yoshio, M. & Amine, K. (2002). Synthesis and electrochemical properties of ZnO-coated $LiNi_{0.5}Mn_{1.5}O_4$ spinel as 5 V cathode material for lithium secondary batteries. *Electrochemical and Solid State Letters,* Vol.5, No.5, (May 2002), pp.A99–A102, ISSN 1099-0062

Sun, Y.; Myung, S.; Kim, M.; Prakash, J. & Amine, K. (2005). Synthesis and characterization of $Li[(Ni_{0.8}Co_{0.1}Mn_{0.1})_{0.8}(Ni_{0.5}Mn_{0.5})_{0.2}]O_2$ with the microscale core-shell structure as the positive electrode material for lithium batteries. *Journal of the American Chemical Society,* Vol.127, No.38, (September 2005), pp. 13411–13418, ISSN 0002-7863

Sun, Y.; Myung, S.; Park, B.; Prakash, J.; Belharouak, I. & Amine, K. (2009). High-energy cathode material for long-life and safe lithium batteries. *Nature Material,* Vol.8, No.4, (April 2009), pp. 320–324, ISSN 1476-1122

Takahashi, K.; Wang, Y. & Cao, G. (2006). Fabrication and Li^+-intercalation properties of V_2O_5-TiO_2 composite nanorod arrays. *Applied Physics A-Materials Science & Processing,* Vol.82, No.1, (January 2006), pp.27–31, ISSN 0947-8396

Thackeray, M.-M.; Kang, S.-H.; Johnson, C.-S.; Vaughey, J.T.; Benedek, R. & Kackney, S.A. (2007). Li_2MnO_3-stabilized $LiMO_2$ (M = Mn, Ni, Co) electrodes for lithium-ion batteries. *Journal of Materials Chemistry,* Vol.17, No.30, (Apirl 2007), pp.3112–3125, ISSN 0959-9428

Vu, A. & Stein, A. (2011). Multiconstituent Synthesis of $LiFePO_4$/C Composites with Hierarchical Porosity as Cathode Materials for Lithium Ion Batteries. *Chemistry of Materials,*, DOI: 10.1021/cm201197j, ISSN 0897-4756

Wang, G.; Yang, L.; Chen, Y.; Wang, J.; Bewlay, S. & Liu, H. (2005). An investigation of polypyrrole-$LiFePO_4$ composite cathode materials for lithium-ion batteries. *Electrochimica Acta,* Vol.50, No.24, (August 2005), pp. 4649–4654, ISSN 0013-4686

Wang, H.-L.; Cui, L.-F.; Yang, Y.-A.; Casalongue, H.-S.; Robinson, J.-T.; Liang, Y.-Y.; Cui, Y. & Dai, H.-J. (2010). Mn_3O_4-graphene hybrid as a high-capacity anode material for lithium ion batteries. *Journal of the American Chemical Society,* Vol.132, No.40, (October 2010), pp.13978–13980, ISSN 0002-7863

Wang, H.; Zhang, W.; Zhu, L. & Chen, M. (2007). Effect of $LiFePO_4$ coating on electrochemical performance of $LiCoO_2$ at high temperature. *Solid State Ionics,* Vol.178, No.1-2, (January 2007), pp. 131–136, ISSN 0167-2738

Wang, Y. & Cao, G. (2008). Developments in nanostructured cathode materials for high-performance lithium-ion batteries. *Advanced Materials,* Vol.20, No.12, (June 2008), pp. 2251–2269, ISSN 0935-9648

Wang, Y.; Wang, Y.; Hosono, E.; Wang, K. & Zhou, H. (2008). The design of a $LiFePO_4$/carbon nanocomposite with a core-shell structure and its synthesis by an in situ polymerization restriction method. *Angewandte Chemie-International Edition,* Vol.47, No.39, (August 2008), pp. 7461–7465, ISSN 1433-7851

Wang, Y.; Wu, M.; Jiao, Z. & Lee, J. (2009). Sn@CNT and Sn@C@CNT nanostructures for superior reversible lithium ion storage. *Chemistry of Materials,* Vol.21, No.14, (July 2009), pp. 3210–3215, ISSN 0897-4756

Winter, M.; Besenhard, J.; Spahr, M. & Novak, P. (1998). Insertion electrode materials for rechargeable lithium batteries. *Advanced Materials,* Vol.10, No.10, (July 1998), pp. 725–763, ISSN 0935-9648

Wu, F. & Wang, Y. (2011). Self-assembled echinus-like nanostructures of mesoporous CoO nanorod@CNT for lithium-ion batteries. *Journal of Materials Chemistry,* Vol.21, No.18, (March 2011), pp. 6636–6641, ISSN 0959-9428

Wu, H.-M.; Belharouak, I.; Abouimrane, A.; Sun, Y.-K. & Amine, K. (2010). Surface modification of $LiNi_{0.5}Mn_{1.5}O_4$ by ZrP_2O_7 and ZrO_2. *Journal of Power Sources,* Vol.195, No.9, (May 2010), pp.2909–2913, ISSN 0378-7753

Wu, P.; Du, N.; Zhang, H.; Yu, J. & Yang, D. (2010). CNTs@SnO_2@C coaxial nanocables with highly reversible lithium storage. *Journal of Physical Chemistry C,* Vol.114, No.51, (December 2010), pp. 22535–22538, ISSN 1932-7447

Wu, Z.; Ren, W.; Wen, L.; Gao, L.; Zhao, J.; Chen, Z.; Zhou, G.; Li, F. & Cheng, H. (2010). Graphene anchored with Co_3O_4 nanoparticles as anode of lithium ion batteries with enhanced reversible capacity and cyclic performance. *ACS Nano,* Vol.4, No.6, (January 2010), pp. 3187–3194, ISSN 1936-0851

Wu, Z.-S.; Ren, W.; Wen, L.; Gao, L.; Zhao, J.; Chen, Z.; Zhou, G.; Li, F. & Cheng, H.-M. (2010). Graphene anchored with Co_3O_4 nanoparticles as anode of lithium ion batteries with enhanced reversible capacity and cyclic performance. *ACS Nano,* Vol.4, No.6, (June 2010), pp. 3187–3194, ISSN 1936-0851

Xia, H.; Meng, Y.-S.; Lu, L. & Ceder, G. (2007). Electrochemcial properties of nonstoichiometric $LiNi_{0.5}Mn_{1.5}O_{4-\delta}$ thin-film electrodes prepared by pulsed laser deposition. *Journal of the Electrochemical Society,* Vol.154, No.8, (May 2007), pp.A737–A743, ISSN 0013-4651

Xiao, J.; Choi, D.; Cosimbescu, L.; Koech, P.; Liu, J. & Lemmon, J.-P. (2010). Exfoliated MoS_2 nanocomposite as an anode material for lithium ion batteries. *Chemistry of Materials,* Vol.22, No.16, (August 2010), pp. 4522–4524, ISSN 0897-4756

Xiong, H.; Shen, W.; Guo, B.; Bo, S.; Cui, W.; Chen, L.; Li, H. & Xia, Y. (2011). Anomalous lithium storage in a novel nanonet composed by SnO_2 nanoparticles and poly(ethylene glycol) chains. *Journal of Materials Chemistry,* Vol.21, No.9, (January 2011), pp. 2845–2847, ISSN 0959-9428

Xue, L.; Wei, Z.; Li, R.; Liu, J.; Huang, T. & Yu, A. (2011). Design and synthesis of Cu_6Sn_5-coated TiO_2 nanotube arrays as anode material for lithium ion batteries. *Journal of Materials Chemistry,* Vol.21, No.9, (January 2011), pp.3216–3220, ISSN 0959-9428

Yamada, H.; Tagawa, K.; Komatsu, M.; Moriguchi, I. & Kudo, T. (2007). High power battery electrodes using nanoporous V_2O_5/carbon composites. *Journal of Physical Chemistry C,* Vol.111, No.23, (June 2007), pp. 8397–8402, ISSN 1932-7447

Yang, S.; Feng, X.; Ivanovici, S. & Mullen, K. (2010). Fabrication of graphene-encapsulated oxide nanoparticles: towards high-performance anode materials for lithium storage. *Angewandte Chemie-International Edition,* Vol.49, No.45, (September 2010), pp. 8408–8411, ISSN 1433-7851

Yang, T.; Zhang, N.; Lang, Y. & Sun, K. (2011). Enhanced rate performance of carbon-coated $LiNi_{0.5}Mn_{1.5}O_4$ cathode material for lithium ion batteries. *Electrochimica Acta*, Vol.56, No.11, (April 2011), pp. 4058–4064, ISSN 0013-4686

Yang, Z.; Du, G.; Guo, Z.; Yu, Xe.; Chen, Z.; Guo, T. & Liu, H. (2011). TiO_2(B)@carbon composite nanowires as anode for lithium ion batteries with enhanced reversible capacity and cyclic performance. *Journal of Materials Chemistry*, Vol.21, No.24, (May 2011), pp. 8591–8596, ISSN 0959-9428

Yan, J.; Sumboja, A.; Khoo, E. & Lee, P. (2011). V_2O_5 loaded on SnO_2 nanowires for high-rate Li ion batteries. *Advanced Materials*, Vol.23, No.6, (February 2011), pp. 746–750, ISSN 0935-9648

Yi, T.-F.; Zhu, Y.-R.; Zhu, X.-D.; Shu, J.; Yue, C.-B. & Zhou, A.-N. (2009). A review of recent developments in the surface modification of $LiMn_2O_4$ as cathode material of power lithium-ion battery. *Ionics*, Vol.15, No.6, (December 2009), pp. 779–784, ISSN 0947-7047

Yi, T.; Shu, J.; Wang, Y.; Xue, J.; Meng, J.; Yue, C. & Zhu, R. (2011). Effect of treated temperature on structure and performance of $LiCoO_2$ coated by $Li_4Ti_5O_{12}$. *Surface & Coatings Technology*, Vol.205, No.13-14, (March 2011), pp. 3885–3889, ISSN 0257-8972

Yu, W.-J.; Hou, P.-X.; Zhang, L.-L.; Li, F.; Liu, C. & Cheng, H.-M. (2010). Preparation and electrochemical property of Fe_2O_3 nanoparticles-filled carbon nanotubes. *Chemical Communications*, Vol.46, No.45, (October 2010), pp. 8576–8578, ISSN 1359-7345

Zhang, H.-X.; Feng, C.; Zhai, Y.-C.; Jiang, K.-L.; Li, Q.-Q. & Fan, S.-S. (2009). Cross-stacked carbon nanotube sheets uniformly Loaded with SnO_2 nanoparticles: A novel binder-free and high-capacity anode material for lithium-ion batteries. *Advanced Materials*, Vol.21, No.22, (January 2009), pp.2299–2304, ISSN 0935-9648

Zhang, M.; Lei, D.; Yin, X.; Chen, L.; Li, Q.; Wang, Y. & Wang, T. (2010). Magnetite/graphene composites: microwave irradiation synthesis and enhanced cycling and rate performances for lithium ion batteries. *Journal of Materials Chemistry*, Vol.20, No.26, (June 2010), pp. 5538–5543, ISSN 0959-9428

Zhang, S.; Du, Z.; Lin, R.; Jiang, T.; Liu, G.; Wu, X. & Weng, D. (2010). Nickel nanocone-array supported silicon anode for high-performance lithium-Ion batteries. *Advanced Materials*, Vol.22, No.47, (December 2010), pp. 5378–5382, ISSN 0935-9648

Zhang, W.-M.; Hu, J.-S.; Guo, Y.-G.; Zheng, S.-F.; & Zhong, L.-S. (2008). Tin-nanoparticles encapsulated in elastic hollow carbon spheres for high-performance anode material in lithium-ion batteries. *Advanced Materials*, Vol.20, No.6, (February 2008), pp. 1160–1165, ISSN 0935-9648

Zhou, X.; Wang, F.; Zhu, Y. & Liu., Z. (2011). Graphene modified $LiFePO_4$ cathode materials for high power lithium ion batteries. *Journal of Materials Chemistry*, Vol.21, No.10, (January 2011), pp. 3353–3358, ISSN 0959-9428

Zhou, Y.; Wang, J.; Hu, Y.; O'Hayre, R. & Shao., Z. (2010). A porous $LiFePO_4$ and carbon nanotube composite. *Chemical Communications*, Vol.46, No.38, (August 2010), pp. 7151–7153, ISSN 1359-7345

Zhou, Z.-B.; Takeda, M.; Fujii, T. & Ue, M. (2005). $Li[C_2F_5BF_3]$ as an electrolyte salt for 4 V class lithium-ion cells. *Journal of the Electrochemical Society*, Vol.152, No.2, (January 2005), pp.A351–A356, ISSN 0013-4651

Zhu, X.-J.; Zhu, Y.-W.; Murali, S.; Stollers, M.-D. & Ruoff, R.-S. (2011). Nanostructured reduced graphene oxide/Fe_2O_3 composite as a high-performance anode material for lithium ion batteries. *ACS Nano,* Vol.5, No.4, (April 2011), pp.3333–3338, ISSN 1936-0851

4

$LiNi_{0.5}Mn_{1.5}O_4$ Spinel and Its Derivatives as Cathodes for Li-Ion Batteries

Liu Guoqiang
School of Material and Metallurgy, Northeastern University, Shenyang, China

1. Introduction

It is well known that lithium-ion batteries are common in consumer electronics. It is one of the most popular types of rechargeable battery for portable electronics, with the best energy densities, no memory effect, and a slow loss of charge when not in use [1, 2]. Beyond consumer electronics, LIBs are also growing in popularity for military, electric vehicle, and aerospace applications. Its excellent properties originate from its materials including cathode, anode and electrolyte and so on. For cathode materials, there are mainly three kinds of materials which have been widely studied and applied commercially, including layered oxide $LiCoO_2$, spinel $LiMn_2O_4$ and olivine $LiFePO_4$. Among the cathode materials, $LiCoO_2$ has been used since the invention of LIB [3], while $LiMn_2O_4$ and $LiFPO_4$ are considered as promising ones due to less toxicity, low cost, more safety and good electrochemical properties [4, 5]. In term of redox energy level, these materials can be charged and discharged at around 4 V, which limits their energy density. The spinel $LiNi_{0.5}Mn_{1.5}O_4$ is becoming a research focus recently. The most remarkable property of spinel $LiNi_{0.5}Mn_{1.5}O_4$ is its discharge voltage plateau at around 4.7 V. In some cases, using $LiNi_{0.5}Mn_{1.5}O_4$ will lead fewer cells at the battery pack level. For example, hundreds of ordinary lithium ion batteries are needed to meet the requirement of electric vehicle (EV) in the state of start-up, accelerate and climb-up [6] because more energy is needed in this case. If the high voltage cells are utilized, the amount of batteries used for EV can decrease greatly. This chapter gives a detailed introduction on $LiNi_{0.5}Mn_{1.5}O_4$ spinel and the latest research advances in this area.

2. Structures of $LiNi_{0.5}Mn_{1.5}O_4$

There are two kinds of crystal structure for spinel $LiNi_{0.5}Mn_{1.5}O_4$, i.e. face-centered spinel ($Fd3m$) and primitive simple cubic crystal ($P4_332$). For $LiNi_{0.5}Mn_{1.5}O_4$ with a face-centered structure ($Fd3m$), the lithium ions are located in the 8a sites of the structure, the manganese and nickel ions are randomly distributed in the 16d sites. The oxygen ions which are cubic-close-packed (ccp) occupy the 32e positions. For $LiNi_{0.5}Mn_{1.5}O_4$ ($P4_332$) with a primitive simple cubic structure, the manganese ions are distributed in 12d sites, and nickel ions in 4a sites. The oxygen ions occupy the 24e and 8c positions, while the lithium ions are located in the 8c sites. In this case, the Ni and Mn ions are ordered regularly [7-9]. Whether $LiNi_{0.5}Mn_{1.5}O_4$ has a structure of face-centered spinel ($Fd3m$) or primitive simple cubic

($P4_332$) depends on its synthetic routes. In synthesizing $LiNi_{0.5}Mn_{1.5}O_4$, annealing process at 700°C after calcination led to the ordering of Ni and Mn ions, making it transformed from face-centered spinel ($Fd3m$) to primitive cubic crystal ($P4_332$). Schematic drawing of the structures of $LiNi_{0.5}Mn_{1.5}O_4$ is shown in Fig. 1 [10].

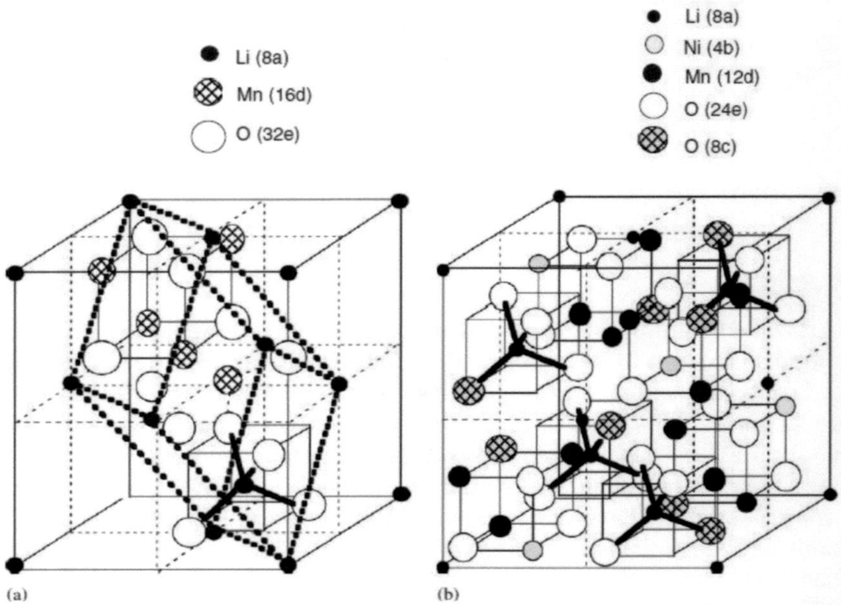

Fig. 1. Schematic drawing of the structures of $LiNi_{0.5}Mn_{1.5}O_4$ spinel lattice: a) face-centered spinel ($Fd3m$) b) primitive simple cubic ($P4_332$) [10]

Infrared spectroscopy is an effective method to distinguish these two structures. Infrared spectra of ordered ($P4_332$) and disordered ($Fd3m$) $LiNi_{0.5}Mn_{1.5}O_4$ exhibit different patterns between 650 and 450 cm^{-1}. At this band range, there are apparent spectra at 588 and 430 cm^{-1} for ordered $LiNi_{0.5}Mn_{1.5}O_4$. The intensity ratio of two bands at 619 and 588 cm^{-1} can be used qualitatively to assess percentage of ordering in spinel which contains both ordered and disordered $LiNi_{0.5}Mn_{1.5}O_4$ [11].

The diffusion path of Li in the spinel structure is a three-dimensional network. Lithium moves from one tetrahedral site to the next through a vacant octahedral site. The activation barriers of migration are greatly influenced by the electrostatic repulsion from the nearest transition metal. Because the distribution of Ni and Mn is different in ordered ($P4_332$) and disordered ($Fd3m$) $LiNi_{0.5}Mn_{1.5}O_4$, the activation barriers for migration of Li will be different from each other. Although the previous studies showed that disordered $LiNi_{0.5}Mn_{1.5}O_4$ exhibited better cycling performance than ordered $LiNi_{0.5}Mn_{1.5}O_4$ at high rates [12-13], a recent study shows that the activation barriers for Li ion transportation in ordered ($P4_332$) $LiNi_{0.5}Mn_{1.5}O_4$ can be as low as around 300 meV according to first-principles calculation, so the ordered $LiNi_{0.5}Mn_{1.5}O_4$ can exhibit good cycle ability at high current rates [14].

In the synthesis of $LiNi_{0.5}Mn_{1.5}O_4$, the high calcination temperature sometimes leads to the reduction of the Mn oxidation state from +4 to +3, which results in the formation of Fd3m structure. When annealed at 700 °C in air after a high-temperature calcination at 1000 °C, the resulting powders does not contain Mn^{3+} [15]. It was reported that $LiNi_{0.5}Mn_{1.5}O_4$ synthesized under O_2 atmosphere has the cubic spinel structure with a space group of $P4_332$ instead of $Fd3m$ [16, 17].

3. Mechanism of high voltage and insertion/deinsertion

Based on the results obtained with the systems $LiMn_{2-y}Ni_yO_4$ and $LiCr_yMn_{2-y}O_4$, Dahn and Sigala [18, 19] previously pointed out that the high voltage originated from the oxidation of nickel and chromium ion. The 4.1 V plateau was related to the oxidation of Mn^{3+} to Mn^{4+} and the 4.7 V plateau to the oxidation of Ni^{2+} to Ni^{4+}. The oxidation of chromium ion could bring about a high voltage of 4.9 V. Yang [20] suggested that a significant amount of Mn^{4+} ion in the spinel framework was essential for electrochemical reaction to occur at around 5 V. His view was supported by Kawai [21] who argued that the presence of manganese was necessary to keep the high voltage capacity because manganese-free spinel oxides, such as $Li_2NiGe_3O_8$, did not show any capacity above 4.5 V. The influence of doping metals including M = Cu [22-24], Co [25], Cr [26-29], Fe [30-32], Al [33, 34], and Zn [35] on the properties of $LiM_{0.5}Mn_{1.5}O_4$ have been investigated. Among these materials, Ni-doped compound $LiM_{0.5}Mn_{1.5}O_4$ displays higher capacity and better cycle ability. For spinel $LiNi_{0.5}Mn_{1.5}O_4$, there is a capacity occurring at 4.6-4.7 V, which can be attributed to a two electron process, Ni^{2+}/Ni^{4+}. While in the 4 V region, the electrode sometimes shows some minor redox behavior, related to the Mn^{3+}/Mn^{4+} couple. When there are more Mn^{4+} and Ni^{2+} in $LiNi_{0.5}Mn_{1.5}O_4$, then the corresponding capacity at 4 V will be less and that at 5 V will be large. [36, 37].

Gao [38] put forward an explanation for the origin of high voltage. As an electron is removed from Mn^{3+}, it is removed from Mn eg (↑) which has an electron binding energy at around 1.5–1.6 eV, and this accounts for the 4.1V plateau. When there are no more electrons left on Mn eg (↑) (all Mn are Mn^{4+}), electrons are removed from Ni eg (↑↓) which has an electron binding energy of about 2.1 eV, and the voltage plateau moves up to 4.7 V because of the increased energy needed to remove electrons.

Terada [39] studied the mechanism of the oxidation reaction during Li deintercalation by measuring the *in situ* XAFS spectra of $Li_{1-x}(Mn,M)_2O_4$ (M=Cr, Co, Ni). It is found from the Ni K-edge XAFS analysis that Ni in $Li_{1-x}Mn_{1.69}Ni_{0.31}O_4$ experiences three distinct valence states during Li deintercalation, Ni^{2+}, Ni^{3+} and Ni^{4+}. The X-ray absorption near-edge structures (XANES) of Mn and M shows that the high voltage (~5 V) in the cathode materials is due to the oxidation of M^{3+} to M^{4+} (M = Cr, Co), and M^{2+} to M^{4+} (M = Ni). The origin of the low voltage (3.9-4.3 V) is ascribed to the oxidation of Mn^{3+} to Mn^{4+}.

Ariyoshi [40] reported that the reaction at *ca.* 4.7 V consisted of two cubic/cubic two-phase, *i.e.* □ $[Ni_{1/2}Mn_{3/2}]O_4$ was reduced to $Li[Ni_{1/2}Mn_{3/2}]O_4$ via $□_{1/2}$ $[Ni_{1/2}Mn_{3/2}]O_4$.The flat voltage at 4.7 V consists of two voltages of 4.718 and 4.739 V. The reaction of $Li[Ni_{1/2}Mn_{3/2}]O_4$ to $Li_2[Ni_{1/2}Mn_{3/2}]O_4$ proceeds into a cubic/tetragonal two-phase reaction with the reversible potential at 2.795 V.

4. Synthesis of $LiNi_{0.5}Mn_{1.5}O_4$

There are mainly two kinds of methods to synthesize electrode materials for lithium ion batteries, i.e. solid-state reaction method and wet chemical method. Solid-state reaction method is simple and suitable for mass production. However, it is difficult to obtain pure products by this method. Some impurities containing nickel oxide usually exist in the products. Among the impurities, $Li_xNi_{1-x}O$ is also related to the loss of oxygen at high temperatures. The capacity of $LiNi_{0.5}Mn_{1.5}O_4$ prepared through solid-state reaction is about only 120 mAh g^{-1} [41]. Fang [42] prepared $LiNi_{0.5}Mn_{1.5}O_4$ by an improved solid-state reaction. He used appropriate amounts of Li_2CO_3, NiO and electrolytic MnO_2 as reactants. After being thoroughly ball-milled, the mixed precursors were heated up to 900 °C, then directly cooled down to 600 °C and heated for 24 h in air. The heating and cooling rates were about 30 °C/min and 10 °C/min, respectively. The product could deliver 143 mAh g^{-1} at 5/7C and still retained 141 mAh g^{-1} after 30 cycles. Fang also synthesized $LiNi_{0.5}Mn_{1.5}O_4$ using a one-step solid-state reaction at 600 °C in air. The prepared product delivered up to 138 mAh g^{-1}, and the capacity retained 128 mAh g^{-1} after 30 cycles [43]. Recently Chen employed a mechanical activated solid state reaction from stoichiometric amount of $Ni(NO_3)_2 \cdot 6H_2O$, MnO_2 and Li_2CO_3 to prepare $LiNi_{0.5}Mn_{1.5}O_4$. Its reversible capacity was about 145 mAh g^{-1} and remained 143 mAh g^{-1} after 10 cycles [44]. Other solid-state reaction reports have also been reported [45-56]. Wet chemical methods make the reactions take place among reactants at the molecular level. It is common that after precursors are obtained by wet method, less energy or lower reaction temperature are needed to turn the precursors into final products. Wet chemical methods include coprecipitation method [57-60], polymer-pyrolysis method [61, 62], ultrasonic-assisted co-precipitation (UACP) method [63, 64], sol-gel method [65-67], radiated polymer gel method [68], sucrose-aided combustion method [69], spray-drying method [70], emulsion drying method [71], composite carbonate process [72], molten salt method [73, 74], mechanochemical process [75], poly (methyl methacrylate) (PMMA)- assisted method [76] ultrasonic spray pyrolysis [77], polymer-assisted synthesis [78], combinational annealing method [79], pulsed laser deposition [80], electrophoretic deposition [81], spin-coating deposition [82], carbon combustion synthesis [83], soft combustion reaction method [84], pulsed laser deposition [85], spray drying and post-annealing [86], rheological method [87], polymer-mediated growth [88], self-reaction method [89], internal combustion type spray pyrolysis [90, 91], a chloride-ammonia co-precipitation method [92], a novel carbon exo-templating method [93], flame type spray pyrolysis [94], self-combustion reaction (SCR) [95] and so on. Comparing to solid-state reaction method, the electrochemical properties of $LiNi_{0.5}Mn_{1.5}O_4$ are much improved by these methods.

As for the impurity phase of $Li_xNi_{1-x}O$ in product, it is believed that they come from the loss of oxygen at high temperatures. The tetravalent manganese (Mn^{4+}) is unstable at high temperatures and can be converted to trivalent (Mn^{3+}) so that oxygen may partially evolve out of the lattice to form $LiNi_{0.5}Mn_{1.5}O_{4-y}$. When x value becomes large, this phase becomes unstable and may decompose into two phases, i.e., $LiNi_{0.5-z}Mn_{1.5}O_{4-y}$ and $Li_xNi_{1-x}O$. The overall reaction process can be depicted as follows:

$$LiNi_{0.5}Mn_{1.5}O_4 \rightarrow qLi_xNi_{1-x}O + LiNi_{0.5-z}Mn_{1.5}O_{4-y} + sO_2$$

To get rid of the impurities, an annealing process after the high temperature treatment is usually necessary. It is acknowledged that impurity $Li_xNi_{1-x}O$ can deteriorate the electrochemical properties of products. However, so far there have not been special researches about how the impurity $Li_xNi_{1-x}O$ affects the electrochemical performances of products. In order to further investigate the effect of impurity $Li_xNi_{1-x}O$. The $LiNi_{0.5}Mn_{1.5}O_4$ compounds were synthesized by solid state reaction method. Fig. 2 (a) and (b) show the XRD patterns of two products. One $LiNi_{0.5}Mn_{1.5}O_4$ (1) was synthesized at 850 ^{0}C for 12 h, and the other $LiNi_{0.5}Mn_{1.5}O_4$ (2) was synthesized at 850 ^{0}C for 12 h and then annealed at 600 ^{0}C for 12 h. The reference material $Li_{0.26}Ni_{0.72}O$ is also illustrated in Fig. 2 (a). It can be seen that there are small peaks at 37.5^0, 43.6^0 and 63.3^0 in the patterns of two products, illustrating that there was a secondary phase $Li_xNi_{1-x}O$. The intensity of the impurity $Li_xNi_{1-x}O$ peaks decreased due to the annealing process.

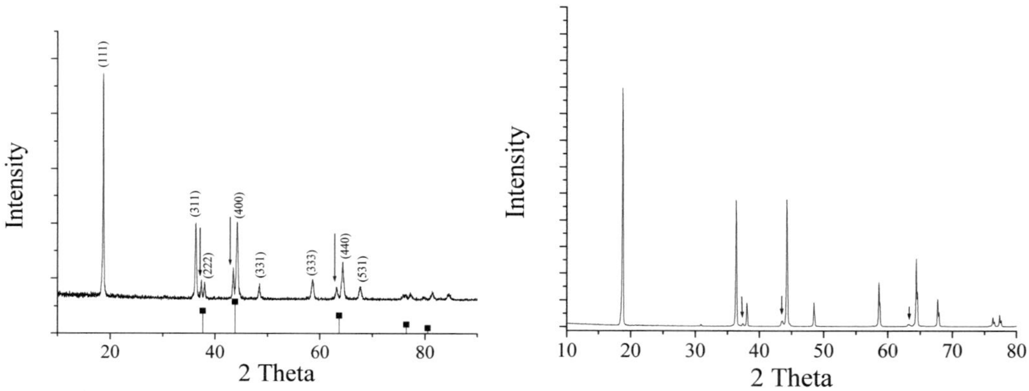

Fig. 2. a) XRD patterns of $LiNi_{0.5}Mn_{1.5}O_4$ (1) and $Li_xNi_{1-x}O$
b) XRD patterns of $LiNi_{0.5}Mn_{1.5}O_4$ (2)

Figure 3 (a) shows the charge-discharge curves of $LiNi_{0.5}Mn_{1.5}O_4$ (1) which was synthesized without annealing process. Its discharge capacities were 121.5 mAh g^{-1} at 0.2 C and 117.6 mAh g^{-1} at 0.7 C, respectively. The cycle performance at 0.7 C was displayed in Fig. 3 (b). It can be found that there is only small capacity decay after 50 cycles. The theoretical capacity of $LiNi_{0.5}Mn_{1.5}O_4$ is about 148 mAh g^{-1}. There is a capacity of about 26 mAh g^{-1} that is not delivered by the sample $LiNi_{0.5}Mn_{1.5}O_4$ (1).

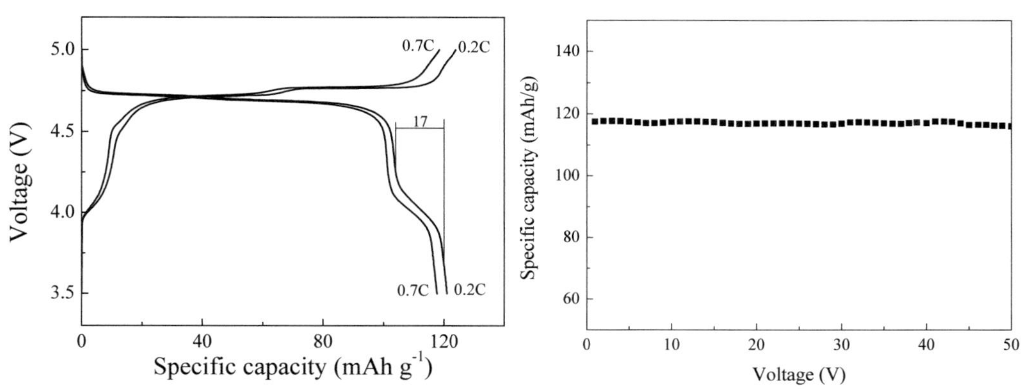

Fig. 3. a) Charge/discharge curves for $LiNi_{0.5}Mn_{1.5}O_4$ (1)
b) Cycle performances of $LiNi_{0.5}Mn_{1.5}O_4$ (1)

Figure 4 (a) illustrates the charge-discharge curves of the sample $LiNi_{0.5}Mn_{1.5}O_4$ (2). This test was conducted at 0.5 C charge current and different discharge current rates. The discharge capacities were 119.5 mAh g^{-1} at 0.5 C and 116.3 mAh g^{-1} at 1 C, respectively. The specific capacity around 4 V was about 13.6 mAh g^{-1}. It is less than that of $LiNi_{0.5}Mn_{1.5}O_4$ (1). The specific capacity of sample $LiNi_{0.5}Mn_{1.5}O_4$ (1) around 4 V was about 17.0 mAh g^{-1}. This proves that there was less amount of Mn^{3+} in sample $LiNi_{0.5}Mn_{1.5}O_4$ (2) than sample $LiNi_{0.5}Mn_{1.5}O_4$ (1). The reason is that there is less oxygen deficiency due to the annealing process.

The cycle performances of sample $LiNi_{0.5}Mn_{1.5}O_4$ (2) are shown in Fig. 4 (b). Its discharge capacities at 2 C and 4 C were 107.5 and 98.5 mAh g^{-1}, respectively. The capacity retention was good for every current rate.

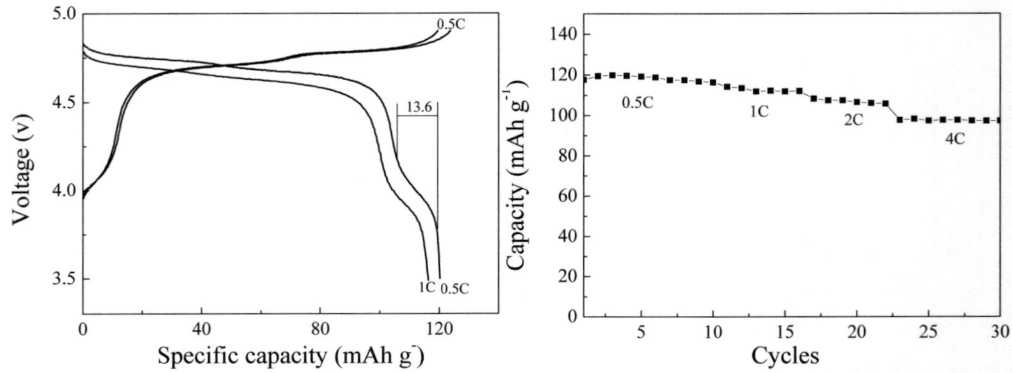

Fig. 4. a) Charge/discharge curves for $LiNi_{0.5}Mn_{1.5}O_4$ (2)
b) Cycle performances of $LiNi_{0.5}Mn_{1.5}O_4$ (2)

The above results demonstrate that the impurity $Li_xNi_{1-x}O$ can reduce the specific capacity of $LiNi_{0.5}Mn_{1.5}O_4$. However, there is no obvious evidence that the impurity $Li_xNi_{1-x}O$ impairs the cycle performances of products. According to previous reports, the $Li_xNi_{1-x}O$ phase can be used as an anode material for lithium ion batteries, exhibiting good electrochemical properties. At 100 mA g^{-1}, its discharge capacity of the first cycle was up to 1480 mAh g^{-1} below 1.5 V [96].

5. Nano-sized $LiNi_{0.5}Mn_{1.5}O_4$ spinels

Nanostructure materials have both advantages and disadvantages for lithium batteries. The advantages include short path lengths for Li^+ transport, short path lengths for electronic transport, higher electrode/electrolyte contact area leading to higher charge/discharge rates, while the disadvantage include an increase in undesirable electrode/electrolyte reactions due to high surface area, leading to self-discharge, poor cycling and calendar life [97, 98].

Usually, nano-sized $LiNi_{0.5}Mn_{1.5}O_4$ can be obtained via wet chemical methods. In this process, the precursor compounds including Li, Ni and Mn salts are mixed homogenously at atomic scale. After further calcination, nano-sized $LiNi_{0.5}Mn_{1.5}O_4$ particles can be obtained at an low temperatures. When sintering temperature continues to go up, the particle size of $LiNi_{0.5}Mn_{1.5}O_4$ increase, and finally they will turn into micro-sized products.

The methods to synthesize nano-sized $LiNi_{0.5}Mn_{1.5}O_4$ include polymer-pyrolysis method [99], hydrothermal synthesis [100], thermal decomposition of acetate [101], composite carbonate process [102] and so on.

Based on the research results that have been reported, the formation temperature of spinel phase is as low as 450 °C, whereas the growth of integrated $LiNi_{0.5}Mn_{1.5}O_4$ crystals takes place at relatively higher calcination temperature. The calcination temperature has significant effects on the structure and morphology of the materials so as to affect their electrochemical performance. The higher calcination temperature leads to higher crystallinity that helps to increase the electrode capacity while it may produce particles with relatively large size and long diffusion distances for lithium ions, which makes lithium ions insertion–extraction difficult. Therefore, with the combination of these two factors, the powders calcined at proper temperature will deliver the highest discharge capacity.

Some researches relative to nano-sized $LiNi_{0.5}Mn_{1.5}O_4$ spinels have been reported. In general, the nanometer particles exhibit a good performance at high rates due to the shortened diffusion paths, whereas at low rates the reactivity towards the electrolyte increases and the cell performance is lowered. Micrometric particles, which are less reactive towards the electrolyte, are a better choice for making electrodes under these latter conditions.

Recently, some improvements have been achieved. Nanometer $LiNi_{0.5}Mn_{1.5}O_4$ with good electrochemical performance over a wide range of rate capabilities by modifying the experimental synthetic conditions has also been reported. For example, Lafont [103] synthesized a nano material $LiMg_{0.05}Ni_{0.45}Mn_{1.5}O_4$ of about 50 nm in size with an ordered cubic spinel phase ($P4_332$) by auto-ignition method. It displayed good capacity retention of 131 mAh g^{-1} at C/10 and 90 mAh g^{-1} at 5C. By using a template method, Arrebola [104] synthesized $LiNi_{0.5}Mn_{1.5}O_4$ nanorods and nanoparticles using PEG 800 (PEG: polyethyleneglycol) as the template. Highly crystalline nanometric $LiNi_{0.5}Mn_{1.5}O_4$ of 70–80 nm was prepared at 800°C. Its electrochemical properties were measured at different charge/discharge rates of C/4, 2C, 4C, 8C and 15C, the capacity values were 121 mAh g^{-1} at 2C to 98 mAh g^{-1} at 15C, and faded slowly on cycling.

Hydrothermal synthesis includes various techniques of crystallizing substances from high-temperature aqueous solutions at high vapor pressures. The method is also particularly suitable for the growth of large good-quality crystals while maintaining good control over their composition. Now it is often used to synthesize nano scale materials including electrode materials for lithium ion batteries. Recently, it is reported that nano $LiNi_{0.5}Mn_{1.5}O_4$ was fabricated by this approach, and the products exhibited good performances. For example, $LiOH \cdot H_2O$, $MnSO_4 \cdot H_2O$, $NiSO_4 \cdot 6H_2O$, $(NH_4)_2S_2O_8$ were used as reactants, and they were dissolved in deionized water in a Teflon-lined stainless steel autoclave. Then, the autoclave was sealed and heated at 180°C for some time. The nano scale products were finally obtained. It delivered 100, 91, 74, and 73 mAh g^{-1} at current densities of 28, 140, 1400, and 2800 mA g^{-1}, respectively. The rate capability of such a nanosized 5 V spinel is better than those of a submicron $LiNi_{0.5}Mn_{1.5}O_4$ [105]. Fig. 5 (a) and (b) show the SEM photographs and charge-discharge curves, respectively.

Besides particle sizes, particle morphology and crystallinity also play a role in properties of materials. Kunduraci [106] synthesized a three dimensional mesoporous network structure with nanosize particles and high crystallinity. This morphology allows easy electrolyte penetration into pores and continuous interconnectivity of particles, yielding high power densities at fast discharges.

At present the electrode materials have reached their intrinsic limit, nano materials provide a new chance to improve their properties. It is no doubt that nano-sized electrode materials including nano $LiNi_{0.5}Mn_{1.5}O_4$ will gradually be applied in future high-energy lithium ion batteries. To realize the commercial application of nano materials, some technical obstacles such as undesirable electrode/electrolyte reactions due to high surface area, self-discharge and poor calendar life, etc have to be solved.

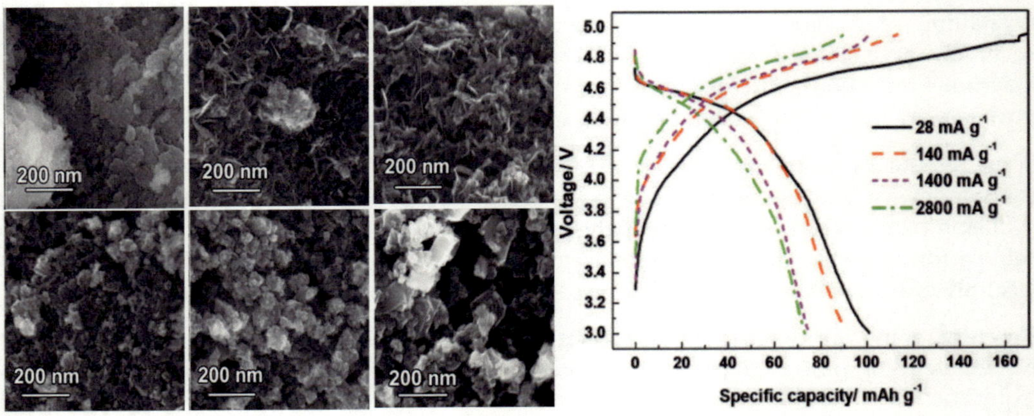

Fig. 5. a) SEM images of the samples synthesized in 1.1 M LiOH at 180°C
b) Charge-discharge curves of products [105], for various hydrothermal reaction times: (a) 0, (b) 2, (c) 6, (d) 12, (e) 24, (f) 336 h

6. Doping elements in $LiNi_{0.5}Mn_{1.5}O_4$ spinels

The structural and electrochemical properties of the $LiNi_{0.5}Mn_{1.5}O_4$ could also be affected by the substitution of other metal ions. Cation doping is considered to be an effective way to modify the intrinsic properties of electrode materials. Taking doping Cu as an example [107], the amount of Cu will affect the lattice parameters, the cation disorder in the spinel lattice, the particle morphology, as well as the electrochemical properties. In situ XAS experiment, the Cu K-edge XANES spectra of $LiCu_{0.25}Ni_{0.25}Mn_{1.5}O_4$ shows that the Cu valence only changes between 4.2 and 4.7 V. Therefore Cu can participate in the charge process in this range may be due to the oxidation of Cu^{2+} to Cu^{3+}. Although the reversible discharge capacity decreases with increasing Cu amount, optimized composition such as $LiCu_{0.25}Ni_{0.25}Mn_{1.5}O_4$ exhibits high capacities at high rates. In addition, the doping with appropriate amount in $LiNi_{0.5}Mn_{1.5}O_4$ can improve electrical conductivity, and help to improve electrochemical performances. For example, the electronic conductivity conductivities of the $LiNi_{0.5}Mn_{1.5}O_4$, $Li_{1.1}Ni_{0.35}Ru_{0.05}Mn_{1.5}O_4$, and $LiNi_{0.4}Ru_{0.05}Mn_{1.5}O_4$ measured from EIS at room temperature are 1.18×10^{-4}, 5.32×10^{-4}, and 4.73×10^{-4} S cm^{-1}, respectively. Although substitution of Ni^{2+} ions with heavier Ru^{4+} ions may reduce the theoretical capacity, the results show that a small doping content does not affect the accessible capacity at low current rates; on the contrary, larger accessible capacity can be obtained due to enhanced conductivity.

Figure 6 (a) and (b) show the scanning electron microscopy and charge-discharge curves. A remarkable cyclic performance at 1470 mA g^{-1} (10 C) charge/discharge rate is achieved for the $LiNi_{0.4}Ru_{0.05}Mn_{1.5}O_4$ synthesized by the polymer-assisted method, which can initially deliver 121 mAh g^{-1} and maintain about 82.6% of the initial capacity at the 500th cycle [108].

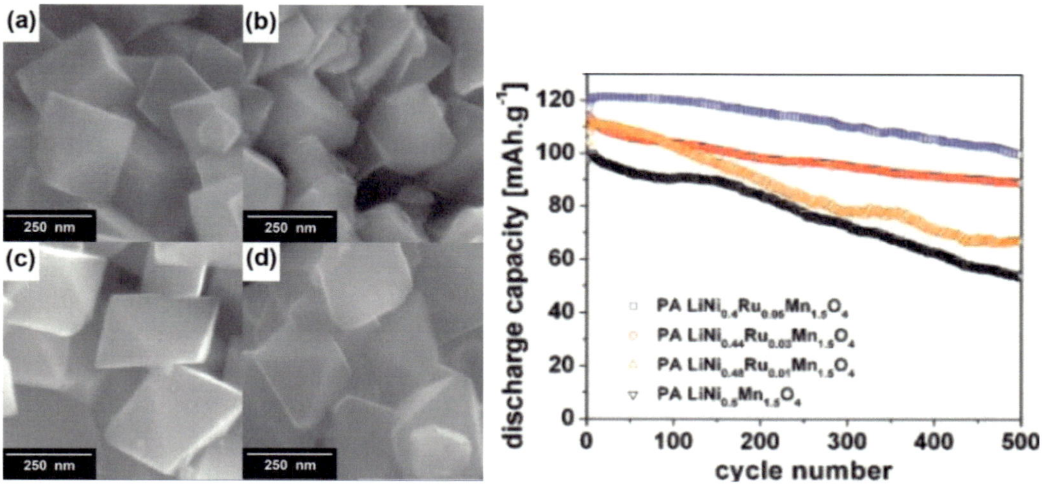

Fig. 6. a) Scanning electron microscopy of the
b) Capacity retention of PAS-$LiNi_{0.5-2x}Ru_xMn_{1.5}O_4$
PA-$LiNi_{0.5-2x}Ru_xMn_{1.5-x}O_4$: (a) x = 0, (b) x = 0.01, (c) x = 0.03, (d) x = 0.05 [126]

Generally speaking, doping elements can substitute for Ni or Mn in $LiNi_{0.5}Mn_{1.5}O_4$. For instance, element Cr will substitute for Ni in $LiNi_{0.5}Mn_{1.5}O_4$ because the ionic radius of Cr^{3+} is 0.615 Å which is close to that of Ni^{2+} (0.69 Å). Partial replacement of Ni in $LiNi_{0.5}Mn_{1.5}O_4$ with Cr is an effective approach to improve the electrochemical properties of $LiNi_{0.5}Mn_{1.5}O_4$ because the bonding energy of Cr–O is stronger than that of Mn–O and Ni–O. The stronger Cr–O bond is in favor of maintaining the spinel structure during cycling. This prevents the structural disintegration of the material. Besides replacing Ni, Mn in $LiNi_{0.5}Mn_{1.5}O_4$ can also be substituted for. In the case of Al doping, the ionic radius of Al^{3+} is 0.62 Å, which is nearly the same as that of Mn^{4+} (0.54 Å), so Al can substitute for Mn in $LiNi_{0.5}Mn_{1.5}O_4$. The strong Al–O bond is also beneficial to improving the electrochemical properties of $LiNi_{0.5}Mn_{1.5}O_4$. Doping with Fe has also achieved good experimental results. The $LiMn_{1.5}Ni_{0.42}Fe_{0.08}O_4$ delivered a capacity of 136 $mAhg^{-1}$ at C/6 rate with capacity retention of 100% in 100 cycles and a remarkably high capacity of 106 $mAhg^{-1}$ at 10 C rate. The material could deliver capacities of 143, 118 and 111 mAh g^{-1} at current densities of 1.0, 4.0 and 5.0 mA cm^{-2} with excellent capacity retention, respectively.

So far, many researches related to doping elements have been reported. These researches include doping Al [109], Fe [110-112], Cu [113], Co [114,115], Ti [116-118], Cr [119-123], Mg [124], Zn [125] and Ru [126].

Cation doping like doping Ru and Fe has achieved some encouraging results, improving the rate capability to a certain extent. Cation doping can improve conductivity, enlarge lattice constants and form stronger M-O bond, etc., which are favorable for the migration of lithium ions and maintaining stable crystal structure. When choosing appropriate element and amount better electrochemical properties can be expected. Perhaps electronic structure of the crystal can provide another theoretical explanation to the role of cation doping.

Besides cation doping, there are some researches relative to the substitution of small amount of F^- for O^{2-} anion [127-129]. In this case it is assumed that O^{2-} and F^- ions are located at the

32e sites. The doped compounds like $LiNi_{0.5}Mn_{1.5}O_{4-x}F_x$ have smaller lattice parameter than $LiNi_{0.5}Mn_{1.5}O_4$ because fluorine substitution changes the oxidation state of transition metal components and more Mn^{3+} ions with larger ionic radius (r = 0.645 Å) will replace Mn^{4+} ions (r = 0.53 Å) for electro-neutrality. The content of fluorine has influence on electrochemical properties of the doped compounds. On one hand, strong Li–F bonding may hinder Li^+ extraction, leading to a lower reversible capacity. On the other hand, fluorine doping makes spinel structure more stable due to the strong M–F bonding, which is favorable for the cyclic stability. According to the previous research report [129], the compound $LiNi_{0.5}Mn_{1.5}O_{3.9}F_{0.1}$ displayed good electrochemical properties of an initial capacity of 122 mAh g^{-1} and a capacity retention of 91% after 100 cycles. In addition, Oh [127] studied the effect of fluorine substitution on thermal stability. He reported that the $LiNi_{0.5}Mn_{1.5}O_4$ electrode had an abrupt exothermic peak at around 238.3 °C (1958 J g^{-1}) when charged to 5.0V, while $Li_\delta Ni_{0.5}Mn_{1.5}O_{3.9}F_{0.1}$ electrodes exhibited smaller exothermic peaks at higher temperatures, i.e., 246.3 °C (464.2 J g^{-1}). So fluorine substitution is advantageous for the thermal stability of $Li_\delta Ni_{0.5}Mn_{1.5}O_{4-x}F_x$ spinel.

7. Surface modification

Although surface modifications applied to high voltage material $LiNi_{0.5}Mn_{1.5}O_4$ are much less than those applied to cathode materials with layer structure like $LiCoO_2$, they are also effective ways to improve the properties of $LiNi_{0.5}Mn_{1.5}O_4$. It is believed that the high surface reactivity of the $LiNi_{0.5}Mn_{1.5}O_4$ with the electrolyte at high operating voltage results in the formation of SEI film, which significantly hinders the insertion/extraction of Li^+ ion, the charge transfer and hence the kinetics of the electrochemical processes. In order to improve its electrochemical behavior, coating the electrode material $LiNi_{0.5}Mn_{1.5}O_4$ with chemically stable compounds has been applied. The coating layer can hinder the formation of SEI film, and protect cathode materials from being attacked by HF. So far, surface modification of 5 V spinels has been limited mainly to Bi_2O_3[130], Al_2O_3[131, 132], ZnO [133-135], Li_3PO_4 [136], SiO_2 [137], Zn [138], Au [139], $AlPO_4$ [140], ZrP_2O_7 [141], BiOF [142] which lead to better cycle performance and rate capability retention. However, the effect of coating the nanometric spinel $LiNi_{0.5}Mn_{1.5}O_4$ with Ag on its rate capability was negative [143]. According to Liu [131], Al_2O_3-modified sample exhibited the best cyclability (99% capacity retention in 50 cycles) with a capacity of 120 mAh g^{-1}, while Bi_2O_3-coated sample exhibited the best rate capability. At a rate of 10C, the Bi_2O_3-coated sample could deliver a capacity of about 90 mAh g^{-1} after 50 cycles. Liu [132] thought that Al_2O_3 reacted with the surface of $LiMn_{1.42}Ni_{0.42}Co_{0.16}O_4$ during the annealing process and formed $LiAlO_2$ that exhibited good lithium-ion conductivity. Therefore, "Al_2O_3" modification layer acts as both a protection shell and as a fast lithium-ion diffusion channel, rendering both excellent cycling performance and good rate capability for the Al_2O_3-coated $LiMn_{1.42}Ni_{0.42}Co_{0.16}O_4$. Similarly, Bi_2O_3 is reduced on the cathode surface during electrochemical cycling to metallic Bi, which is an electronic conductor, rendering both excellent rate capability and good cycling performance for the Bi_2O_3-coated $LiMn_{1.42}Ni_{0.42}Co_{0.16}O_4$. In addition, the microstructure of the surface modification layer plays an important role in determining the electrochemical performances of the active material. Some experimental results indicate that the surface modifications neither change the bulk structure nor cause any change in the cation disorder of the spinel sample. In addition, electrolyte is easy to decompose on the surface of the 5 V spinel cathodes because of the higher operating voltage, resulting in the formation of thick

SEI layers. The Al_2O_3 coating is the most effective in suppressing of the development of the SEI layer. Thin SEI layer allow lithium-ion conduction. Fig. 7 shows the TEM images and rate capabilities of 2 wt % Al_2O_3-coated $LiMn_{1.42}Ni_{0.42}Co_{0.16}O_4$.

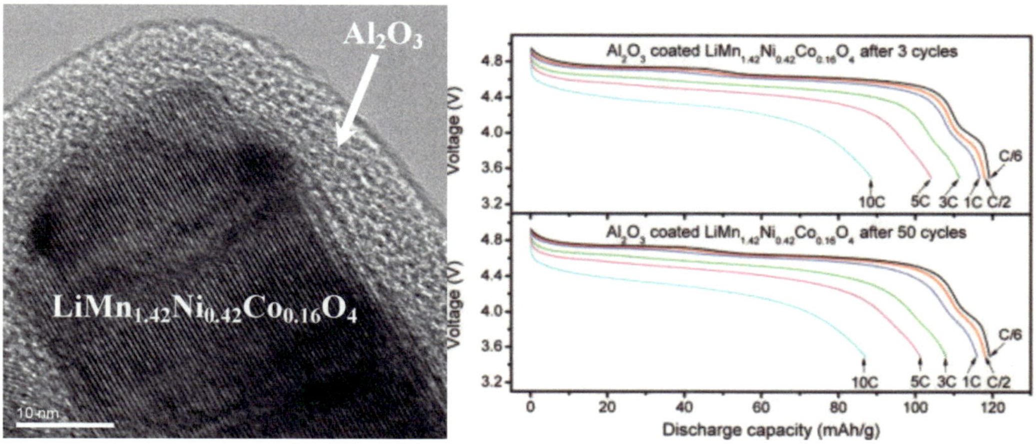

Fig. 7. a) TEM images of Al_2O_3-coated $LiMn_{1.42}Ni_{0.42}Co_{0.16}O_4$
b) Discharge profiles illustrating rate capabilities [132]

Besides the above mentioned coating layers, coating carbon should also be a good choice because it is a better conductor. Recently, it was reported that the carbon-coated material $LiNi_{0.5}Mn_{1.5}O_4$ was synthesized by a sol–gel method. The XRD patterns demonstrate that the spinel structure is not affected after coating the $LiNi_{0.5}Mn_{1.5}O_4$ powder with carbon. The lattice parameter was 0.8178 nm for pristine $LiNi_{0.5}Mn_{1.5}O_4$, while lattice parameters of $LiNi_{0.5}Mn_{1.5}O_4$ coated with different amount of carbon varied from 0.8171 to 0.8177 nm. The carbon layer was consecutive, and the thickness range of carbon layer was from approximate 10 to 20 nm. The carbon coating made the powders coarser and more agglomerated. The conductive carbon layer not only avoided the direct contact between the active cathode material and the electrolyte, but also provided pathways for electron transfer. Accordingly, the electrochemical properties of $LiNi_{0.5}Mn_{1.5}O_4$ were also improved duo to carbon layer, for example, when the $LiNi_{0.5}Mn_{1.5}O_4$ was modified with optimal 1wt.% sucrose, its discharge capacity could reach 130 mAh g^{-1} at 1 C discharge rate with a high retention of 92% after 100 cycles and a high 114 mAh g^{-1} at 5 C discharge rate [144].

8. Fabrication advanced LI-Ion batteries

The 5V material $LiNi_{0.5}Mn_{1.5}O_4$ has also been considered in new lithium-ion battery system. The $LiNi_{0.5}Mn_{1.5}O_4/Li_4Ti_5O_{12}$ (LNMO/LTO) cell is a good example [145-154].

The graphite anode with brittle layer structure can suffer from exfoliation when lithium ion inserts into its structure in electrolyte, which deteriorates the properties of batteries. Also, it is believed that the operating potential plateau of the carbon anode is close to that of metal lithium so that "dendrite" could still be unavoidable. And the solid electrolyte interface (SEI) layer on the carbon electrode, which is usually formed at the potential below 0.8V versus Li^+/Li and accompanied over time with active lithium loss, an increase in impedance and a decrease in specific capacity, limits the lifetime and rate capability of the lithium-ion batteries. Furthermore, there are some other drawbacks, such as thermal stability concerns, and the bad compatibility with propylene carbonate-based electrolytes and some functional electrolytes, i.e. the flame-retardant electrolytes containing phosphates or phosphonate. The spinel $Li_4Ti_5O_{12}$ (LTO) has been considered as a zero-strain insertion material duo to its excellent reversibility and structural stability in the charge–discharge process. So it is a promising alternative anode material. In addition, its structure remains nearly unchanged in PC-containing electrolyte, which makes batteries safer than those with graphite anode [155, 156]. In recent years, there have been some researches about full lithium-ion cells. Many 2V lithium-ion battery systems have been studied, such as $LiCoO_2$/LTO, $LiMn_2O_4$/LTO, $LiFePO_4$/LTO, $LiNi_{1/3}Co_{1/3}Mn_{1/3}O_2$/LTO, etc. Although these batteries exhibit good cycleability, rate capability and stability associated with safety, the low voltage indicates that the battery has low energy density.

Because the operating voltage of spinel $LiNi_{0.5}Mn_{1.5}O_4$ can reach 4.7V, the full batteries can output a voltage of over 3 V if using $LiNi_{0.5}Mn_{1.5}O_4$ as cathode and $Li_4Ti_5O_{12}$ as anode respectively. Fig. 8 (a) and (b) show the charge-discharge curves and cyclic voltammograms of LNMTO/LTO batteries. LNMTO represents Ti-doped compound $LiNi_{0.3}Mn_{1.2}Ti_{0.3}O_4$. It can be seen that this battery displays a discharge voltage profile at around 3.2V.

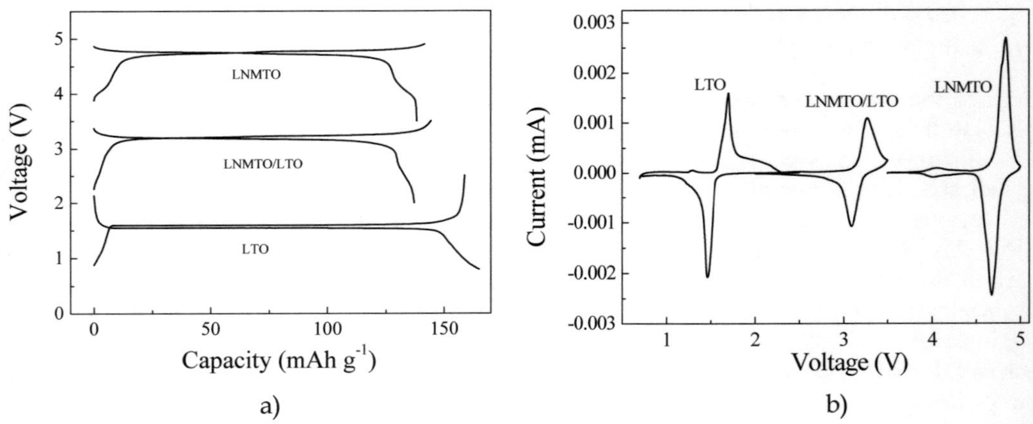

Fig. 8. a) Charge-discharge curves; b) Cyclic voltammograms

The properties of new battery $LiNi_{0.5}Mn_{1.5}O_4/Li_4Ti_5O_{12}$ depended on both cathode and anode materials. Nano scale $Li_4Ti_5O_{12}$ displays good electrochemical properties, and will be applied in this battery. Because cathode $LiNi_{0.5}Mn_{1.5}O_4$ and anode $Li_4Ti_5O_{12}$ have different specific capacities, there are two ways to fabricate the full batteries. One is LNMO-limited cell, another is LTO-limited cell. The experiment results indicate that the LNMO/LTO cell system with the capacity limited by LTO has the better cycling performance than that limited by LNMO [145].

Figure 9 show that the 3V $LiNi_{0.5}Mn_{1.5}O_4/Li_4Ti_5O_{12}$ lithium-ion battery with electrolyte (1M $LiPF_6$/EC +DMC (1:1)) exhibit perfect cycling performance. A LTO-limited cell showed high capacity retention of 85% after 2900 cycles.

In addition, Arrebola [157] have tried to combine $LiNi_{0.5}Mn_{1.5}O_4$ spinel and Si nanoparticles to fabricate new Li-ion batteries. Because Si composite could deliver capacities as high as 3850 (with super P) and 4300 (with MCMC) mAh g^{-1}, this $LiNi_{0.5}Mn_{1.5}O_4$/Si cell was expected to have higher capacity. At present, the battery could deliver a capacity of around 1000 mAh g^{-1} after 30 cycles with good cycling properties. Xia [158] reported that the properties of $LiNi_{0.5}Mn_{1.5}O_4$/ (Cu-Sn) cell which has an average working voltage at 4.0 V.

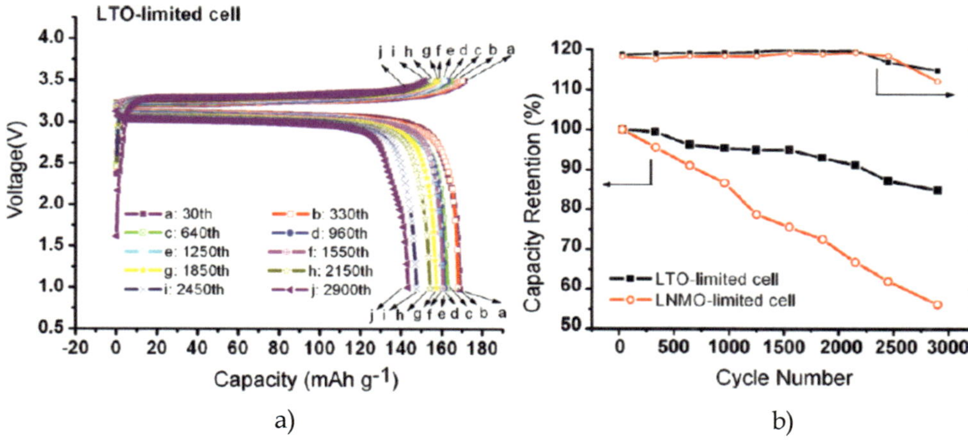

Fig. 9. a) Charge-discharge curves; b) Capacity retention [145]

9. Summary

In recent years, spinel $LiNi_{0.5}Mn_{1.5}O_4$ has become a research focus, and great advance has been achieved. This material has the merits of $LiMn_2O_4$ such as inexpensive and environmentally benign, and it has a prominent advantage of 4.7 discharge plateau which is much higher than other cathode materials. The spinel $LiNi_{0.5}Mn_{1.5}O_4$ has two possible structures, i.e., face-centered spinel (*Fd3m*) and primitive simple cubic crystal (*P4₃32*). According to the previous studies, the spinel with Fd3m structure exhibits better cycling performance than spinel with $P4_332$ structure at high rates. However, recent research demonstrates that the activation barrier for Li ion transportation in ordered ($P4_332$) $LiNi_{0.5}Mn_{1.5}O_4$ is the lowest one, so the ordered $LiNi_{0.5}Mn_{1.5}O_4$ can exhibit the best cycle ability at high current rates. Further studies are needed to solve this disagreement. The microstructure and surface are the key factors affecting its electrochemical properties. Doping elements can improve electrochemical properties, such as doping Ru, Fe and so on. It is well known that the doped elements can make crystal structure more stable, which is in favor of Li insertion/deinsertion. However, the transportation of Li ion in structure is also affected by electronic structure of materials. So far there has not been satisfactory and scientific explanation for this aspect. Making surface modification on the surface of $LiNi_{0.5}Mn_{1.5}O_4$, can not only protect electrode materials from attacking of HF which generates from electrolyte decomposing, but also suppress the development of the SEI layer. This also helps to improve the electrochemical properties of spinel $LiNi_{0.5}Mn_{1.5}O_4$. The Li insertion/deinsertion is affected by particle morphology and size which depend on synthesis methods as well. Nano materials can lead to higher charge/discharge rates. New lithium-ion battery system can be put into practice when combining $LiNi_{0.5}Mn_{1.5}O_4$ and other anode materials such as $Li_4Ti_5O_{12}$. This system has exhibited excellent electrochemical properties.

10. References

[1] M. Winter and J. Brodd, Chem. Rev. 104, 4254, 2004
[2] Lithium Ion technical handbook, Gold Peak Industries Ltd. Taiwan, 2009
[3] Hitachi Maxell KK, Patent JP63211565-A, 1988
[4] Nagaurat, M. Yokokawa, T. Hashimoto, Patent GB2196785-A, 1988
[5] A.K. Padhi, K.S. Nanjundaswamy, J.B.Goodenough, Journal of The Electrochemical Society 14, 1188, 1997
[6] S. Patoux, C. Bourbon, Le Cars, Patent FUS2007292760-A1, 2007
[7] A.V.S. Walter, B. Scrosati, Advances in lithium-ion batteries Kluwer Academic/Plenum, New York, 2002
[8] G.A. Nazri, Lithium ion battery science and technology, Springer, USA, 2008
[9] S. Patoux, L. Daniel, C. Bourbon, H. Lignier, C. Pagano, F.L. Cras, S. Jouanneau, S. Martinet J Power Sources 189, 344, 2009
[10] N. Amdouni, K. Zaghib, F. Gendron, A. Mauger, C.M. Julien, J Magn Magn Mater 309, 100, 2007
[11] M. Kunduraci, J. F. Al-Sharab, G. G. Amatucci, Chem. Mater., 18, 3585, 2006
[12] J.H. Kim, C.S. Yoon, S.T. Myung, J. Prakash and Y.K. Sun, Electrochemical and Solid-State Letters, 7 (7), A216, 2004
[13] J.H. Kim, S.T. Myung, C.S. Yoon, S.G. Kang and Y.K. Sun, Chem. Mater., 16, 906, 2004
[14] X.H. Ma, B. Kang, G. Ceder, Journal of the Electrochemical Society, 157(8), A925, 2010
[15] Y.Idemoto, H. Narai, N. Koura, *J. Power Sources*, 125, 119, 2003
[16] T.Ohzuku, K. Ariyoshi, S. Yamamoto, *J. Ceram. Soc. Jpn.* 110, 501, 2002
[17] J.H. Kim, S.T. Myung, C.S. Yoon, S.G. Kang, Y.K. Sun, *Chem. Mater.* 16, 906, 2004
[18] Q.M. Zhong, A. Bonakdarpour, M.J. Zhang, Y. Gao, J.R. Dahn JR, Journal of the electrochemical society, 144, 205, 1997
[19] C. Sigala, D. Guyomard, A. Verbaere, Y. Piffard, M. Tournoux, Solid State lonics, 81, 167, 1995
[20] S.H. Yang, R.L. Middaugh, Solid State Ionics, 139, 13, 2001
[21] H. Kawai, M. Nagata, H. Tukamoto, A.R. West, Journal of Power Sources, 81–82, 67, 1999
[22] Y. Ein-Eli, S.H. Lu, M.A. Rzeznik, Journal of The Electrochemical Society, 145, 3383, 2004
[23] Y. Ein-Eli, W.F. Howard, S.H. Lu, S. Mukerjee, J. McBreen, J.T. Vaughey, M.M. Thackeray Journal of The Electrochemical Society, 1451, 1238, 1998
[24] N. Biskup, J.L. Martinez, P. Diaz-Carrasco, J. Morales, Journal of Applied Physics, 100, 093908, 2006
[25] H. Shigemura, M. Tabuchi, H. Kobayashi, H. Sakaebe, A. Hirano, H. Kageyama, Journal of Materials Chemistry, 12, 1882, 2002
[26] K. Oikawa, T. Kamiyama, F. Izumi, D. Nakazato, H. Ikuta, M. Wakihara, Journal of Solid State Chemistry, 146, 322, 1999
[27] C. Sigala, A. Verbaere, J.L. Mansot, D. Guyomard, Y. Piffard, M. Tournoux, Journal of Solid State Chemistry, 132, 372, 1997
[28] M.N. Obrovac, Y. Gao, J.R. Dahn, Physical Review B, 57, 5728, 1998
[29] M. Aklalouch, J.M. Amarilla, R.M. Rojas, I. Saadoune, J.M. Rojo, Journal of Power Sources, 185, 501, 2008
[30] H. Shigemura, H. Sakaebe, H. Kageyama, H. Kobayashi, A.R. West, R. Kanno, S. Morimoto , S. Nasu, M. Tabuchi, Journal of The Electrochemical Society, 148, A730, 2006
[31] K. Amine, H. Tukamoto, H. Yasuda, Y. Fujita, Journal of Power Sources, 68, 604, 1997
[32] A. Eftekhari, Journal of Power Sources, 132, 240, 2004
[33] D. Song, H. Ikuta, T. Uchida, M. Wakihara, Solid State lonics, 117, 151, 1999

[34] J.S. Kim, J.T. Vaughey, C.S. Johnson, M.M. Thackeray, Journal of The Electrochemical Society, 150, A1498, 2003
[35] Y.J. Lee, S.H. Park, C. Eng, J.B. Parise, C.P. Grey, Chem. Mater. 14, 194, 2005
[36] Y. Shin, A. Manthiram, Electrochimica Acta, 48, 3583, 2003
[37] Y.E. Eli, J.T. Vaughey, M.M. Thackeray, S. Mukerjee, X.Q. Yang and McBreen, Journal of The Electrochemical Society, 146 (3), 908, 1999
[38] Y. Gao and K. Myrtle, Physical Review B 54, 16670, 1996
[39] Y. Terada, K. Yasaka, F. Nishikawa, T. Konishi, M. Yoshio and I. Nakai, Journal of Solid State Chemistry 156, 286, 2001
[40] K. Ariyoshi, Y. Iwakoshi, N. Nakayama, and T. Ohzuku, Journal of The Electrochemical Society 151, A296, 2004
[41] Y. Idemoto, H. Narai, N. Koura, Journal of Power Sources, 119–121, 125, 2003
[42] H.S. Fang, Z.X. Wang, X.H. Li, H.J. Guo, W.J. Peng, Journal of Power Sources, 153, 174 2006
[43] H.S. Fang, Z.X. Wang, X.H. Li, H.J. Guo, W.J. Peng, Materials Letters, 60, 1273, 2006
[44] Z.Y. Chen, S. Ji, V. Linkov, J.L. Zhang, W. Zhu, Journal of Power Sources, 189, 507, 2009
[45] G.Q. Liu, L. Qi, L. Wen, Rare Metal Materials and Engineer, 35, 299, 2006
[46] Y. Ji, Z.X. Wang, Z.L. Yin, H.J. Guo, W.J. Peng, X.H. Li, Chinese Journal of Inorganic Chemistry, 23, 597, 2007
[47] W, Qiu, T. Li, H. Zhao, J. Liu, Patent CN1321881-C, 2008
[48] Z.Y. Chen, S. Ji, H.L. Zhu, S. Pasupathi, B. Bladergroen, V. Linkov, South African Journal of Chemistry-Suid-Afrikaanse Tydskrif Vir Chemie, 61, 157, 2008
[49] Z. Wang, X. Li, W. Peng, R.Yuan, X. Jiang, Patent CN101148263-A, 2008
[50] H.S. Fang, L.P. Li, G.S. Li, Chinese Journal of Inorganic Chemistry, 167, 223, 2007
[51] Y. Takahashi, H. Sasaoka, R. Kuzuo, N. Kijima, J. Akimoto, Electrochemical and Solid State Letters, 9, A203, 2006
[52] Z.X. Wang, H.S. Fang, Z.L. Yin, X.H. Li, H.J. Guo, W.J. Peng, Journal of Central South University of Technology, 15, 1429, 2006
[53] Z.Y. Chen, J. Xiao, H.L. Zhu, Y.X. Liu, Chinese Journal of Inorganic Chemistry 21: 1417-1421, (2005)
[54] Fang HS, Wang ZX, Li XH, Yin ZL, Guo HJ, Peng WJ, Chinese Journal of Inorganic Chemistry, 22, 311, 2006
[55] B. Zhang, Z.X. Wang, H.J. Guo, Transactions of Nonferrous Metal Society of China, 17, 287, 2007
[56] T. Ohzuku, K. Ariyoshi, S. Yamamoto, Journal of The Ceramic Society of Japan, 110, 501, 2002
[57] Q. Sun, Z.X. Wang, X.H. Li, H.J. Guo, W.J. Peng, 17, Transactions of Nonferrous Metal Society of China, S917-S922, 2007
[58] J. Xiao, L.Y. Zeng, Z.Y. Chen, H. Zhao, Z.D. Peng, Chinese Journal of Inorganic Chemistry 22, 685, 2006
[59] Y.K. Fan, J.M. Wang, X.B. Ye, J.Q. Zhang, Materials Chemistry and Physics 103, 19, 2007
[60] G.Q. Liu, Y.J. Wang, Q. Lu, Electrochimica Acta 50, 1965, 2005
[61] L.F. Xiao, Y.Q. Zhao, Y.Y. Yang, X.P. Ai, H.X. Yang, Y.L. Cao, J Solid State Electrochem 12, 687, 2008
[62] L.H. Yu, Y.L. Cao, H.X. Yang, X.P. Ai. J Solid State Electrochem. 10, 283, 2006
[63] T.F. Yi, Y.R. Zhu, Electrochimica Acta 53, 3120, 2008
[64] T.F. Yi, X.G. Hu, Journal of Power Sources 167, 185, 2007
[65] L.J. Fu, H. Liu, C. Li, Y.P. Wu, E. Rahm, R. Holze, H.Q. Wu, Progress in Materials Science 50, 881, 2005

[66] H. Liu, Y.P. Wu, E. Rahm, R. Holze, H.Q. Wu, J Solid State Eletrochem 8, 450, 2004
[67] T.F. Yi, C.Y. Li, Y.R. Zhu, J. Shu, R.S. Zhu, J Solid State Electrochem 13, 913, 2009
[68] H.Y. Xu, S. Xie, N. Ding, B.L. Liu, Y. Shang, C.H. Chen, Electrochim Acta 51, 4352, 2006
[69] M.G. Lazarraga, L. Pascual, H. Gadjov, D. Kovacheva, K. Petrov, J.M. Amarilla, R.M. Rojas, M.A. Martin-Luengo and J.M. Rojo, Mater. Chem. 14, 1640, 2004
[70] H.M. Wu, J.P. Tu, X.T. Chen, D.Q. Shi, X.B. Zhao, G.S. Cao, Electrochimica Acta 51, 4148, 2006
[71] D. Kovacheva, B. Markovsky, G. Salitra, Y. Talyosef, M. Gorova, D. Aurbach, Electrochimica Acta 50, 5553, 2005
[72] J.C. Arrebola, A. Caballero, L. Hernan, J. Morales, Electrochemical and Solid-State Letters 8, A641, 2005
[73] L. Wen, Q. Lu, G.X. Xu, Electrochimica Acta 51, 4388, 2006
[74] J.H. Kim, S.T. Myung, Y.K. Sun, Electrochim. Acta 49, 220, 2004
[75] J.M. Amarilla, R.M. Rojas, F. Pico, L. Pascual, K. Petrov, D. Kovachevab, Journal of Power Sources 174, 1212, 2007
[76] A. Caballero, M. Cruz, L. Hernan, M. Melero, J. Morales, Journal of The Electrochemical Society 152, A552, 2005
[77] J.C. Arrebola, A. Caballero, L. Hernán and J. Morales, European Journal of Inorganic Chemistry 21, 3295, 2008
[78] M. Kunduraci, G.G. Amatucci, Electrochimica Acta 53, 4193, 2008
[79] S.H. Yang, R.L. Middaugh, Solid State Ionics 139, 13, 2001
[80] H. Xia, L. Lu, M.O. Lai, Electrochimica Acta 54, 5986, 2009
[81] A. Caballero, L. Hernan, M. Melero, J. Morales, R. Moreno, B. Ferrari, Journal of Power Sources 158, 583, 2006
[82] J.C. Arrebola, A. Caballero, L. Hernan, M. Melero, J. Morales, E.R.Castellon, Journal of Power Sources 162, 606, 2006
[83] L. Zhang, X.Y. Lv, Y.X. Wen, F. Wang, H.F. Su, Journal of Alloys and Compounds, 480, 802, 2009
[84] Z.Q. Zhao, J.F. Ma, H. Tian, L.J. Xie, J. Zhou, P.W. Wu, Y.G. Wang, J.T. Tao, X.Y. Zhu, Journal of The American Ceramic Society 88, 3549, 2005
[85] H. Xia, Y.S. Meng, L. Lu, G. Ceder, Journal of The Electrochemical Society 154, A737, 2007
[86] D.C. Li, A. Ito, K. Kobayakawa, H. Noguchi, Y. Sato, Electrochimica Acta 52, 1919, 2007
[87] Z.Q. He, L.Z. Xiong, X.M. Wu, W.P. Liu, S. Chen, K.L. Huang, Chinese Journal of Inorganic Chemistry 23, 875, 2007
[88] J.C. Arrebola, A. Caballero, L. Hernan, J. Morales, European Journal of Inorganic Chemistry 21, 3295, 2008
[89] K. Takahashi, M. Saitoh, M. Sano, M. Fujita, K. Kifune, Journal of The Electrochemical Society 151, A173, 2004
[90] M. Kojima, I. Mukoyama, K. Myoujin, T. Kodera, T. Ogihara, Electroceramics in Japan XI 388, 85, 2009
[91] I. Mukoyama, K. Myoujin, T. Nakamura, H. Ozawa, T. Ogihara, M. Uede, Electroceramics in Japan X 350, 191, 2007
[92] L. Zhang, X.Y. Lv, Y.X. Wen, F. Wang, H.F. Su, Journal of Alloys and Compounds 480, 802, 2009
[93] M.W. Raja, S. Mahanty, R.N. Basu, Solid State Ionics 180, 1261, 2009
[94] M. Yamada, B. Dongying, T. Kodera, K. Myoujin, T. Ogihara, Journal of The Ceramic Society of Japan 117, 1017, 2009
[95] W.F. Fan, M.Z. Qu, G.C. Peng, Z. L. Yu, Chinese Journal of Inorganic Chemistry 25, 124, 2009

[96] Z.C Li, C.W. Wang, X.L. Ma, L.J. Yuan, J.T. Sun, Mater. Chem. Phys. 91, 36, 2005
[97] U. Lafont, C. Locati, E.M. KelderSolid State Ionics, 177, 3023, 2006
[98] U. Lafont, C. Locati, W.J.H. Borghols, A. Łasinska, J. Dygas, A.V. Chadwick, Journal of Power Sources 189, 179, 2009
[99] L.F. Xiao, Y.Q. Zhao, Y.Y. Yang, X.P. Ai, H.X. Yang, Y.L. Cao, J Solid State Electrochem 12, 687, 2008
[100] X.K. Huang, Q.S. Zhang, J.L. Gan, H.T. Chang, and Y.Yang, *Journal of The Electrochemical Society,* 158, A139, 2011
[101] X. Fang, Y. Lu, N. Ding, X.Y. Feng, C. Liu, C.H. Chen, Electrochimica Acta 55, 832, 2010
[102] Y.S. Lee, Y.K. Sun, S. Ota, T. Miyashita, M. Yoshio, Electrochemistry Communications 4 989, 2002
[103] U. Lafont, C. Locati, E.M. Kelder, Solid State Ionics 177, 3023, 2006
[104] J.C. Arrebola, A. Caballero, M. Cruz, L. Hernán, J. Morales and E.R. Castellón, Adv. Funct. Mater. 16, 1904, 2006
[105] X.K. Huang, Q.S. Zhang, J.L. Gan, H.T. Chang, Y. Yang, *Journal of The Electrochemical Society,* 158, 139, 2011
[106] M. Kunduraci, J.F. Al-Sharab and G.G. Amatucci, Chem. Mater. 18, 3585, 2006
[107] M.C.Yang, B.Xu, J.H. Cheng, C.J. Pan, B.J. Hwang, and Y.S. Meng, Chem. Mater. 23, 2832, 2011
[108] H.L. Wang, T.A. Tan, P. Yang, M. O. Lai, and L. Lu, J. Phys. Chem. C 115, 6102, 2011
[109] M. Kunduraci, J.F. Al-Sharab and G.G. Amatucci, Chem. Mater. 18, 3585, 2006
[110] S.H. Yang, R.L. Middaugh, Solid State Ionics 139, 13, 2001
[111] B. Leon, J.M. Lloris, C.P. Vicente, J.L. Tirado, Electrochemical and Solid State Letters 9, A96, 2006
[112] R. Alcantara, M. Jaraba, P. Lavela, J.M. Lloris, C.P. Vicente, J.L. Tirado, Journal of The Electrochemical Society 152, A13, 2005
[113] J. Liu, A. Manthiram, Journal of Physical Chemistry C 113, 15073, 2009
[114] Y.E. Eli, J.T. Vaughey, M.M. Thackeray, S. Mukerjee, X.Q. Yang and J. McBreencJournal of The Electrochemical Society 146, 908, 1999
[115] R. Alcantara, M. Jaraba, P. Lavela and J.L. Tirado, Journal of The Electrochemical Society 151, A53, 2004
[116] A. Ito, D. Li, Y. Lee, K. Kobayakawa, Y. Sato, Journal of Power Sources 185, 1429, 2008
[117] J.H. Kim, S.T. Myung, C.S. Yoon, I.H. Oh and Y.K. Sun, Journal of The Electrochemical Society 151, A1911, 2004
[118] G.Q. Liu, W.S. Yuan, G.Y. Liu, Y.W. Tian, Journal of Alloys and Compounds 484, 567, 2009
[119] R. Alcantara, M. Jaraba, P. Lavela, J.L. Tirado, P. Biensan, J.P. Peres, Chemistry of Materials 15, 2376, 2003
[120] S.B. Park, W.S. Eom, WIl. Cho, H. Jang, Journal of Power Sources 159, 679, 2006
[121] M. Aklalouchb, J.M. Amarilla, R.M. Roja, I. Saadouneb, J.M. Rojo Journal of Power Sources 185, 501, 2008
[122] K.J. Hong, Y.K. Sun, Journal of Power Sources 109, 427, 2002
[123] G.Q. Liu, H.W. Xie, Y.W. Tian, Materials Research Bulletin, 42, 1955, 2007
[124] M. Aklalouch, R.M. Rojas, J.M. Rojo, I. Saadoune, J.M. Amarilla, Electrochimica Acta 54, 7542, 2009
[125] C. Locati, U. Lafont, L. Simonin, F. Ooms and E.M. Kelder, Journal of Power Sources 174, 847, 2007
[126] R. Alcantara, M. Jaraba, P. Lavela, Chemistry of Materials 16, 1573, 2004
[127] J. Liu, A. Manthiram, Journal of The Electrochemical Society 156, A66, 2009

[128] H.L. Wang, H. Xia, M.O. Lai, L. Lu, Electrochemistry Communications 11, 1539, 2009
[129] S.W. Oh, S.H. Park, J.H. Kim, Y.C. Bae, Y.K. Sun, Journal of Power Sources 157, 464, 2006
[130] J. Liu, A. Manthiram, Journal of the Electrochemical Society 156, A833, 2009
[131] J. Liu, A. Manthiram, Journal of the Electrochemical Society 156, S13, 2009
[132] J. Liu and A. Manthiram, Chem. Mater. 21, 1695, 2009
[133] Y.K. Sun, Y.S. Lee, M. Yoshio and K. Aminec, Electrochemical and Solid-State Letters 5, A99, 2002
[134] Y.K. Sun, C.S. Yoon and I.H. Oh, Electrochimica Acta 48, 503, 2003
[135] Y.K. Sun, Y.S. Lee, M. Yoshio, K. Amine, Journal of The Electrochemical Society 150, L11, 2003
[136] Y. Kobayashi, H. Miyashiro, K. Takei, H. Shigemura, M. Tabuchi, H. Kageyama and T. Iwahori, Journal of The Electrochemical Society 150, A1577, 2003
[137] Y. Fan, J. Wang, Z. Tang, W. He and J. Zhang, Electrochimica Acta 52, 3870, 2007
[138] R. Alcantara, M. Jaraba, P. Lavela, J.L. Tirado, Journal of Electroanalytical Chemistry 566, 187, 2004
[139] J. Arrebola, A. Caballero, L. Hernan, J. Morales, E.R. Castellon, J.R.R. Barrado, Journal of The Electrochemical Society 154, A178, 2007
[140] J. Arrebola, A. Caballero, L. Hernan, J. Morales, E.R. Castellon, Electrochemical and Solid State Letters 8, A303, 2005
[141] J.Y. Shi, C.W. Yi, K. Kim, Journal of Power Sources, 195, 6860, 2010
[142] H.M. Wu, I. Belharouak, I, A. Abouimrane, et al. Journal of Power Sources, 195, 2909, 2010
[143] H.B. Kang, S.T. Myung, K. Amine, et al. Journal of Power Sources, 195, 2023, 2010
[144] T.Y. Yang, N.Q. Zhang, Y.Lang, K.N. Sun, Electrochimica Acta, 56, 4058, 2011
[145] H.F. Xiang, X. Zhang, Q.Y. Jin, C.P. Zhang, C.H. Chen and X.W. Ge, Journal of Power Sources183, 355, 2008
[146] H.F. Xiang, Q.Y. Jin, R. Wang, C.H. Chen, X.W. Ge, Journal of Power Sources 179, 351, 2008
[147] K. Ariyoshi, S. Yamamoto, T. Ohzuku, Journal of Power Sources 119, 959, 2003
[148] H.M. Wu, I. Belharouak, H. Deng, A. Abouimrane, Y.K. Sun, K. Amine, Journal of The Electrochemical Society 156, A1047, 2009
[149] M. Imazaki, K.Ariyoshi, T.Ohzuku, Journal of The Electrochemical Society 156, A780, 200
[150] T. Amazutsumi, K. Ariyoshi, K. Okumura, T. Ohzuku, Electrochemistry 75, 867, 2007
[151] K. Ariyoshi, R. Yamatoa, Y. Makimura, T. Amazutsumi, Y. Maeda, T. Ohzukua, Electrochemistry 76, 46, 2008
[152] T. Ohzuku, K. Ariyoshi, S. Yamamoto, Y. Makimura, Chemistry Letters 12, 1270, 2001
[153] Y. Maeda, K. Ariyoshi, T. Kawai, T. Sekiya, T. Ohzuku, Journal of The Ceramic Society of Japan 117, 1216, 2009
[154] S.J. Niu, M. Chen, J.M. Jin, Y. Sun, Electronic Components and Materials 26, 49, 2007 (in Chinese)
[155] K. Zaghib, M. Simoneau, M. Armand, M. Gauthier, Journal of Power Sources 81, 300, 1999
[156] K. Zaghib, M. Armand, M. Gauthier, Journal of The Electrochemical Society 145, 3135, 1998
[157] J.C. Arrebola, A. Caballero, J.L. Gómez-Cámer, L. Hernán, J. Morales, L. Sánchez, Electrochemistry Communications 11, 1061, 2009
[158] Y.Y. Xia, T. Sakai, T. Fujieda, M. Wada, H. Yoshinaga, Electrochemical and Solid State Letters 4, A9, 2001

5

New Developments in Solid Electrolytes for Thin-Film Lithium Batteries

Inseok Seo and Steve W. Martin
*Department of Materials Science and Engineering,
Iowa State University, Iowa,
USA*

1. Introduction

Research on lithium-ion secondary batteries began in the 1980s because of the growing demand for power sources for portable electronic devices. After the early 1990s, the demands for higher capacities and even smaller sizes energy systems significantly increased. Further, the explosive growth in the use of limited fossil fuels and their associated environmental issues and economical aspects are major concerns. Hence, the enormous growth in the demand for low-cost, environmentally friendly energy sources over the past decade has generated a significant need for high energy density portable energy sources.

The enormous growth in portable consumer electronic devices such as mobile phones, laptop computers, digital cameras, and personal digital assistants over the past decade has generated a large interest in compact, high-energy density and lightweight batteries. As power requirements become more demanding, batteries are also expected to provide higher energy densities. In recent decades, $LiCoO_2$ and $LiMn_2O_4$ cathode materials and graphite anodes have been developed and are common in most lithium batteries(Kenji et al.; Peled et al., 1996; Peng et al., 1998; Wang et al., 2002; Xu et al., 2002). However, graphite-based materials are less attractive in terms of capacity when compared to lithium metal, 372 vs. 3800 mAh/g, respectively, in spite of graphite's higher cyclability and safer operation than lithium metal anodes (Tarascon and Armand, 2001). Even still, lithium-based batteries have become enormously important batteries due to their relatively high capacity and low weight. A comparison of many different anode (bottom) and cathode (top) combinations are shown in Figure (1)(Tarascon and Armand, 2001). Further, lithium is very lightweight and has a high electrochemical equivalency and these properties make lithium an attractive battery anode. Therefore, rechargeable lithium batteries are attractive for numerous reasons: high voltages, high energy densities, wide operating temperature ranges, good power density, flat discharge characteristics, and excellent shelf life. However, as shown in Figure (1) the 10-fold increase in capacity of Li metal over graphite has prompted continued effort to develop rechargeable lithium batteries based upon lithium metal anodes for use in a wide variety of applications.

Although the implementation of lithium metal as the anode material in lithium batteries is attractive, electrolytes with high ionic conductivity that are stable in contact with metallic lithium are still lacking. Around ten years ago, lithium batteries with lithium metal anodes using liquid electrolytes, which show the highest ionic conductivities, failed because of serious safety issues(Crowther and West, 2008). Lithium metal anodes tend to form dendrites during charging and discharging processes due to plating-out reactions between lithium metal and liquid electrolytes(Crowther and West, 2008). For these reasons, lithium-based solid-state electrolytes instead of liquid electrolytes are now of great interest and many researchers have been examining them in solid-state batteries because solid electrolytes do not have the aforementioned safety issues and show a smaller temperature dependence to the ionic conductivity compared to some liquid electrolytes. In addition, with the recent surge in interest of various kinds of portable electronic devices and electric and hybrid-electric vehicles, the importance of portable energy sources like secondary batteries has increased. It is widely recognized that all-solid-state energy devices show promise towards improving the safety and reliability of lithium batteries(Hayashi et al., 2009).

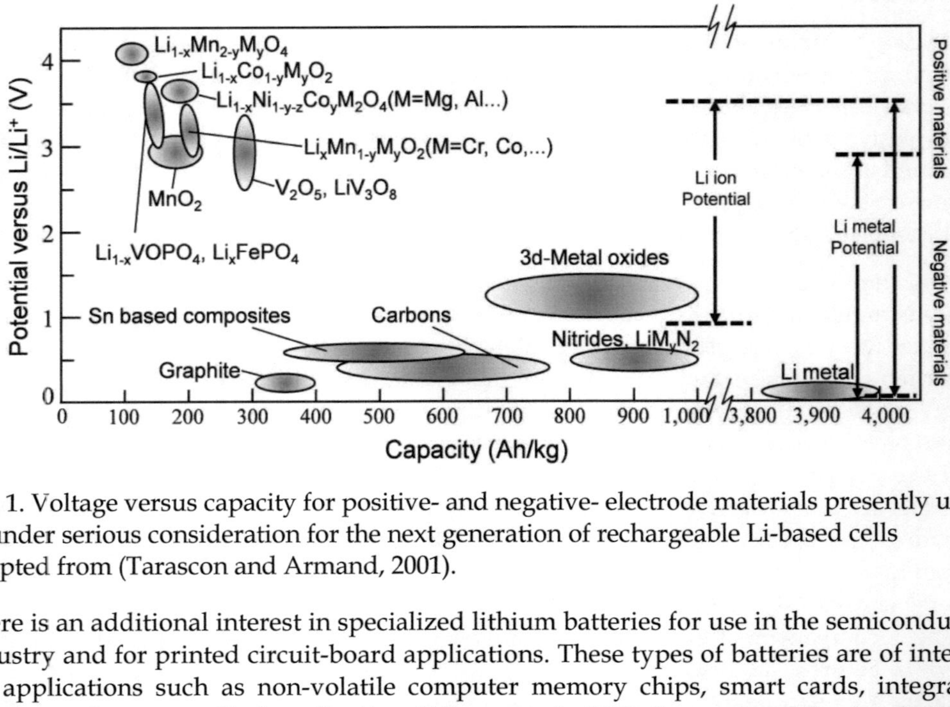

Fig. 1. Voltage versus capacity for positive- and negative- electrode materials presently used or under serious consideration for the next generation of rechargeable Li-based cells adopted from (Tarascon and Armand, 2001).

There is an additional interest in specialized lithium batteries for use in the semiconductor industry and for printed circuit-board applications. These types of batteries are of interest for applications such as non-volatile computer memory chips, smart cards, integrated circuits, and some medical applications(Albano et al., 2008; Souquet and Duclot, 2002). In addition, as the increasing tendency of many advanced technologies is towards miniaturization, the future development of batteries is aiming at smaller dimensions with higher power densities. The development of new technologies and miniaturization in the microelectronics industry has reduced the power and current requirements of small power electronic devices such as smart cards and other CMOS circuit applications(Albano et al., 2008; Souquet and Duclot, 2002). Therefore, developing improved solid-state thin-film batteries will allow better compatibility with microelectronic processing and components.

Therefore, solid-state lithium secondary batteries have attracted much attention because the replacement of conventional liquid electrolytes with an inorganic solid electrolyte may improve the safety and reliability of lithium batteries utilizing high capacity lithium metal anodes(Jones and Akridge, 1992).

Although solid-state batteries have many potential advantages over competitive batteries, solid electrolytes must have higher Li^+ ionic conductivity for them to succeed in commercial applications. Solid electrolytes are a key material of all-solid-state energy storage devices and have been extensively studied in the fields of materials science(Scholz and Meyer, 1994), polymer science(Croce et al., 2001; Fauteux et al., 1995; Song et al., 1999), and electrochemistry(Scholz and Meyer, 1994). Much research has been devoted to the preparation of solid electrolytes made of various materials including ceramics(Abe et al., 2005; Jak et al., 1999a; Jak et al., 1999b), glasses(Iriyama et al., 2005; Lee et al., 2002; Lee et al., 2007; Takada et al., 1995) and glass ceramics(Hayashi et al., 2010; Minami et al., 2011; Ohtomo et al., 2005).

Among these materials for electrolytes, amorphous or glassy materials often have superior ionic conductivities over corresponding crystalline materials because they can form over a wide range of compositions, have isotropic properties, do not have grain boundaries, and can form thin-films easily(Angell, 1983; Martin, 1991). Because of their more open disordered structure, amorphous materials typically have higher ionic conductivities than the corresponding crystalline material(Angell, 1983; Martin, 1991). In addition, single ion conduction can be realized because glassy materials belong to decoupled systems in which the mode of ion conduction relaxation is decoupled from the mode of structural relaxation(Kanert et al., 1994; Patel and Martin, 1992). For these reasons, amorphous or glassy materials are thus among the more promising candidates of solid electrolytes because of their properties of single ion conduction and high ionic conductivities.

Oxide-based electrolytes are currently widely studied because of their stability in air, easy preparation, and their long shelf life(Cho et al., 2007; Jamal et al., 1999). However, they show a critical disadvantage which is their low ionic conductivity. Even still, so-called "LiPON" films formed from sputtering Li_3PO_4 in N_2 atmospheres are currently one of the primary solid-state thin-film electrolytes in use because of these above mentioned advantages(Bates et al., 1993; Bates et al., 2000a; Bates et al., 2000b; Dudney, 2000; Neudecker et al., 2000; West et al., 2004; Yu et al., 1997). However, this easily prepared material has a relatively low Li^+ ion conductivity of ~10^{-6} S/cm at 25 °C as compared to sulfide-based materials whose Li^+ conductivities are in the range of 10^{-3} S/cm at 25 °C(Hayashi et al., 2003; Komiya et al., 2001; Minami et al., 2006; Mizuno et al., 2005).

Because lithium containing thio-materials show higher ionic conductivities than corresponding oxide materials, much research has been conducted on the use of the thio-materials as solid electrolytes. Recently, sulfide materials have been investigated such as SiS_2(Aotani et al., 1994; Hayashi et al., 2002; Hirai et al., 1995; Kennedy, 1989), GeS_2(Haizheng et al., 2004; Kawamoto and Nishida, 1976; Pradel et al., 1985), P_2S_5(Hayashi et al., 2005; Mercier et al., 1981a; Mizuno et al., 2005; Murayama et al., 2004), and B_2S_3(Hintenlang and Bray, 1985; Wada et al., 1983). Among these sulfide materials, GeS_2 is particularly attractive as a base material because it is less hygroscopic(Yamashita and Yamanaka, 2003), more oxidatively stable and enables a more electrochemically stable matrix for lithium-ion conduction to be prepared(Xia et al., 2009). While much research has been done on ion-conducting bulk sulfide glasses prepared by melt-quenching, only a few studies of thin-film ion conducting sulfides have been reported because of the difficulty in preparing them. For example, while Kim(Kim et al., 2005) et al. and Itoh et al.(Itoh et al., 2006) reported on the $Li_2S + GeS_2$ bulk glass system, detailed characterizations of thin-films in this system have not been reported so far. Several thin-film techniques such as pulsed laser deposition (PLD)(Jin et al., 2000; Tabata et al., 1994), radio frequency (RF) sputtering(Bates et al., 1993; Bates et al., 2000a; Bates et al., 2000b; Nakayama et al., 2003; Neudecker et al., 2000; Yu et al., 1997), e-beam evaporation(Bobeico et al., 2003; Wu et al., 2000), physical vapor deposition (PVD)(Kong et al., 2001; Narayan et al., 1992), and chemical vapor deposition (CVD)(Chhowalla et al., 2001) have been used to produce thin-films. Among these techniques, sputtering techniques have been shown to produce high quality thin-films. Furthermore, of the few reports that do exist on sulfide thin-films in the open literature, most show that the films tended to be strongly oxidized either during sputtering, caused possibly by leakage of the RF chamber, or by exposure to air after sputtering(Yamashita et al., 1996a). In addition, these thin-films were found to be Li deficient compared to that of the targets from which they were made(Yamashita et al., 1996a). Therefore, although sulfide films may have good potential in thin-film batteries, sulfide thin-films produced so far appear to be less than optimized and for this reason have found limited applications.

In order to investigate sulfide-based thin-films more extensively, lithium thio-germanate thin-films were carefully sputtered under well-controlled conditions in this study. Since GeS_2-based materials are typically more stable in air than other sulfide materials, GeS_2-based thin-film electrolytes for Li-ion thin-film batteries were grown by RF magnetron sputtering in Ar atmospheres. The starting materials, GeS_2 and Li_2S, and the target materials, Li_2GeS_3, Li_4GeS_4, and Li_6GeS_5, were characterized by X-ray diffraction to verify the phase purity of the targets used to produce thin-films. Further structural characterization of the starting materials, target materials, and their thin-films sputtered by RF sputtering in Ar atmospheres was conducted by Raman and IR spectroscopy to verify purity, contamination, and to examine the structures between targets and their thin-films. The surface morphology and the thickness of the thin-films were characterized by field emission scanning electron microscopy (FE-SEM).

The starting materials, target materials, and thin-films were carefully analyzed by x-ray photoelectron spectroscopy (XPS). To minimize contamination of the films produced in this work, every experimental step was performed carefully and in particular, the RF sputtering conditions were optimized to obtain consistency between target and thin-film compositions and to specifically produce films with near stoichiometric lithium concentrations. The starting materials, Li_2S and GeS_2, the target materials, and thin-films were characterized by XPS for compositional and chemical shift analysis. In order to determine if a maximum conductivity in the $nLi_2S + GeS_2$ system exists, the Li_2S content ranged from $n = 1$ to $n = 4$, 50 mol % to 80 mol %. Ionic conductivities of the thin-films were characterized by impedance spectroscopy. The ionic conductivities were measured over the temperature range from -25 °C to 100 °C in 25 °C increments and over the frequency range from 0.1 Hz to 1 MHz. Before we turn to a detailed description of this work, we first give an overview of the different electrodes that can be used with thin-film electrolytes and of the research progress to date on the chemistry and composition of thin-film electrolytes.

2. Electrodes

2.1 Anodes

As mentioned above in the introduction section, the development of advanced all-solid-state lithium-ion batteries with high energy densities is strongly desired because current lithium-ion batteries using liquid electrolytes potentially have safety issues(Machida et al., 2002; Machida et al., 2004). Because the battery performance strongly depends on the quality of the electrode materials, the electrode materials are very important in battery system. Although many different electrode materials have been developed for the conventional lithium-ion battery which used liquid electrolytes, many of the anode materials developed so far are not suitable for the solid-state lithium-ion batteries. Therefore, in this section, anode and cathode materials which are suitable for the solid-state lithium-ion battery are reviewed.

2.1.1 Graphite/carbon

Graphite/carbon materials have been commonly used as anode materials for the commercial lithium-ion battery using liquid electrolytes because graphite/carbon materials have many advantages including (1) a good cyclability, (2) a relatively large specific capacity of ~370 mAh/g, and (3) a low anode electrode potential of ~0.2 V compared to the Li/Li+ electrode(Buiel and Dahn, 1999; Wu et al., 2003). Although carbon materials have some advantages for conventional Li-ion batteries, not all carbon materials are suitable for all-solid-state lithium-ion batteries with inorganic solid electrolytes. The reason for this is that during charge and discharge processes, the electrochemical lithium insertion into the anode materials, carbon materials, are not completely reversible in solid-state lithium-ion batteries with an inorganic electrolyte. In order to improve the performance of batteries, metallic lithium is very attractive compared to the graphite/carbon materials because metallic lithium has around ten times higher capacity than that of graphite/carbon materials.

2.1.2 Lithium silicide

Lithium silicide ($Li_{4.4}Si$) is a good candidate as an anode material for all-solid-state lithium-ion batteries because $Li_{4.4}Si$ has a large theoretical specific capacity of ~4000 mAh/g, has a high negative potential close to that of lithium metal, and Si is very abundant and is a non-toxic material (Armand and Tarascon, 2008; Lee et al., 2001). However, $Li_{4.4}Si$ has a severe volume expansion of over 300% for the $Li_{4.4}Si$ phase during charge and discharge processes. Thus, in its common form, the material shows poor cyclability compared to graphite and has barriers for commercial application (Kubota et al., 2008). Four different lithium silicides, $Li_{4.4}S$, $Li_{3.25}Si$, $Li_{2.33}Si$, and $Li_{1.71}Si$, as intermetallic phases have been reported in Li-Si system(Sharma and Seefurth, 1976).

2.1.3 Lithium metal

Lithium metal as an anode material has high energy density and it has been recognized as the best candidate for lithium-ion batteries(Tarascon and Armand, 2001). While dendrite formation during cycling is found with liquid electrolytes, lithium metal does not form dendrites in all solid-state lithium-ion batteries. In solid-state batteries, the major challenges are interface resistances and electrochemical stability at the contact area between the anode and electrolyte. Further, extensive effort is demanded before lithium metal is applied to commercial all solid state lithium ion batteries.

2.2 Cathode

While anode materials play an important role in supporting the lithium source, the cathode materials also plays an important role in supporting the reducible/oxidizable ion for secondary lithium-ion batteries. There are key requirements for good cathode materials to be used successfully in rechargeable lithium-ion batteries(Whittingham, 2004). The cathode materials should react with lithium metal in a reversible manner, with a high free-energy of formation and react very rapidly both on insertion and removal. In addition, these materials need to be a good electrical conductors, be electrochemically stable, have a low cost, and need to be environmentally safe (Whittingham, 2004). Until now, many cathode materials have been studied. In this section, some of the more representative cathode materials are reviewed.

2.2.1 Vanadium pentoxide (V_2O_5)

Vanadium pentoxide, V_2O_5, has been studied for three decades(Dickens et al., 1979; Whittingham, 1976). V_2O_5 as a cathode material is an alternative because of its low cost, plentiful resources, and greater safety compared to commercial cathodes such as $LiCoO_2$ and $LiNiO_2$(Wang and Cao, 2008). The main disadvantages of the V_2O_5 material are its low capacity, low conductivity, and poor structural stability(Li et al., 2007). Recently, the V_2O_5-based material, polyaniline (PAN)-V_2O_5 composites have been extensively studied to improve conductivity, cyclabilty, and coulombic efficiency of the electrode materials used in lithium batteries(Malta and Torresi, 2005; Pang et al., 2005).

2.2.2 Lithium cobalt oxide ($LiCoO_2$)

Lithium cobalt oxide, $LiCoO_2$, cathode material was discovered by John Goodenough in 1980 when he worked at Oxford University(Mizushima et al., 1980). Research on $LiCoO_2$ material has been widely done because of its high energy density and good cyclability(Wang et al., 1999) and relatively high theoretical capacity of 272 mAh/g. $LiCoO_2$ cathode material is attractive because of its high energy density and reversible lithium-ion intercalation(Chiang et al., 1998; Kumta et al., 1998). Furthermore, $LiCoO_2$ material has a layer structure which can be suitable for the accommodation of the large changes of the lithium contents. Therefore, it can be cycled more than 500 times with 80-90 % capacity retention (Patil et al., 2008). Thin-film $LiCoO_2$ cathode materials also show good power density when discharged between 3.0V and 4.2V(Kim et al., 2000) because of the layered $LiCoO_2$ structure. Amatucci et al.(Amatucci et al., 1996) reported that $LiCoO_2$ can be reversibly form the Li-ion and CoO_2. In addition, the preparation of $LiCoO_2$ is very facile over other comparable materials. For these reasons, it became the most common cathode material in lithium batteries.

However, cobalt is relatively expensive compared to other elements such as Ni, Mn, and V. In order to make it cheaper and improve the reversible capacity, Yonezawa et al.(Yonezawa et al., 1998) and Huang et al.(Huang et al., 1999) applied doping materials such as fluorine, magnesium, aluminum, nickel, copper or tin. If LiF is doped, the reversible capacity improved compared to pure $LiCoO_2$(Yonezawa et al., 1998). If Al is incorporated partially to substitute for cobalt, the working and open voltages increased. Huang et al. reported that the reversible capacity of $LiAl_{0.15}Co_{0.85}O_2$ reached up to 160 mAh/g without volume change after 10 cycles(Huang et al., 1999). Especially, self-discharge effects of the thin-film batteries using $LiCoO_2$ cathodes are negligible (Dudney, 2005). Thus, thin-film batteries using $LiCoO_2$ cathode can hold full charge for three years(Dudney, 2005).

2.2.3 Lithium manganese oxide ($LiMn_2O_4$)

Lithium manganese oxide, $LiMn_2O_4$, is an attractive cathode material and has been widely studied because the material has advantages from ecological and economical perspectives as well as easy preparation (Kang and Goodenough, 2000; Lee et al., 2004; Liu and Shen, 2003; Ohzuku et al., 1991). However, $LiMn_2O_4$ has a serious drawback. Before cycling, the structure of the $LiMn_2O_4$ is cubic. Then, on cycling, the spinel structure is destroyed due to a cubic-tetragonal phase transition induced by Jahn-Teller distortion (David et al., 1987; Gummow et al., 1993; Gummow and Thackeray, 1994). For this reason, batteries with $LiMn_2O_4$ cathodes show capacity loss and poor cyclability(Gummow et al., 1994; Myung et al., 2000). Pure $LiMn_2O_4$ has been improved by doping. If chromium is doped into $LiMn_2O_4$, it can form $Li_{1+x}Mn_{0.5}Cr_{0.5}O_2$ and the doped $Li_{1+x}Mn_{0.5}Cr_{0.5}O_2$ reveals improved capacity and cyclability(Sigala et al., 1995). It can be assumed that Mn plays an important role to stabilize the structure of the chromium oxide. However, chromium materials are toxic and expensive. Therefore, in order to fabricate successfully stabilized layer structural framework, the doping of other elements into $LiMn_2O_4$ has been studied.

2.2.4 Lithium nickel oxide ($LiNiO_2$)

Because $LiNiO_2$ is cheaper than $LiCoO_2$ and the redox potential is higher than that of $LiCoO_2$, The $LiNiO_2$ material has become an attractive as a cathode material for Li-ion batteries(Campbell et al., 1990). The structure of the $LiNiO_2$ is layered similar to $LiCoO_2$(Zhecheva and Stoyanova, 1993). The layered $LiNiO_2$ structure has a wide homogeneity range, $Li_xNi_{2-x}O_2$ ($0.6 < x < 1$)(Bronger et al., 1964). Upon cycling, the capacity of the materials fades because Ni^{2+} ions migrate to Li^+ sites. The appearance of Ni^{2+} in the Li^+ sites obstructs Li^+ diffusion and the lithium-ion transfer during cycling(Li et al., 1992). For this reason, the $LiNiO_2$ battery shows poor cycle performance compared to $LiCoO_2$(Dahn et al., 1991). $LiNiO_2$ has some drawbacks such as being unstable in the overcharge state as well as easy decomposition at high temperature. Furthermore, lithium oxide contents in the $LiNiO_2$ decrease when heat treatment is performed due to the volatility of Li_2O. The Li deficient defect structure results in gradual collapse of oxide structure during cycling and the specific charge decreases during cycling of the $LiNiO_2$ electrode(Hirano et al., 1995). In order to improve the performance of the $LiNiO_2$ structure, many researchers have studied this material using doping elements such as Co, Ti, Mn, Al, Mg, Fe, Zn, Ga, Sb, and S(Chang et al., 2000; Chowdari et al., 2001; Cui et al., 2011; Gao et al., 1998; Nishida et al., 1997; Park et al., 2005; Park and Sun, 2003; Pouillerie et al., 2000; Reimers et al., 1993). The reversible capacity of $LiNi_{0.75}Ti_{0.125}O_2$ reached up to 190 mAh/g(Gao et al., 1998). Therefore, from this point of view, the doped $LiNiO_2$ can be a good candidate as a cathode material for secondary Li-ion batteries; however, safety is a serious concern.

2.2.5 Lithium iron phosphate ($LiFePO_4$)

Lithium iron phosphate, $LiFePO_4$, was reported by John Goodenough's research group in 1996, as a cathode material for rechargeable lithium batteries(Padhi et al., 1997). Conventional cathode materials, $LiCoO_2$ and $LiNiO_2$, have drawbacks such as the high cost, toxicity, safety issues, and electrochemically instability. $LiFePO_4$ is a promising candidate for secondary lithium-ion batteries because of its relatively high energy density, low cost, good safety, and high thermal stability compared to conventional cathode materials(Padhi et al., 1997). However, there is a key barrier in that $LiFePO_4$ has an intrinsically low electrical conductivity of 10^{-9} to 10^{-10} S/cm(Andersson et al., 2000; Barker et al., 2003; Padhi et al., 1997). Therefore, early studies on $LiFePO_4$ showed that it was not the best cathode material over other conventional cathode materials. These problems were resolved later by reducing the particle size, doping using cations of materials such as Al, Nb, and Zr, and coating the $LiFePO_4$ particles using a conductive carbon material(Huang et al., 2001; Prosini et al., 2001; Shi et al., 2003; Yamada et al., 2001). By using these methods, greatly improved electrochemical response and full capacity of $LiFePO_4$ was obtained with prolonged cycle life. Recently, $LiFePO_4$ can be used up to 90 % of its theoretical capacity, 165 mAh/g, and at high rate capabilities(Yamada et al., 2001). Thus, optimized $LiFePO_4$ is a good candidate as a cathode material for the solid-state thin-film batteries.

3. Electrolytes

Among the three components, anode, cathode, and electrolyte, in a battery system, battery performance strongly depends on the performance of the electrolyte. The basic requirements of an appropriate electrolyte for lithium-ion batteries are high ionic conductivity, electrochemical and thermal stability, and good performance at low and high temperatures. Because liquid electrolytes have a higher ionic conductivity compared to polymer and solid electrolytes, they have been widely used. However, liquid electrolytes have strong ionic conductivity temperature dependence and can have safety issues related to the flammability of the organic liquid. Recently, solid-state lithium secondary batteries have attracted much attention because the replacement of conventional liquid electrolyte with an inorganic solid electrolyte may improve the safety and reliability of lithium batteries utilizing high capacity lithium metal anodes(Jones and Akridge, 1992). There are two types of solid state electrolytes; one is thin-film electrolytes grown by RF sputtering(Bates et al., 1993; Bates et al., 2000a; Bates et al., 2000b; Dudney et al., 1999; Neudecker et al., 2000; Seo and Martin, 2011a, b, c; Yu et al., 1997) or PLD(Jin et al., 2000; Tabata et al., 1994) etc. and the other is bulk electrolytes fabricated by typically using melting processes. All-solid-state thin-film batteries using inorganic amorphous electrolytes such as LiPON have been reported and LiPON shows excellent cyclability, over 50,000 cycles, at room temperature(Bates et al., 1993; Bates et al., 2000a; Bates et al., 2000b; Dudney et al., 1999; Neudecker et al., 2000; Yu et al., 1997). However, bulk type batteries using bulk electrolytes have an advantage of improving cell capacity by the addition of large amounts of active materials to the cell. For both the thin-film and bulk batteries, Li_2S based electrolytes are promising because of their high ionic conductivities compared to oxide electrolytes. In the next section, we report recent data of the thin-film electrolytes for the solid-state batteries.

3.1 Polymer electrolytes

Polymer electrolytes for use in lithium batteries were rapidly developed in the 1970s(Fenton et al., 1973). It was found that these materials could offer a safer battery than liquid electrolytes which are corrosive, flammable, or toxic. In recent decades, polymer electrolytes have been widely studied including polyethyleneoxide (PEO)(Appetecchi et al., 2003), polyacrylonitrile (PAN)(Yu et al., 2001), polymethylmethacrylate (PMMA)(Rajendran and Uma, 2000), and polyvinylidenefluoride (PVDF)(Saunier et al., 2004). However, the ionic conductivity of these polymer electrolytes is still too low at room temperature for commercial batteries. In addition, the primary concerns with these electrolytes involve their reactivity with a lithium metal as an anode. Their reactivity with lithium poses safety concerns because lithium dendrites can grow towards the cathode and ultimately short-circuit the cell. For this reason, gel-type polymer electrolytes were developed and the electrolytes show improved ionic conductivity compared to conventional polymer electrolytes, but they still have lower ionic conductivities than those of liquid electrolytes. The gel-type polymer electrolytes have ionic conductivity of ~10^{-4} S/cm at room temperature. In the gel-type polymer electrolytes, the liquid element can be trapped in polymer matrix so the leakage problems associated with liquid electrolytes can be resolved. The low conductivity of the polymer/gel polymer electrolytes can be overcome by introducing inorganic ceramic particles to form a composite material that is more "solid". These materials have conductivities two or three orders of magnitude lower than aqueous electrolytes. However, thin polymer films on the order of 100 μm thick can compensate for their diminished conductivities(Birke et al., 1999). There is also the opportunity of increasing the operating temperature of the cell to around 90 °C. Therefore, optimized thin polymer electrolytes can be promising electrolytes for the thin-film solid-state batteries.

3.2 Solid-state electrolytes

While commercial cells will continue to be fabricated using organic polymeric electrolytes due to their ease of fabrication and low cost, solid-state electrolytes will also attract attention for their possible use in special applications. Solid-state electrolytes are attractive because they provide a hard surface that is capable of suppressing side reactions and inhibiting dendritic growth of lithium that is capable of short-circuiting a cell(Schalkwijk and Scrosati, 2002). However, one disadvantage of these electrolytes is their potential to form cracks or voids if there is poor adhesion to the electrode materials. In order to successfully fabricate all-solid-state lithium batteries with good performance, the design of the electrodes and electrolytes are important.

A number of candidate materials have been investigated for use as solid electrolytes in batteries. The most attractive candidates to date are glassy materials. These electrolytes have many advantages over their crystalline counterparts such as physical isotropy, absence of grain boundaries, good compositional flexibility, and good workability. The anisotropy and grain boundaries present in crystalline materials lead to resistive loss, decreasing cell efficiency, as well as chemical attack, raising safety concerns. A number of different systems have been explored and are discussed specifically below.

3.2.1 Oxide glasses

Oxide glass electrolytes for solid-state batteries have been widely studied because they have the primary advantage of being relatively stable in air allowing for ease of fabrication. However, oxide glasses have received less attention for their use as electrolyte materials because they exhibit very low ionic conductivities and high activation energies. The best of the oxide materials appears to be those glasses with mixed formers such as SiO_2 and B_2O_3(Lee et al., 2002; Nogami and Moriya, 1982; Zhang et al., 2004). These glasses have a conductivity at room temperature on the order of ~10^{-7} S/cm. These materials might prove promising if produced into thin-films. However, chemistries with a higher ionic conductivity are more desirable. Sulfide materials, discussed in more detail below, are of interest for this reason. In terms of conductivity, it is clear that oxide glasses have significantly lower conductivities than those of sulfide materials(Boukamp and Huggins, 1978; Elmoudane et al., 2000; Ito et al., 1983; Mercier et al., 1981b; Murayama et al., 2004).

3.2.2 Oxinitride glasses

The most commercially viable material in this category is the lithium phosphorus oxy-nitride (LiPON) glass. This material was first discovered and reported in the 1980s by Marchand(Marchand et al., 1988) and Larson(Larson and Day, 1986). These materials were not thin-films but bulk glass materials. In addition, their properties were not fully characterized until Oak Ridge National Laboratory (ORNL) reported LiPON thin-films(Bates et al., 1993; Bates et al., 2000a; Bates et al., 2000b; Dudney, 2000; Dudney, 2005; Dudney et al., 1999; Neudecker et al., 2000; Yu et al., 1997). The LiPON thin-films were deposited using a high purity lithium phosphate, Li_3PO_4, target by RF magnetron sputtering technique in nitrogen atmosphere. It was found that the resulting thin-film with a typical composition of $Li_{2.9}PO_{3.3}N_{0.36}$ contained 6 at % nitrogen. This additional nitrogen was found to enhance the ionic conductivity at room temperature from ~10^{-8} S/cm in the starting Li_3PO_4 target to a value of ~10^{-6} S/cm. Furthermore, these films were found to be highly stable in contact with metallic lithium. It is believed that a thin passivating layer of Li_3N is formed between the lithium and electrolyte which prevents lithium dendrite growth, but allows ion conduction. The nitrogen was found to substitute for oxygen and form 2 and 3-coordinated nitrogen groups, effectively crosslinking the structure. The schematic of the structural units of LiPON are shown Figure (2). This crosslinking is believed to decrease the electrostatic energy of the overall network, allowing for faster ion conduction. Thin-film batteries comprised of Li-$LiCoO_2$ cells and Li-$LiMn_2O_4$ cells have been fabricated using the LiPON electrolyte at ORNL(Bates et al., 1993; Bates et al., 2000a; Bates et al., 2000b; Dudney, 2000; Dudney, 2005; Dudney et al., 1999; Neudecker et al., 2000; Park et al., 2007; Yu et al., 1997). Nam et al. reported LiPON using V_2O_5 cathode materials(Jeon et al., 2001). These types of batteries are being commercialized and target for applications in implantable medical devices, CMOS-based integrated circuits, and RF identification (RFID) tags for inventory control and anti-theft protection.

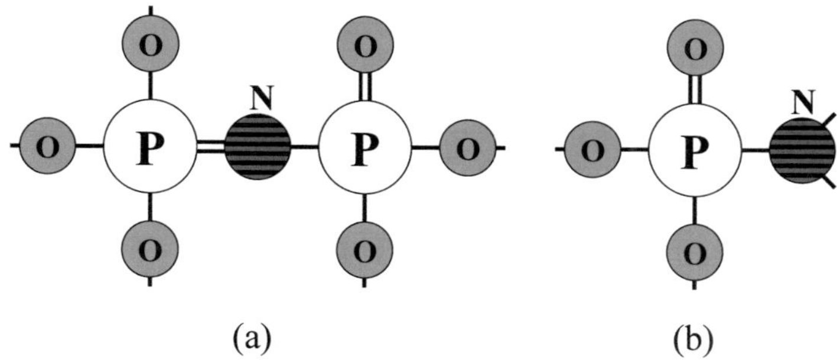

Fig. 2. Proposed structural units present in LIPON sputtered films. (a) two-coordinated bridging nitrogen unit and (b) three-coordinated nitrogen unit.

3.2.3 Sulfide glasses

Sulfide glasses were first reported on in the 1980s(Kennedy, 1989; Mercier et al., 1981a; Zhang and Kennedy, 1990). These glasses were based on SiS_2, P_2S_5, and B_2S_3 and are doped with an alkali sulfide such as lithium sulfide. It was found that these materials have exceptional conductivities, ~ 10^{-3} S/cm, at room temperature. This is attributed primarily to the larger ionic radius of sulfur and its high atomic polarizability(Kim et al., 2006). This is believed to create weaker covalent bonds between the sulfur and the lithium ions. As a result, the potential energy barrier that must be overcome is decreased, and lithium-ion conduction is facilitated. Unfortunately, these glasses have not been widely used because they are highly reactive in air/moisture and corrosive with silica containers. A high quality, low O_2 and H_2O, glovebox is absolutely necessary to fabricate such glasses without contamination.

The thin-films related to the Li_2S-GeS_2-Ga_2S_3 glass system have recently been prepared using RF sputtering(Yamashita and Yamanaka, 2003; Yamashita et al., 1996b). Successfully deposited films were produced using this method, however, the authors reported that the ionic conductivities of the thin-films were diminished compared to that of the target material. This was a result of the films being deficient in lithium and enriched in germanium from the XPS composition analysis of the films. In addition, the thin-films were contaminated by O_2 during sputtering due to the leakage of the RF chamber. Although the ionic conductivities of the sulfide materials are higher than oxide materials, if the sulfide thin-films exhibit lithium deficiency and contamination, there are few benefits of sulfide materials.

3.2.4 Oxy-sulfide glasses

Efforts to combine the advantages of oxide and sulfide glasses have resulted in the research of a class of oxy-sulfide materials(Hayashi et al., 1996; Kondo et al., 1992; Takada et al., 1996). It was found that adding a small amount, approximately 5 mole %, of different lithium metal oxides to a base sulfide glass, improved the conductivity over 10^{-3} S/cm(Minami et al., 2008; Ohtomo et al., 2005). Furthermore, the stability of the structure was observed to improve from thermal analysis results. Structural analysis of these materials has demonstrated that the oxygen typically occupies a bridging anion site, leaving sulfur at the non-bridging sites for lithium mobility. Some solid-state batteries have been fabricated using these oxy-sulfide compositions, and initial results appear to indicate good electrochemical stability (Hayashi et al., 2010; Minami et al., 2011; Ohtomo et al., 2005).

3.2.5 Thio-nitride glasses

Lithium nitride (Li_3N) materials are somewhat attractive as a solid electrolyte because of their high ion conductivity, 2-4 × 10^{-4} S/cm at 25 °C(Lapp et al., 1983). The single crystal Li_3N has an impressive high ionic conductivity of 1.2 × 10^{-3} S/cm at room temperature(Rabenau, 1982). However, it is impossible to use Li_3N itself as an electrolyte in secondary batteries because Li_3N decomposes at low voltage (Yonco et al., 1975). For this reason, thio-nitride glasses have been studied with high ionic conduction(Iio et al., 2002; Sakamoto et al., 1999). The motivation behind these materials comes from the fact that doping nitrogen into oxide systems improved the ionic conductivity(Unuma and Sakka, 1987; Wang et al., 1995c). Furthermore, doping of nitrogen into oxide glasses has been found to improve the hardness and chemical durability (Sakka, 1986).

4. All-solid-state thin-film batteries

4.1 History of thin-film batteries

All-solid-state thin-film batteries were reported first by Hitachi Co. Ltd in Japan in 1982. The TiS_2 cathode material was prepared by chemical vapor deposition (CVD), a $Li_{12}Si_3P_2O_{20}$ electrolyte was grown by radio frequency (RF) sputtering, and a lithium metal anode material was deposited by a vacuum evaporation(Kanehori et al., 1986; Miyauchi et al., 1983). NTT Co. also reported thin-film batteries using a $Li_{12}Si_3P_2O_{20}$ electrolyte with $LiCoO_2$ or $LiMnO_2$ cathode materials grown by RF sputtering(Ohtsuka et al., 1990; Ohtsuka and Yamaki, 1989; Yamaki et al., 1996). The performance of the thin-film batteries was not as good as current thin-film batteries.

In 1980s, Union Carbide Corporation and Eveready Battery Co., Ltd. in USA developed thin-film batteries using sulfide glass electrolytes, $Li_4P_2S_7$ or Li_3PO_4-P_2S_5, and Li metal anode or LiI anode(Akridge and Vourlis, 1986, 1988). They improved the battery performance in 1990s to reach over 1000 cycle performance between 1.5V and 2.8V and 10 to 135 $\mu A/cm^2$(Jones and Akridge, 1996). Bellcore Co., Ltd. also developed thin-film batteries using a $LiMnO_2$ cathode, lithium metal as an anode, and lithium borophosphate (LiBP) or lithium phosphorus oxynitride (LiPON) glass as an electrolyte(Shokoohi et al., 1991). The cell showed over 150 cycles with 3.5~4.3 V and 70 $\mu A/cm^2$.

Recently, Bates and Dudney et al. at Oak Ridge National Laboratory (ORNL) reported significant progress on LiPON-based thin-film batteries which were produced by an RF sputtering technique(Bates et al., 1993; Bates et al., 2000a; Bates et al., 2000b; Dudney et al., 1999; Wang et al., 1995a; Wang et al., 1995b; Wang et al., 1995c; Yu et al., 1997). In order to fabricate LiPON thin-film batteries, the metallic anode was produced by vacuum evaporation and anode and cathodes were produced by RF sputtering. The LiPON thin-film batteries are very stable in air compared to lithium oxide or sulfide based batteries in spite of LiPON's low ionic conductivity of ~10^{-6} (S/cm). The LiPON thin-films reported by ORNL showed very good performance between 2-5 V and over 10,000 cycles. Furthermore, ORNL reported also a Li-free thin-film battery with an in-situ plated Li anode on copper electrode (Neudecker et al., 2000).

LiPON is now known as a standard electrolyte for the thin-film batteries and it has been widely studied by a number of research groups. Park et al. in Korea reported "mesa-type" all-solid-state LiPON thin-film battery using a $LiMn_2O_4$ cathode(Park et al., 1999). Baba et al. in Japan reported also LiPON thin-film batteries using a $Li_xV_2O_5$ anode material and V_2O_5 or $LiMn_2O_4$ cathode materials produced by RF sputtering(Baba et al., 2001; Baba et al., 1999; Komaba et al., 2000).

Jourdaine et al. in France reported thin-film batteries produced by RF sputtering. They successfully fabricated the cell using metallic lithium as an anode, Li_2O-B_2O_3-P_2O_5 or Li_2O-B_2O_3 glasses as electrolytes, and V_2O_5-TeO_2 or V_2O_5-P_2O_5 as cathodes, respectively(Jourdaine et al., 1988).

4.2 Thin-film techniques

There are many vapor deposition techniques that can be employed in order to produce thin-film materials. These include simple heating of a source material, laser-induced vaporization, or bombarding the material with energetic ions. All of these techniques are performed under vacuum and rely on the kinetic theory of gases in order to understand their behavior.

4.2.1 Pulsed laser deposition (PLD)

Pulsed laser deposition (PLD) involves using a laser beam to vaporize the surface of a target material(Chrisey and Hubler, 1994). One of the most common lasers used is the KrF excimer laser, operating at 248 nm with the following parameters: a pulse on the order of 25 ns, a power density of 2.4×10^8 W/cm^2, and a repetition rate of 50 Hz. In general, the PLD process can be divided into four stages(Chrisey and Hubler, 1994). First, the laser beam is focused onto the target material. The elements in the target are rapidly heated to their evaporation temperature where there are sufficiently high flux densities over a short pulse duration. This ablation process involves many complex physical phenomena such as collisional, thermal and electronic excitation, exfoliation and hydrodynamics. Second, the ablated target elements move towards the substrate according to the laws of gas-dynamics. In the third stage, the high energy atoms bombard the substrate surface where a collision region is formed between the incident flow and the sputtered atoms. A film begins to grow after a thermalized region develops and when the condensation rate is higher than the rate of sputtered atoms. Finally, nucleation and growth of a thin-film occurs on the substrate. This step depends on many factors such as the density, energy, ionization degree, and the temperature of the substrate. PLD has some advantage over other techniques in that the stoichiometry of the target can be retained in the deposition film and many different materials can be deposited, and can be easily handled compared to other techniques such as CVD and ion implantation techniques(Bao et al., 2005; Kaczmarek, 1996). On the other hand, it has some disadvantages such as the deposition of droplets(Yoshitake et al., 2001), the splashing or the particulates deposition on the thin-film, and lower energy density and lower deposition rate compared to other techniques(Willmott and Huber, 2000).

4.2.2 Radio frequency (RF) sputtering

Sputtering is a technique whereby energetic ions from a plasma are used to bombard a target (which is the cathode of the discharge), and ejecting atoms into the plasma. These atoms then impinge upon the substrate (the anode) and form a coating. Additionally, a magnet can be added to these two setups in order to enhance the deposition rates. RF magnetron sputtering is a reliable technique used to deposit many different types of films, including electrically insulating samples. A high-voltage RF source at a frequency of typically 13.56 MHz is used to ionize a sputtering gas which produces the plasma(Yamashita et al., 1999). The ionized gas then bombards the target where multiple collisions take place, releasing atoms of the target material into the plasma. These atoms condense upon the substrate which is placed in front of the target(Nalwa, 2002). A permanent magnet is added to the sputtering gun in order to enhance the deposition rate. This is done by the trapping of electrons from a Hall effect near the target surface(Nalwa, 2002). This magnet creates lines of magnetic flux that are perpendicular to the electric field or parallel to the target surface. This static magnetic field retains secondary electrons in that region which drift in a cycloidal path on the target and increase the number of collisions that occur.

While many different thin-film deposition techniques could be used in this research, RF magnetron sputtering (RFMS) has been chosen as the technique of choice. The most important reasons for selecting RFMS as the technique of choice are given here(Dudney et al., 1999; Souquet and Duclot, 2002). First, there is no need in the project to produce thick films. To produce a protective barrier for lithium metal anodes, a layer is needed and only needs to be thick enough so that it does not have large numbers of pin holes that will lead to failure of the anode. A layer 50 to 5000 Å is thought to be thick enough. Such layers can easily be produced by RFMS. Secondly, in the thin-film lithium battery research, there is no need for thick films and films 500 to 10,000 Å are thick enough, which are again attainable with sputtering techniques. In addition, sputtering can be done within the confines of a sealed glovebox, can be used with multiple targets and film chemistries, can be used to produce very uniform films of high compositional integrity, and produces films with excellent adherence to the substrate. Finally, it is possible to deposit insulator films through RF reaction sputtering at rates higher than those of DC methods(Davidse, 1967).

4.2.3 Chemical Vapor Deposition (CVD)

Chemical Vapor Deposition (CVD) process is related to transform gaseous molecules, precursor, by chemical reactions in the thin-film or power on the substrate(Mount, 2003). CVD processing is usually used to apply various fields such as integrated circuits, optoelectronic devices and sensors, micro-machines, and fine metal and ceramic powders. CVD has many advantages compared to physical vapor deposition (PVD) techniques such as sputtering and molecular beam evaporation. While PVD processes may not give complete coverage due to a shadowing effect, CVD can be allowed to coat thin-films of three dimensional structures with large aspect ratios. The deposition rates of the CVD are several times higher than that of PVD. In addition, ultra high vacuum is not needed and high purity film can be produced by CVD process. However, there are some disadvantages of the CVD process. High temperatures of the deposition temperature, over 600 ^{o}C, are not suitable for already grown thin-films on substrates. CVD precursors are sometimes dangerous and toxic and many precursors for CVD, for example metal organic chemical vapor deposition (MOCVD) precursors, are very expensive.

4.3 Recent results for the lithium thio-germanate thin-film electrolytes

4.3.1 X-ray diffraction of the starting materials and targets

In this study, GeS_2 and Li_2S as starting materials were used to synthesize the target material. To verify the phase purity, XRD pattern of GeS_2 glass powder, Li_2S crystalline powder and three target materials are shown in Figure (3). While GeS_2 glass powder is verified to be amorphous, Li_2S powder shows several sharp peaks. The XRD pattern of the Li_2S powder closely matches the JCPDS data(Cunningham et al., 1972). From the JCPDS data, it is verified that the system and space group of Li_2S powder are face-centered cubic and Fm-3m, respectively (Cunningham et al., 1972).

The Li_2GeS_3 target shows an XRD amorphous pattern without dominant peaks because the melt-quenching technique combined with its 50% GeS_2 glass former composition are sufficient to make this phase amorphous on cooling during preparation. The XRD patterns of the Li_4GeS_4 and Li_6GeS_5 targets, on the other hand, are polycrystalline and show sharp peaks because the Li_4GeS_4 and Li_6GeS_5 target contain only 33 % and 25% of the GeS_2 glass former, respectively, which are not sufficient to vitrify these melts on quenching.

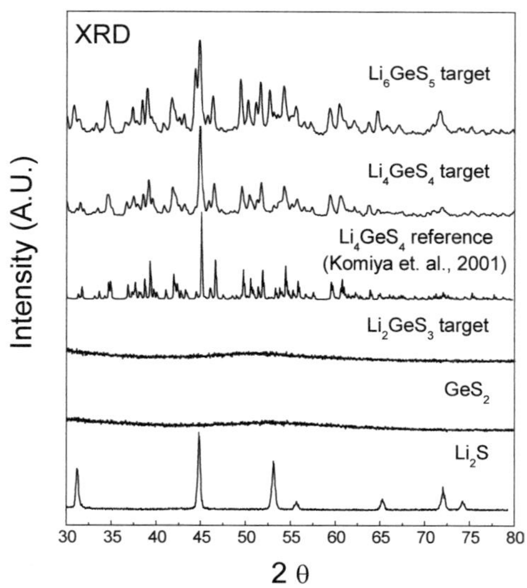

Fig. 3. XRD patterns of crystalline Li_2S, GeS_2 glass, and target materials

To verify the XRD pattern of the Li_4GeS_4 target material which was quenched on a brass plate in the glovebox, and the reference data(Komiya et al., 2001) of Li_4GeS_4 is also shown in Figure (3). The XRD pattern of our experimental Li_4GeS_4 target material shows slightly broader peaks than those of the reference data(Komiya et al., 2001). A possible reason is that the Li_4GeS_4 target was quenched more quickly on a brass plate. This rapid quenching presumably produces a more defective crystal structure than typical slow cooled or solid-state reaction prepared samples. However, the XRD pattern of the Li_4GeS_4 target material still appears to closely match the reported reference pattern. Murayama et al.(Murayama et al., 2002) reported that the structure of Li_4GeS_4 is related to that of γ-Li_3PO_4 and is comprised of hexagonal close-packed sulfide ions with germanium ions distributed over the tetrahedral sites. In this structure, the Li^+ ions are located in both octahedral and tetrahedral sites. Murayama et al.(Murayama et al., 2002) suggested that the distribution of Li^+ ions in the LiS_4 tetrahedra, the interstitial tetrahedral sites, and the LiS_6 octahedra sites forms conduction pathways in the crystal. For this reason, the Li_4GeS_4 material shows higher ionic conductivity than oxide materials.

While XRD data of Li_4GeS_4 do not show peaks related to those of Li_2S, XRD data of Li_6GeS_5 show peaks related to those of Li_2S. This suggests that the XRD pattern for Li_6GeS_5 agrees well with the expectation that it is composed of equi-molar mixture of Li_4GeS_4 and Li_2S. The Li_4GeS_4 and Li_6GeS_5 targets are crystalline as shown in Figure (3). The fact can also be seen from the Raman spectra in Figure (4). The Li_2GeS_3, Li_4GeS_4, and Li_6GeS_5 thin-films were not characterized by XRD because our standard XRD system is not sensitive enough to examine such thin-films as are reported here.

4.3.2 Raman spectroscopy

Starting materials, GeS_2 and Li_2S, targets, and thin films were characterized by Raman spectroscopy in order to analyze their purity and to determine their chemical structure and are shown in Figure (4). In the Raman spectrum of GeS_2, a strong main peak appears at ~340 cm^{-1} that agrees well with that of literature(Cernosek et al., 1997) and is assigned to the

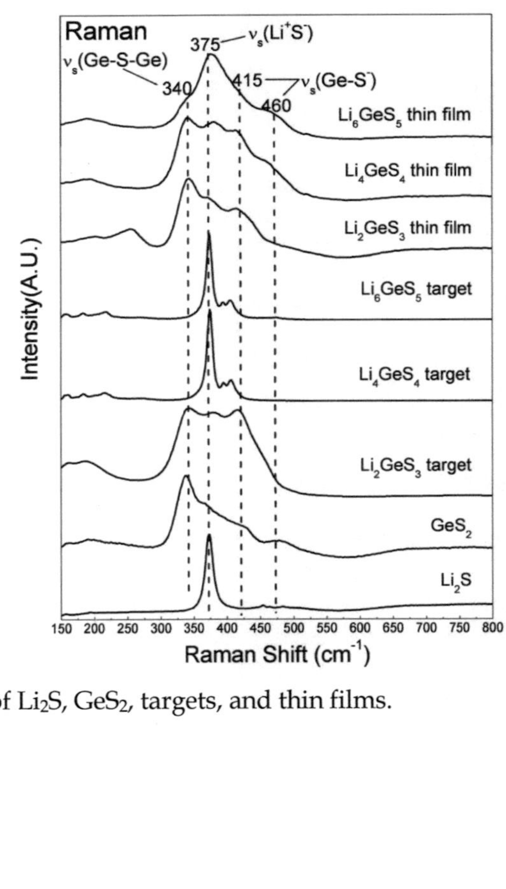

Fig. 4. Raman spectra of Li_2S, GeS_2, targets, and thin films.

symmetric stretching of bridging sulfur, S (BS), (Ge-S-Ge) in the $GeS_{4/2}$ tetrahedra. The Raman spectrum of Li_2S shows a single strong peak at ~375 cm^{-1} which is assigned to Li^+-S^- stretching modes. The Raman spectrum of the Li_2S is sharper than that of GeS_2 glass because Li_2S is crystalline while the GeS_2 is glassy.

In the spectrum of the Li_2GeS_3 target, there are three dominant peaks at 340, 375 and 415 cm^{-1}. The peak at 340 cm^{-1} is found in GeS_2 and is assigned to bridging sulfur (Ge-S-Ge) bonding. The peak at 375 cm^{-1} is found in the Raman spectrum of Li_2S and for this reason is assigned to Li^+S^- ionic bonding. The peak at 415 cm^{-1} is assigned to non-bridging sulfur (NBS) ≡Ge-S^- ionic bonding. While there are three peaks in the Raman spectrum of Li_2GeS_3 target, the Raman spectra of the Li_4GeS_4 and Li_6GeS_5 targets show only one dominant peak at 375 cm^{-1}. The strong main Raman peak in the both Li_4GeS_4 and Li_6GeS_5 target materials appears at 375 cm^{-1} which is at the same peak position of Li_2S. This indicates that the 375 cm^{-1} peak in both of the target materials was related to that of the Li_2S component. The narrowing of the Raman peaks in spectra of Li_6GeS_5 and Li_4GeS_4 compounds compared to that of Li_2GeS_3 arises from the polycrystalline structure of the former compound and the glassy structure of the latter.

The Raman spectrum of the Li_2GeS_3 thin-film shows three dominant peaks at 340, 375, and 415 cm^{-1}. The peak at 340 cm^{-1} coincindes with GeS_2 main peak position and is assigned to the BS (Ge-S-Ge) mode. The 375 cm^{-1} peak is assigned to Li^+-S^- modes and the 415 cm^{-1} peak is assigned to NBS (Ge-S^-) modes.

Among the three peaks, the peak at 340 cm^{-1} has the highest intensity. This is due to the high fractions (50%) of GeS_2 glass former in Li_2GeS_3. The Raman spectrum of the Li_4GeS_4 thin-film also shows three peaks at 340, 375 and 415 cm^{-1}, like the spectrum of the Li_2GeS_3 thin-film, and another broader peak of lower intensity at 460 cm^{-1}. The intensities of the peaks at 375 and 415 cm^{-1} are higher than those in the spectrum of the Li_2GeS_3 thin-film. This is consistent with the increased Li_2S content in the Li_4GeS_4 compared to Li_2GeS_3 which would increase the concentration of both Li^+S^- and Ge-S^- NBS modes. The Raman spectrum of the Li_6GeS_5 thin-film which has an even higher Li_2S content compared to the other thin-films only has one dominant peak at 375 cm^{-1} which is assigned to the Li^+S^- vibrational mode. This indicates that the Li_6GeS_5 thin-film contains the highest Li_2S content compared to the other two thin-films. There are three low intensity peaks at 340, 415 and 460 cm^{-1} in the spectrum of Li_6GeS_5. As described above, the peak at 340 cm^{-1} is assigned to the bridging sulfur (Ge-S-Ge bonding) and the peaks at 415 and 460 cm^{-1} are assigned to modes of the NBS (Ge-S^-). The peak at 460 cm^{-1} is assigned to the 1 NBS bonding and the peak is not present significantly in thin-films. The peak at 415 cm^{-1} is assigned to 2 NBS and the peak is present in thin-films. On the other hand, the peak at 340 cm^{-1} is assigned to 0 NBS and the peak is present in thin-films. The Raman spectra of all other thin-films do not show sharp peaks, but rather show broad peaks compared to those of crystalline targets (Li_4GeS_4 and Li_6GeS_5) and are consistent with the films being amorphous. As the Li_2S content increases in the targets (Li_2, Li_4, and Li_6), the Li_2S content in the thin-film increases. It can be concluded that although the previous reported literature showed Li_2S deficiency in GeS_2-based thin-films after sputtering compared to that of target,(Yamashita et al., 1996a) the amount of Li_2S in the thin-films in this study increases with the increase of Li_2S in the target and are consistent with the Li_2S content in the targets.

4.3.3 Infrared (IR) spectroscopy

To further characterize the starting materials, Li_2S crystalline powder and GeS_2 glass powder, targets, and their thin-films were characterized by infrared spectroscopy. Attention is focused on the far-IR region (900 to 100 cm^{-1}) in order to evaluate the nature of the chemical bonding in the materials, as well as the mid-IR region (4000 to 400 cm^{-1}) in order to determine how these materials might be contaminated by oxygen and/or moisture before and/or after processing. However, due to the lack of any significant O or OH contamination in the films and the very thin dimension observed, the mid-IR spectra are not shown here. However, in the far-IR region, strong absorptions were observed and arise from the framework structure species Li, Ge, and S.

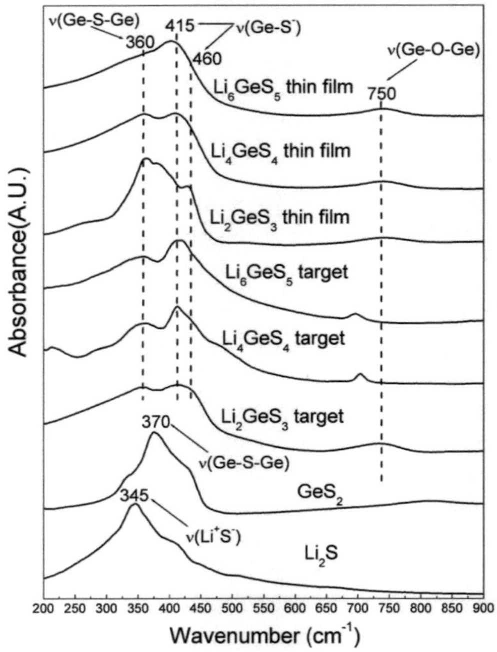

Fig. 5. Infrared spectra of Li_2S, GeS_2, targets, and thin-films

The IR spectra of polycrystalline Li_2S, glassy GeS_2, targets and thin films are shown in Figure (5). The IR peak in the far-IR spectrum of Li_2S at ~345 cm^{-1} is assigned to the ionic bonding of Li^+S^- and the strong peak at ~370 cm^{-1} in the spectrum of glassy GeS_2 is assigned to the BS, ν(Ge-S-Ge, BS) mode of the $GeS_{4/2}$ tetrahedra.(Zhou et al., 1999) It is possible that the broad peak in the IR spectrum of GeS_2 can be deconvoluted into two additional peaks, one centered at ~325 cm^{-1} and the other centered at ~435 cm^{-1}. These two additional peaks also arise from vibrational modes of the $GeS_{4/2}$ tetrahedra. (Frumarova et al., 1996) The shift in wavenumbers can be due to the presence of compressive stress in the film which is expected for films deposited by RF sputtering. In the IR spectra of the GeS_2, there is one broad and low intensity peak at ~800 cm^{-1}.(Zhou et al., 1999) This peak is assigned to the preparation and handling giving rise to a Ge-O bonding mode. It can be assumed that GeS_2 might be slightly contaminated by oxygen during IR sample measurement. In the IR spectra of both the starting materials, there is no peak at ~ 1500 cm^{-1} (O-H vibration mode) or at ~3500 cm^{-1} (O-H stretching mode) so this suggests that these two starting materials are not significantly contaminated by oxygen or moisture.

The IR spectra of Li_2GeS_3, Li_4GeS_4 and Li_6GeS_4 targets show dominant peaks at ~360 cm^{-1} and 415 cm^{-1} and a low intensity peak at ~750 cm^{-1}. The IR peak at ~360 cm^{-1} is assigned to the BS, ν(Ge-S-Ge, BS) mode and the 415 cm^{-1} peak corresponds to the vibration stretch of Ge with two non-bridging sulfur atoms. The broad IR peak at ~360 cm^{-1} can be deconvoluted into two peaks one centered at 345 cm^{-1} corresponding to Li^+S^- mode and the other centered at ~360 cm^{-1} corresponding to Ge-S-Ge mode. In addition, one low intensity peak which is assigned to oxide impurities, ν(Ge-O-Ge) appears at ~750 cm^{-1}. It is possible that contamination occurs when the target materials are melted in the glovebox because a background level of several ppm O_2 exists in the glovebox. Another possibility is that the oxygen comes from the GeS_2, its spectrum in Figure (6) shows that there is a low intensity peak at ~ 750 cm^{-1} assigned to GeO_2.

To the best of our knowledge, the IR spectra of the thio-germanate based thin-film materials have not been reported in the open literature. In this research, in order to characterize the thin-films by IR spectroscopy, the Li_2GeS_3, Li_4GeS_4 and Li_6GeS_5 thin-films were deposited directly on the top side of pressed CsI pellets that provided a mid- and far-IR transparent support for the films. The Li_2GeS_3, Li_4GeS_4 and Li_6GeS_5 thin-films were deposited directly on the pressed CsI pellets and the IR spectra were then collected in transmission. The intense peak at ~360 cm^{-1} can be deconvoluted into two peaks one centered at 345 cm^{-1} corresponding to the Li^+S^- mode and the other centered at ~360 cm^{-1} corresponding to the Ge-S-Ge mode as described above and the intensity of this peak decreases with added Li_2S. In addition, one low intensity peak which is assigned to oxide impurities, ν(Ge-O-Ge) appears at ~750 cm^{-1}. A new band appears at 445 cm^{-1} as a result of the formation of non-bridging sulfurs -Ge-S^--Li^+ (NBS). This NBS band was reported at ~450 cm^{-1} in the IR spectra of binary $xNa_2S + (1-x)GeS_2$ glasses.(Barrau et al., 1980) The NBS band at 445 cm^{-1} diminishes as another NBS band at 415 cm^{-1} grows stronger with further additions of Li_2S and this suggests that the number of NBS per Ge increases with the addition of Li_2S. Indeed, it is expected from the compositions that these would be two NBS in Li_2GeS_3 and four NBS in Li_4GeS_4 and Li_6GeS_5.

4.3.4 Surface morphology and thickness of the thin-film

In order to determine the sputtering rate, the thickness of the thin-films were measured in the cross-section direction by FE-SEM as shown in Figure (6-a). A Ni adhesion layer (~120 nm) is used to improve the adhesion between the Si wafer and the thin-film. The Ni adhesion layer is also very useful for Raman spectroscopy. In particular, when one characterizes the films using micro-Raman spectroscopy, the dominant silicon peak, ~520 cm^{-1}, appears in Raman spectra unless a barrier layer is used. Therefore, the Ni adhesion layer also acted to prevent the appearance of the peak from the silicon substrate. Furthermore, it has been found that Ni is chemically stable in contact with the lithium thio-germanate thin-film electrolytes.(Bourderau et al., 1999) The sputtering power and pressure of 50 W and 25 mtorr (~3.3 Pa) were used, respectively, and the total thickness of the thin- film after two hours of sputtering was ~1.3 μm which gives a sputtering rate of ~5 nm/minute.

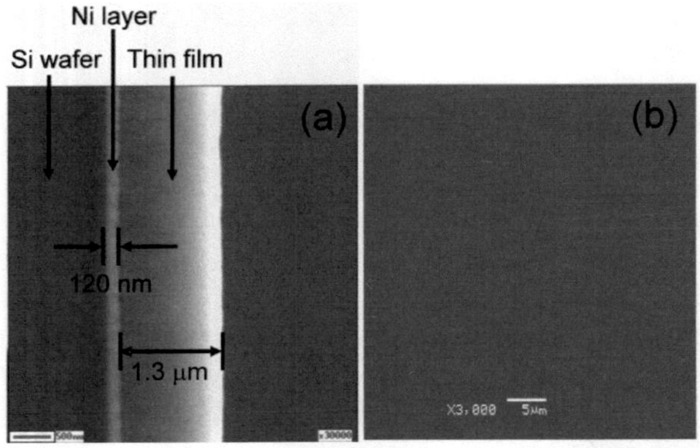

Fig. 6. FE-SEM images of cross-sectional view (a) and top view (b) of Li_4GeS_4 thin film grown on a Ni/Si substrate in an Ar atmosphere.

Figure (6-b) shows the surface morphology of the thin-films produced in an Ar atmosphere. The thin-film surface is mirror-like without any defects or cracks. This suggests that the thin-film electrolytes are homogeneous and have a flat surface morphology. The smooth surface enables the thin-films to decrease the contact resistance between thin-film and the electrodes.

4.3.5 Impedance analysis

In order to measure the ionic conductivity of the thin-films, they were deposited on a single crystal Al_2O_3 substrate. The sapphire substrates were loaded into a d.c. sputtering chamber in the glovebox and covered by the stainless steel mask with two 2 mm × 10 mm slits at 2 mm apart parallel to each other to produce two 2 mm × 10 mm parallel electrodes 2 mm apart on the sapphire substrate. Au electrodes of ~100 nm thickness were sputtered for 20 min. at a sputtering rate of 5 nm/min. through the mask. Lastly, the substrate then was loaded into the RF magnetron sputtering chamber to grow the thin-film electrolytes.

The conductivities of the thin-films were determined from the resulting complex impedance spectra. The semicircle at high frequency represents the response of the thin-film materials to an applied electric field. Thus, the d.c. resistance can be calculated from the semicircle plot. The ionic conductivities of the thin-films can be calculated from the measured d.c. resistance, R, the thickness of the electrolyte t, its area A, and t/A is the cell constant.

The ionic conductivities of the Li_6GeS_5 thin-film grown in Ar atmosphere at various temperatures from -25 °C to 100 °C with 25 °C increments are shown in Figure (7). The ionic conductivities of the Li_6GeS_5 thin-film in Ar atmosphere at 25 °C and at 100 °C are 1.7×10^{-3} S/cm and 3.0×10^{-2} S/cm, respectively. As the temperature increases from -25 °C to 100 °C, the ionic conductivity continually increased and was found to be stable over this temperature range. This thin-film appears to be stable wider temperature range compared to liquid electrolytes (Guyomard and Tarascon, 1995).

Figure (8) shows a Nyquist plot of the complex impedance for the Li_6GeS_5 thin-film grown in an Ar atmosphere over the temperature ranges from 25 °C to 100 °C with 25 °C increments.

The frequency increases for each point from right to left starting at 0.1 Hz and finishing at 1 MHz. The spike at low frequencies represents polarization of the Li ions due to the use of Au blocking electrodes. The d.c resistance can be calculated from the semicircle as shown in Figure (8).

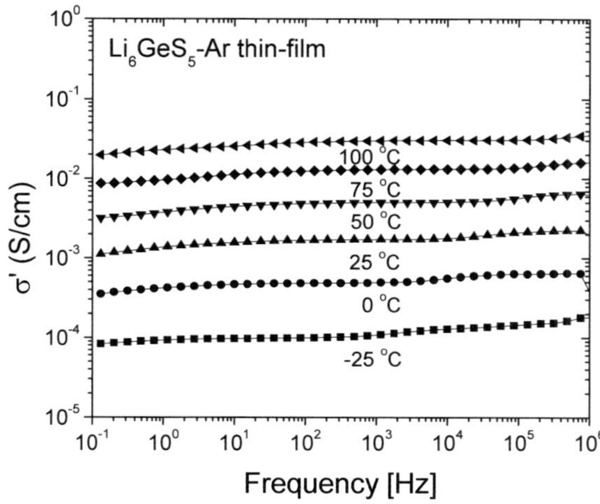

Fig. 7. The ionic conductivity of Li_6GeS_5 thin-film grown in Ar atmospheres over the temperatures from -25 °C to 100 °C with 25 °C increments.

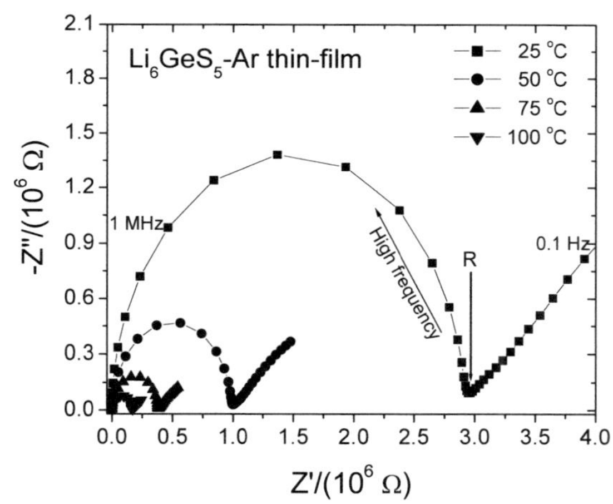

Fig. 8. Nyquist plot of the complex impedance for the Li_6GeS_5 thin-film in Ar atmosphere over the various temperatures from 25 °C to 100 °C.

The ionic conductivities were calculated from the resistance and cell constant relations and are listed in Table (1). The d.c. ionic conductivities of the Li_2GeS_3, Li_4GeS_4, Li_6GeS_5 and Li_8GeS_6 thin-films are also shown in Table (1). The ionic conductivities of all four thin-films were characterized with the same temperature ranges, from -25 °C to 100 °C with 25 °C increments, and same frequency ranges, from 0.1Hz to 1 MHz. As shown in Table 1, the ionic conductivities of the Li_4GeS_4 thin-film are higher than those of Li_2GeS_3 thin-film at each temperature. The reason for the Li_4GeS_4 thin-film having a higher ionic conductivity than the Li_2GeS_3 thin-film is that the Li_4GeS_4 thin-film contains a higher Li_2S content. The ionic conductivities at room temperature and 100 °C of the Li_4GeS_4 thin-film are 7.5×10^{-4} S/cm and 1.3×10^{-2} S/cm, respectively. As the temperature increased, the ionic conductivities increase without decreasing and hence the thin-films are stable over wide temperature ranges.

Temp.	Li_2GeS_3 (S/cm)	Li_4GeS_4 (S/cm)	Li_6GeS_5 (S/cm)	Li_8GeS_6 (S/cm)
-25 °C	4.0×10^{-6}	4.6×10^{-5}	9.7×10^{-5}	2.6×10^{-6}
0 °C	2.5×10^{-5}	2.2×10^{-4}	4.8×10^{-4}	1.5×10^{-5}
25 °C	1.1×10^{-4}	7.5×10^{-4}	1.7×10^{-3}	7.3×10^{-5}
50 °C	3.8×10^{-4}	2.2×10^{-3}	5.0×10^{-3}	1.9×10^{-4}
75 °C	1.1×10^{-3}	5.8×10^{-3}	1.3×10^{-2}	5.7×10^{-4}
100 °C	2.9×10^{-3}	1.3×10^{-2}	3.0×10^{-2}	1.4×10^{-3}

Table 1. D.c. ionic conductivities over the temperatures from -25 °C to 100 °C at 25 °C increments for $nLi_2S + GeS_2$, n = 1, 2, 3, and 4, thin-films grown in Ar atmosphere

The ionic conductivities of the Li_6GeS_5 thin-film at room temperature and 100 °C are 1.7×10^{-3} S/cm and 3.0×10^{-2} S/cm, respectively. The ionic conductivities of the Li_6GeS_5 thin-film increase with increasing temperatures. As n increases in $nLi_2S + GeS_2$ from 1 to 3, the ionic conductivities increase at all temperatures. For the n = 3 thin-films, the ionic conductivity was measured to be $>10^{-3}$ S/cm at 25 °C which is very high compared to the ionic conductivity of oxide thin-films which are $\sim 10^{-6}$ S/cm at 25°C.

To determine if a maximum Li^+ ionic conductivity occurs for this series of materials, a Li_8GeS_6 thin-film, n = 4 in $nLi_2S + GeS_2$, was prepared and the ionic conductivities were analyzed over the same temperature and frequency ranges. The ionic conductivities of the Li_8GeS_6 thin-film at 25 °C and 100 °C are 7.3×10^{-5} S/cm and 1.4×10^{-3} S/cm, respectively. While the d.c. ionic conductivities of the thin-films from n = 1 to 3 in $nLi_2S + GeS_2$ increased with n, the d.c. ionic conductivity of the thin-film for n = 4, Li_8GeS_6, decreased and this is caused by the activation energy increasing, see discussion below.

In this series of films, the n = 3 composition, Li_6GeS_5, is the optimized composition with the highest ionic conductivity in the $nLi_2S + GeS_2$ system, n = 1, 2, 3, and 4. Although the n = 4 composition, Li_8GeS_6, thin-film showed lower ionic conductivities than those of the n = 1(Li_2GeS_3), 2(Li_4GeS_4), and 3(Li_6GeS_5) compositions, the ionic conductivities of all four of these thin-films are significantly higher than that of LiPON. In addition, all compositions n = 1, 2, 3, and 4 of the sulfide thin-film electrolytes are very stable over wide temperature ranges compared to liquid or polymer electrolytes. Therefore, Li-ion batteries using these sulfide thin-film electrolytes are promising for use in solid-state lithium-ion batteries.

For all thin-films, the ionic conductivities were found to follow an Arrhenius law, $\sigma_{d.c.}(T) = \sigma_o \exp(-\Delta E_a/RT)$, over the measured temperature ranges. The Arrhenius plots of the d.c. ionic conductivities of the thin-films over the temperature range from -25 °C to 100 °C with 25 °C increments are shown in Figure (9) and are compared to that of LiPON. The activation energies of conduction, ΔE_a, were calculated from the slope of the Arrhenius plots. The ionic conductivities of the thin-films at room temperature, the activation energies, and pre-exponential factors are listed in Table (2).

The ionic conductivities of the all-sulfide thin-films higher than that of LiPON(Yu et al., 1997). In the case of the Li_6GeS_5 thin-film, the ionic conductivity at 25 °C is approximately three orders of magnitude higher than that of LiPON(Yu et al., 1997). The composition dependence of the ionic conductivities of all thin-films, n = 1, 2, 3, and 4 in $nLi_2S + GeS_2$ system, at 25 °C and their activation energies are shown in Figure (10) to show how they depend upon Li_2S content. The thin-films showed that as Li_2S content increases, the ionic conductivities increase up to n = 3, 75 % Li_2S.

In addition, while the conductivity of the bulk sulfide glasses are less than that of the thin-films, the ionic conductivities of the sulfide bulk glasses(Kim et al., 2006) over the range from 35% to 50 % the ionic conductivities also increased. It is significant to note that the ionic conductivity decreased and the activation energies increased for the thin-films at n = 4.

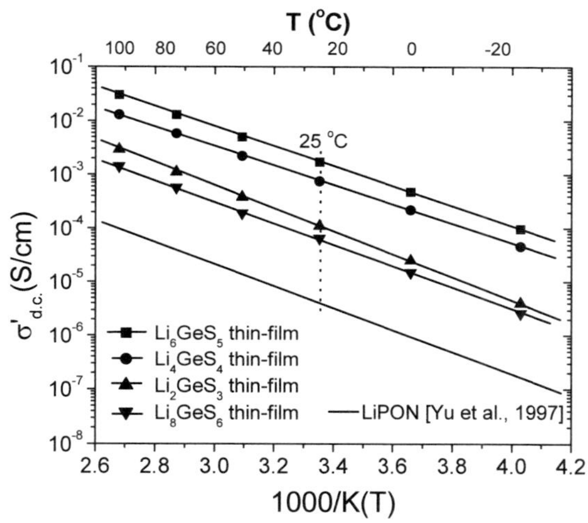

Fig. 9. Arrhenius plot of the ionic conductivities at various temperatures for Li_2GeS_3, Li_4GeS_4, Li_6GeS_5, and Li_8GeS_6 thin-films in Ar atmosphere and comparison of ionic conductivities between thin-films and LiPON (Yu et al., 1997).

Composition	$\sigma_{25°C}$ (S/cm)	ΔE_a (kJ/mol) (± 0.05)	$\log_{10}[\sigma_o$ (S/cm)] (± 0.005)
Li_2GeS_3-Ar thin-film	1.1 (±0.05) × 10^{-4}	40.2	3.096
Li_4GeS_4 -Ar thin-film	7.5 (±0.05) × 10^{-4}	34.5	2.951
Li_6GeS_5 -Ar thin-film	1.7 (±0.05) × 10^{-3}	35.0	3.382
Li_8GeS_6 -Ar thin-film	7.0 (±0.05) × 10^{-5}	38.1	2.763

Table 2. Ionic conductivities at room temperature and activation energies for $nLi_2S + GeS_2$ (n = 1, 2, 3, and 4) thin-films in Ar atmosphere

Further, the effective basicity of the counter and charge compensating anion in the structure of these materials is also expected to change significantly with n. In the n = 1, 2, 3, and 4 films, the structure is expected to consist of increasing numbers of sulfurs possessing a single negative charge, and recent XPS studies of these same films show that these films are comprised of the nominal structures shown in Figure (11). In these structures, the average charge on the sulfur is expected to change from -2/3, -4/4, -6/5 to -8/6. At Li_2S, the formal charge of the sulfur is expected to -2/1. Hence, while increasing the number of Li^+ is important to increasing the ionic conductivity, the negative charge density on the sulfur increases by a factor of 2 in this series and as a result the columbic binding energy of these increasingly basic sulfurs will increase as well. It appears that for the n = 3 composition, the larger number of Li is still important because the appearance of the full -2/1 negatively charged Li_2S unit does increase the conductivity activation energy, 35 kJ/mol for n = 3 versus 34.5 kJ/mol for n = 2, but the conductivity is still higher, presumably because of the composition (n) dependence of the pre-exponential factor.

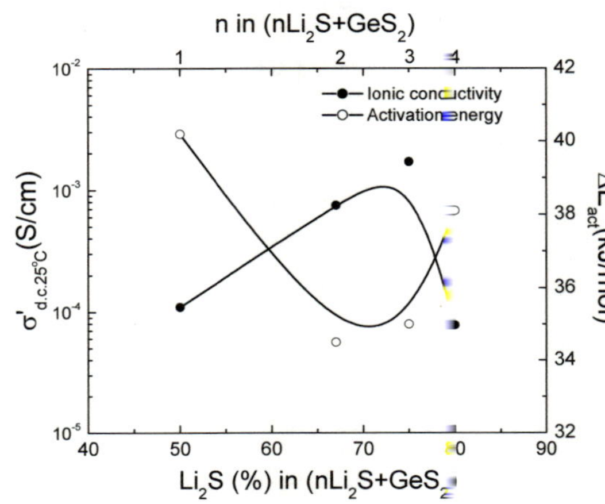

Fig. 10. Ionic conductivities and activation energies of the thin films n = 1, 2, 3, and 4 in $nLi_2S + GeS_2$ system at 25 °C.

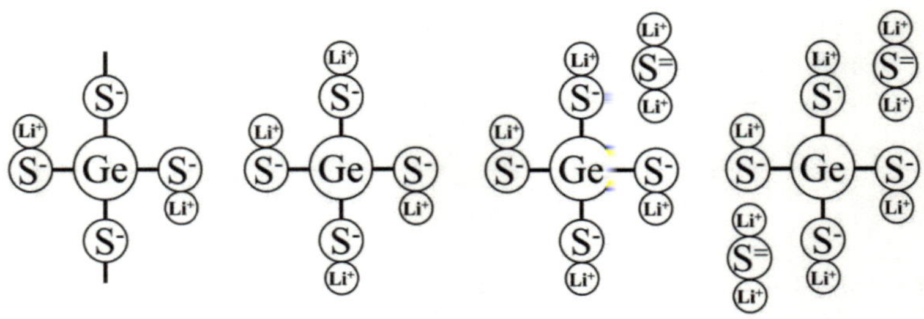

(b) Li_2GeS_3 **(c) Li_4GeS_4** **(d) Li_6GeS_5** **(a) Li_8GeS_6**

Fig. 11. Atomic structure of the four nominal compositions with n (n = 1, 2, 3, and 4).

We have shown in our other studies of these thin-films that the fraction of NBS $Ge\text{-}S^-$ increases with n in these series. For n = 1, the fraction of bridging sulfurs S-Ge-S and non-bridging sulfurs $Ge\text{-}S^-$ are 1/3 and 2/3, respectively. At n = 2, these fractions are 0 and 1, respectively. For n = 3 and n = 4, these units are expected to be connected to non-bridging sulfur units and new $S^{=}$ units. Hence, the fraction of sulfur in non-bridging Ge-S^{-} units and $S^{=}$ units are expected to be 4/5 and 1/5, respectively. In Li_8GeS_6 these fractions change to 4/6 and 2/6, respectively. The increase in the fraction of Li^{+} ions bound to $S^{=}$ units increases from 0 for Li_2GeS_3 and Li_4GeS_4 to 2/6 (33 %) and 4/8 (50 %) for Li_6GeS_5 and Li_8GeS_6, respectively. Due to the high binding energy expected for Li^{+} ions about these $S^{=}$ ions it is therefore not surprising to see that the activation energy passes through a minimum at the n = 2 composition and increases for n = 3 and 4. Such a maximum in conductivity and minimum activation energy have been observed in other high alkali glass forming system where the anionic basicity of the host network increases significantly in the high alkali modifier range(Martin and Angell, 1984).

4.3.6 X-ray photoelectron spectroscopy (XPS) analysis

4.3.6.1 Analysis of the Li_2S and GeS_2 starting materials

In order to verify the purity of the starting materials, the Li_2S and GeS_2 were examined by XPS. In the case of the commercially purchased Li_2S material, Table (3) shows that the concentration of C and O were ~12 % and ~21 % (±3 %), respectively, and as such relatively high.

At %	Li1s	Ge2p3	S2p	C1s	O1s	Comments
	44.7	-	22.9	11.7	20.7	As-prepared
Li_2S	66.1	-	33.9	-	-	Ignoring C and O
	66.7	-	33.3	0.0	0.0	Expected values
	-	34.2	59.2	6.6	0.0	As-prepared
GeS_2	-	35.7	64.3	-	0.0	Ignoring C
	-	33.3	66.7	0.0	0.0	Expected values

Table 3. XPS compositional analysis of the Li_2S and GeS_2 starting materials.

One possibility is that the Li_2S was slightly contaminated on the surface in the glovebox because the glovebox contained several ppm level of oxygen. Another possible reason for this can also include the "see-through" effect due to the double-sided tape used to adhere the powder to the XPS sample holder. The ratio of Li to S, however, 1.95 : 1.00 is very close to the expected value of 2 : 1.

From Table (3), while the Li_2S shows relatively high O content, the GeS_2 material was not contaminated by oxygen due in part to the fact that GeS_2 material is less hygroscopic than other sulfide materials, but also due to the fact that this material was prepared from high purity starting materials, Ge and S (99.9999%), in the very controlled conditions of our laboratory. GeS_2 contains a small percent of C, presumably surface C, see Table (3), and after ignoring C, the compositional data of GeS_2 agrees well with expected values.

Deconvoluted S2p core XPS spectra for crystalline Li_2S (a) and glassy GeS_2 (b) starting materials are shown in Figure (12). The binding energies for sulfur in Li_2S and GeS_2 are at

160.7 eV and 163.2 eV ±0.2 eV, respectively. The reason for the difference in the S2p binding energies between Li_2S and GeS_2 is that S in Li_2S is the fully ionic $S^=$ sulfide anion and the S in the GeS_2 is the fully covalent BS, ≡Ge-S-Ge≡, and hence the binding energy of the covalent BS is higher than that of the ionic sulfide. For the S2p spectra of sulfur species, there is a doublet consisting of $S2p_{3/2}$ and $S2p_{1/2}$ spin-orbit coupled electrons in the intensity ratio of 2:1. The S2p core peaks of Li_2S show one doublet. This doublet arises from the Li_2S bonding and means that only the Li_2S bonding exists. This result agrees well with the literature data(Foix et al., 2001).

If Li_2S was significantly contaminated by oxygen, the deconvoluted S2p spectra would be expected to show additional peaks related to sulfite SO_3^{2-} (166 eV) and/or sulfate SO_4^{2-} (172 eV) contamination(Volynsky et al., 2001). Both the Li_2S and GeS_2 materials do not show significant peaks at 166 eV and 172 eV and suggests that these materials are of high purity. The deconvoluted S2p core peaks of the GeS_2 also show as expected only one doublet arising from the single bridging sulfur structure, ≡Ge-S-Ge≡.

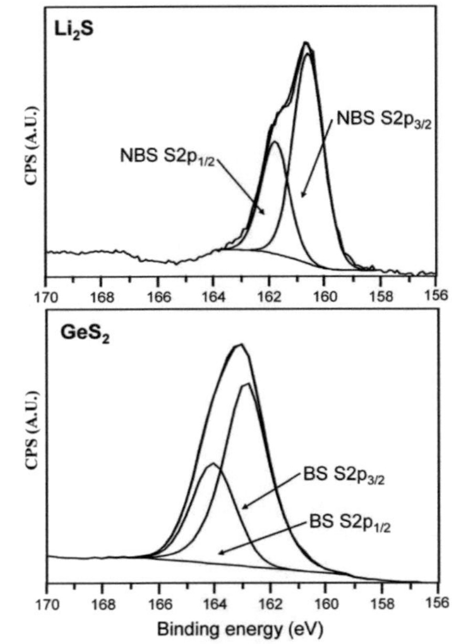

Fig. 12. Deconvoluted S2p core XPS spectra for Li_2S and GeS_2 powder.

4.3.6.2 Target material analysis

After the target materials for RF sputtering were made using $nLi_2S + GeS_2$, n = 1, 2 and 3, their compositions were determined by XPS. The compositional data of the three target materials are shown in Table (4). The target materials show C and O contents and therefore the Li, Ge and S contents are lower than the expected values. If C and O elements are ignored, the compositional data of Li, Ge and S for all three target materials nearly match with the expected values. Although the XPS compositional data are different between the collected and expected data, the differences are within the confidence error limit, ± 3%. Considering the ± 3% error of the XPS data, the small Li deficiency can be ignored.

Ar etching treatments were not performed to remove surface C and O because the target materials were in the form of powders and Ar etching does not work well for powders that do not have large flat smooth surfaces. The deconvoluted S2p spectra of the three target materials are shown in Figure (13).

At %	Li1s	Ge2p3	S2p	C1s	O1s	Comments
Li_2GeS_3 target	26.1	16.4	41.3	10.3	5.9	As-prepared
	30.4	19.1	48.1	-	-	Ignoring C and O
	33.3	16.7	50.0	0	0	Expected values
Li_4GeS_4 Target	36.5	9.6	40.2	8.6	5.1	As-prepared
	42.3	11.1	46.6	-	-	Ignoring C and O
	44.4	11.2	44.4	0.0	0.0	Expected values
Li_6GeS_5 target	40.4	8.0	37.2	6.5	7.9	As-prepared
	47.2	9.3	43.5	-	-	Ignoring C and O
	50.0	8.3	41.7	0	0	Expected values

Table 4. XPS compositional analysis of the Li_2GeS_3, Li_4GeS_4 and Li_6GeS_5 target materials.

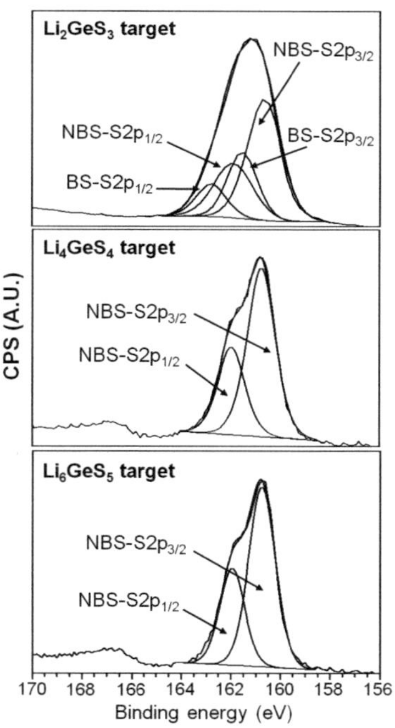

Fig. 13. Deconvoluted S2p core XPS spectra of the Li_2GeS_3 (a), Li_4GeS_4 (b), Li_6GeS_5 (c) target materials.

From the spectra in Figure (13), the binding energies of the NBS, $\equiv Ge\text{-}S^{-}Li^{+}$ and BS, $\equiv Ge\text{-}S\text{-}Ge\equiv$ can be obtained. By comparing these XPS spectra to that of the standard materials, it can be determined that the sulfur spectra do not contain peaks related to sulfite

SO_3^{2-} or sulfate SO_4^{2-} anions which would be shifted to significantly higher binding energies due to their S^{+4} and S^{+6} oxidation states, respectively. This suggests that although the target materials show some oxygen content as shown in Table (5), the contamination may only be on the surface. The low oxygen content is associated with the high purity of the starting materials as well as the fact that the target materials were made in a N_2 filled high quality glovebox. As shown in Figure (13), while two doublets appear in the deconvoluted S2p core peaks of the Li_2GeS_3 target material indicating that there are two chemically distinct surface species, the Li_4GeS_4 and Li_6GeS_5 targets show only one doublet indicating a single chemical species for sulfur.

Materials	E_b S2p$_{3/2-1/2}$ (eV)	Ratio
GeS_2	162.8 - 164.0	100 % BS
Li_2GeS_3 target	160.9 - 162.1 161.7 - 162.9	65.9 % NBS 34.1 % BS
Li_4GeS_4 target	160.8 - 162.0	100 % NBS 0% BS
Li_6GeS_5 target	160.7 - 161.9	100 % NBS 0 % BS
Li_2S	160.5 – 161.7	100 % NBS

Table 5. The XPS binding energies and the ratio of NBS to BS for the starting and target materials.

The binding energies of sulfur in the target compositions and the NBS and BS ratios of the three target materials are shown in Table (5). In order to compare the chemical shifts, the binding energies of the GeS_2 and Li_2S are also listed in Table (5).

While the binding energy of S in GeS_2 shows the highest value due to its BS structure, the binding energy of S in Li_2S shows the lowest value. The binding energies of the target materials are similar to one another and, as expected, are between that of GeS_2 and Li_2S. In the S XPS spectrum of the Li_2GeS_3 target, the low energy doublet is assigned to the NBS and the other higher energy doublet is assigned to the BS. For the Li_2GeS_3 target material, the ratio of the NBS to BS is 65.3 % to 34.7 %. The expected ratio of NBS to BS in the Li_2GeS_3 target composition agrees well with that calculated from the composition of 67 % to 33 %.(Foix et al., 2002) Theoretically, the ratio of the NBS to BS in the Li_4GeS_4 target should be 100 % and 0 %, respectively.

As shown in Table (5), the Li_4GeS_4 target material shows 100 % NBS to 0 % BS ratio. Additionally and as expected, the Li_6GeS_5 target material shows 100 % NBS and 0 % BS. As described above, Li_2S consists of only the $S^=$ anion whereas Li_4GeS_4 consists of 100 % NBS. From the composition, it is expected that the Li_6GeS_5 target should be composed of an equimolar mixture of Li_2S and Li_4GeS_4. However, the XPS spectra data shown in Table (5) shows that Li_6GeS_5 consists of only 100 % NBS and 0 % BS. Strictly speaking, Li_6GeS_5 should consist of Li_4GeS_4 which has 100 % NBS and Li_2S which has 100 % ionic sulfur, $S^=$. While the binding energies of the $S^=$ anion and the NBS are very close, the resolution of our XPS instrument appears to be insufficient to differentiate the chemical shift of $S^=$ anion and the NBS unit, $\equiv$Ge-S$^-$Li$^+$.

4.3.6.3 Lithium thio-germanate thin-film analysis

After sputtering thin-films on Ni-coated Si substrates in Ar atmospheres, they were characterized by XPS to determine their compositions and chemical shifts. It was found that a Ni protective layer on the Si was necessary to prevent reaction of the Si with the Li which produces highly Li deficient films. This is described below in the experimental section. The compositional data of all of the thin-films sputtered in Ar atmospheres using the three different conditions are shown in Table (6).

At %	Li1s	Ge2p3	S2p	C1s	O1s	Comments
Li_2GeS_3 thin-film n = 1	27.2	8.5	37.1	18.6	8.6	As-prepared
	32.6	15.9	47.8	0.0	3.7	Etching for 1 min.
	31.7	16.1	48.1	0.0	4.1	Etching for 5 min.
	33.3	16.7	50.0	0.0	0.0	Expected values
Li_4GeS_4 thin-film n = 2	31.0	5.5	32.1	18.3	13.1	As-prepared
	40.6	12.6	41.3	0.0	5.5	Etching for 1 min.
	41.9	12.9	40.5	0.0	4.7	Etching for 5 min.
	44.4	11.2	44.4	0.0	0.0	Expected values
Li_6GeS_5 thin-film n = 3	35.9	4.9	33.2	14.7	11.3	As-prepared
	43.7	8.9	41.8	0.0	5.6	Etching for 1 min.
	44.6	11.1	41.2	0.0	3.1	Etching for 5 min.
	50.0	8.3	41.7	0.0	0.0	Expected values

Table 6. XPS compositional analysis of the Li_2GeS_3, Li_4GeS_4 and Li_6GeS_5 thin-film grown on Ni-coated Si substrates in an Ar atmosphere.

For the as-prepared thin-films, the C and O contents are slightly higher than those of the targets and the Li, Ge and S contents are slightly lower than their expected values. It is assumed that this arises due to the intrinsically higher chemical reactivity of the surface of thin-films compared to bulk materials. In order to obtain more accurate compositional data of the thin-films, Ar etching was performed on the thin-film surfaces for 1 min. and 5 min. at a rate of ~1 nm/min. As shown in Table (6), after Ar etching for 1 min. the C content in the thin-films reduced to 0 % and the O content decreased significantly. Although some O content still exists in the thin-films, the Li, Ge, and S contents in the thin-films are very close to their expected values. In order to examine deeper profiles of the thin-film, Ar etching for 5 min. was performed at ~1 nm/ min. etching rate. After Ar etching for 5 min. was performed, the compositional data are almost the same compared to the data obtained after Ar etching for 1 minute. This suggests that the thin-films show high uniformity and quality except for the top 1 nm of the surface. Previous literature reported(Yamashita, M., et al., 1996a) that thio-germanate thin-films produced from Li_2S + Ga_2S_3 + GeS_2 by sputtering showed severe Li deficiency. While these ternary thin-films showed as high as ~30 to 40 % Li deficiency compared to the Li in target composition, the thin-films produced in this study only show ~3-5 % Li deficiency. The compositions of the thin-films in this study are consistent with those of the target and it is therefore assumed that the sputtering conditions reported here are optimized and the thin-film compositions are reliable. In order to determine the fractions of NBS and BS in the thin-films, the deconvoluted S2p core peaks for the Li_2GeS_3, Li_4GeS_4, and Li_6GeS_5 as-prepared thin-films (without Ar etching treatment) are shown in Figure (14).

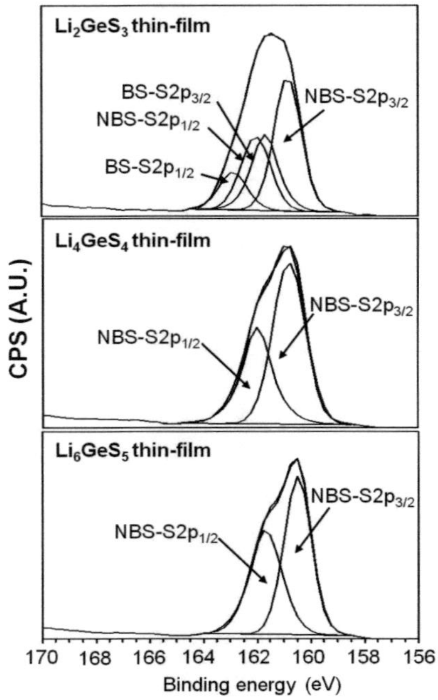

Fig. 14. Deconvoluted S2p core peaks for the Li_2GeS_3, Li_4GeS_4, and Li_6GeS_5 thin-films grown in Ar atmospheres.

While the XPS spectra of the Li, Ge, and S species are unchanged in binding energy with and without Ar etching, Ar etching could reform the chemistry of the Ar sputtered surface. For this reason, it is believed that a better representation of the bonding chemistries, the chemical speciation, of these thin-films are therefore found in the as-prepared surfaces of the thin-films. For example, Foix et al. reported the fractions of NBS and BS in lithium thio-germanate and thio-arsenate bulk glasses and to do so they broke the glasses in the glove box and they characterized the newly exposed broken surface of the glasses without Ar etching(Foix et al., 2001; Volynsky et al., 2001).

In addition, Atashbar et al. reported the XPS deconvoluted data of TiO_2 thin-films without Ar etching(Atashbar et al., 1998). These approaches suggest that although accurate compositional data could be obtained from the Ar etched surface, the data could also be to use the XPS deconvoluted structural analysis is also obtained from the as-prepared surface without Ar etching. The fractions of the NBS and BS in the thin-films were calculated from Figure (14) and are shown in Table (7). In Figure (14) as described above, the Li_2GeS_3 thin-film shows two doublets. The doublet on the low energy side (lower binding energy) is attributed to the NBS and the other doublet on the high energy side (higher binding energy) is associated to BS. The ratios of the NBS and BS in the Li_2GeS_3 thin-film are 64.4 % NBS and

35.6 % BS, respectively. Although the ratios are not exactly the same as the expected values, 67 % NBS and 33 % BS, the differences between those of the Li_2GeS_3 thin-film and expected values are within the error range of ± 3%.

In addition, the ratios of the NBS and BS in the Li_2GeS_3 target and thin-film are very close. This suggests that the target compositions and thin-film compositions are quite consistent. While the deconvoluted S2p core spectra of the Li_2GeS_3 thin-film show two doublets, the deconvoluted S2p core spectra of the Li_4GeS_4 and Li_6GeS_5 thin-films show only one doublet. As described above, the Li_4GeS_4 and Li_6GeS_5 targets also show only one doublet from the NBS. In agreement with these Li_4GeS_4 and Li_6GeS_5 targets, the two thin-films show only one doublet arising from only NBS structures.

Recently, a few XPS studies of Ge-S thin-films have been reported in the literature but the analyzes were very brief.(Gonbeau et al., 2005; Mitkova et al., 2004) However, in this study, the compositions and chemical shifts of the Li-Ge-S thin-films have been thoroughly investigated(Gonbeau et al., 2005; Mitkova et al., 2004). As shown in Figure (14), the spectrum for GeS_2 shows a higher binding energy than those of Li_2S and the thin-films because the GeS_2 is assigned to the BS as described above. As the Li_2S content increases, the binding energy of the thin-films shifts to lower values than that of GeS_2. While the binding energies of the thin-films are similar to one another, with the S peak for Li_2GeS_3 being broader than that for Li_4GeS_4 and Li_6GeS_5 due to the presence of both BS and NBS, the binding energies of the thin-films slightly shifted to lower values. As expected, the binding energy of the Li_2S shows the lowest binding energy of the materials studied here.

Thin-films	E_b S2p$_{3/2-1/2}$ (eV)	NBS : BS	Expected values
Li_2GeS_3 as-prepared	160.9 - 162.1 161.5 - 162.7	65.4 % NBS 34.6 % BS	66.7 % NBS 33.3 % BS
Li_4GeS_4 as-prepared	160.8 - 162.0	100 % NBS 0 % BS	100 % NBS 0 % BS
Li_6GeS_5 as-prepared	160.7 - 161.9	100 % NBS 0 % BS	100 % NBS 0 % BS

Table 7. The XPS binding energies (E_b) and fractions of NBS to BS of the Li_2GeS_3, Li_4GeS_4 and Li_6GeS_5 thin-films.

5. Conclusion

For the first time, lithium thio-germanate thin-film electrolytes for the solid-state lithium-ion batteries grown by RF sputtering were characterized thoroughly by XRD, FE-SEM, Raman, IR, impedance spectroscopy, and XPS. From the XRD pattern, the Li_2GeS_3 (n = 1) target was amorphous and the Li_4GeS_4 (n = 2) and Li_6GeS_5 (n = 3) targets were crystalline as expected from compositions. The Li_6GeS_5 target appears to be consistent with an equi-molar mixture of Li_2S and Li_4GeS_4. FE-SEM of the thin films deposited on Ni-coated Si substrates shows a mirror-like surface without cracks or pits. The Raman spectra of all of the thin-films do not show sharp peaks, rather they show much broader peaks compared to those of crystalline targets (Li_4GeS_4 and Li_6GeS_5) and are consistent with the thin-films being amorphous. This shows that RF sputtering can be used to extend the formation range of amorphous materials from ~50 to ~75 mole % Li_2S.

The Raman and IR spectra also showed the structural and compositional consistency between targets and the thin-films and that the Li_2S content of thin-films increased as expected with Li_2S addition in the targets. These results suggest that the thin-films did not show significant Li deficiency as seen in previous reports after sputtering.

The ionic conductivities of the thin-films at 25 °C obtained are the highest reported for Li^+ ion in a glassy materials and are at least two orders of magnitude higher than those of commercial LiPON thin-film electrolytes. The thin-films materials are stable over wide temperature ranges, so that it can be said that the lithium-ion batteries based on these sulfides materials are very stable over wide temperature ranges and are very promising to apply to commercial products. The purpose of the XPS work was to provide information on the compositional data and the structures of lithium thio-germanate thin-films by means of XPS studies. High purity starting materials were used and targets were produced under well-calibrated and optimized conditions.

For the first time, highly reproducible compositions and chemical shifts of the starting materials, targets, and thin-films of $nLi_2S + GeS_2$ materials were determined by XPS. Although the as-prepared thin-films contained C and O on the surface, the thin-films showed that the C was completely removed and O content decreased significantly after Ar etching for 1 min. This suggests that the thin-films were contaminated by C only at the top 1 nm of the surface and the thin-films contained low oxygen contents in the interior of the film. After Ar etching, the compositions of the thin-films were very close to those expected. Therefore, the thin-films produced by sputtering are very close to their corresponding target materials. Thio-germanate thin-film materials have not been as widely studied as their oxide materials because of the difficulties in preparation. However, in this study, the lithium thio-germanate thin-films were successfully prepared and the compositional data and the chemical shifts were carefully characterized by XPS.

By successfully making thin-films of high quality and high conductivity, they can be applied as thin-film electrolytes for solid-state thin-film batteries. Further extensive effort for solid-state full battery fabrication, however, is needed before this thin-film electrolyte is put to practical use.

6. Acknowledgement

This research was supported by NSF under grant number DMR-0312081 and this research support is gratefully acknowledged. The authors thank Mr. James Anderegg who helped collect all of the XPS spectra and assisted with the experimental details to load and examine the samples without contamination.

7. References

Abe, T., et al., (2005), Lithium-Ion Transfer at the Interface Between Lithium-Ion Conductive Ceramic Electrolyte and Liquid Electrolyte-A Key to Enhancing the Rate Capability of Lithium-Ion Batteries: *Journal of the Electrochemical Society*, Vol. 152, No. 11, pp. A2151-A2154, 00134651.

Akridge, J. R., and H. Vourlis, (1986), Solid-State Batteries Using Vitreous Solid Electrolytes: *Solid State Ionics*, Vol. 18-9, pp. 1082-1087, 0167-2738.

Akridge, J. R., and H. Vourlis, (1988), Performance of Li/TiS2 solid state batteries using phosphorous chalcogenide network former glasses as solid electrolyte: *Solid State Ionics*, Vol. 28-30, No. Part 1, pp. 841-846, 0167-2738.

Albano, F., et al., (2008), A fully integrated microbattery for an implantable micro-electromechanical system: *Journal of Power Sources*, Vol. 185, No. 2, pp. 1524-1532, 0378-7753.

Amatucci, G. G., et al., (1996), CoO_2, the end member of the Li_xCoO_2 solid solution: *Journal of the Electrochemical Society*, Vol. 143, No. 3, pp. 1114-1123, 0013-4651.

Andersson, A. S., et al., (2000), Lithium extraction/insertion in $LiFePO_4$: an X-ray diffraction and Mossbauer spectroscopy study: *Solid State Ionics*, Vol. 130, No. 1-2, pp. 41-52, 0167-2738.

Angell, C. A., (1983), Fast ion motion in glassy and amorphous materials: *Solid State Ionics*, Vol. 9-10, No. DEC, pp. 3-16, 0167-2738.

Aotani, N., et al., (1994), Synthesis and electrochemical properties of lithium ion conductive glass, $Li_3PO_4Li_2SSiS_2$: *Solid State Ionics*, Vol. 68, No. 1-2, pp. 35-39, 0167-2738.

Appetecchi, G. B., et al., (2003), Investigation of swelling phenomena in poly(ethylene oxide)-based polymer electrolytes-III. Preliminary battery tests: *Journal of the Electrochemical Society*, Vol. 150, No. 3, pp. A301-A305, 0013-4651.

Armand, M., and J. M. Tarascon, (2008), Building better batteries: *Nature*, Vol. 451, No. 7179, pp. 652-657, 0028-0836.

Atashbar, M. Z., et al., (1998), XPS study of Nb-doped oxygen sensing TiO_2 thin films prepared by sol-gel method: *Thin Solid Films*, Vol. 326, No. 1-2, pp. 238-244, 0040-6090.

Baba, M., et al., (2001), Fabrication and clcctrochemical characteristics of all-solid-state lithium-ion rechargeable batteries composed of $LiMn_2O_4$ positive and V_2O_5 negative electrodes: *Journal of Power Sources*, Vol. 97-8, pp. 798-800, 0378-7753.

Baba, M., et al., (1999), Fabrication and electrochemical characteristics of all- solid-state lithium-ion batteries using V_2O_5 thin films for both electrodes: *Electrochemical and Solid State Letters*, Vol. 2, No. 7, pp. 320-322, 1099-0062.

Bao, Q., et al., (2005), Pulsed laser deposition and its current research status in preparing hydroxyapatite thin films: *Applied Surface Science*, Vol. 252, No. 5, pp. 1538-1544, 0169-4332.

Barker, J., et al., (2003), Lithium iron(II) phospho-olivines prepared by a novel carbothermal reduction method: *Electrochemical and Solid State Letters*, Vol. 6, No. 3, pp. A53-A55, 1099-0062.

Barrau, B., et al., (1980), Glass formation, structure and ionic conduction in the Na_2S-GeS_2 system: *Journal of Non-Crystalline Solids*, Vol. 37, No. 1, pp. 1-14, 0022-3093.

Bates, J. B., et al., (1993), Fabrication and characterization of amorphous lithium electrolyte thin films and rechargeable thin-film batteries: *Journal of Power Sources*, Vol. 43, No. 1-3, pp. 103-110, 0378-7753.

Bates, J. B., et al., (2000a), Thin-film lithium and lithium-ion batteries: *Solid State Ionics*, Vol. 135, No. 1-4, pp. 33-45, 0167-2738.

Bates, J. B., et al., (2000b), Preferred Orientation of Polycrystalline $LiCoO_2$ Films: *Journal of The Electrochemical Society*, Vol. 147, No. 1, pp. 59-70,

Birke, P., et al., (1999), A first approach to a monolithic all solid state inorganic lithium battery: *Solid State Ionics*, Vol. 118, No. 1-2, pp. 149-157, 0167-2738.

Bobeico, E., et al., (2003), P-type strontium-copper mixed oxide deposited by e-beam evaporation: *Thin Solid Films*, Vol. 444, No. 1-2, pp. 70-74, 0040-6090.

Boukamp, B. A., and R. A. Huggins, (1978), Fast ionic conductivity in lithium nitride: *Materials Research Bulletin*, Vol. 13, No. 1, pp. 23-32, 0025-5408.

Bourderau, S., et al., (1999), Amorphous silicon as a possible anode material for Li-ion batteries: *Journal of Power Sources*, Vol. 81, pp. 233-236, 0378-7753.

Bronger, W., et al., (1964), Zur Kenntnis Der Niccolate Der Alkalimetalle: *Zeitschrift Fur Anorganische Und Allgemeine Chemie*, Vol. 333, No. 4-6, pp. 188-200, 0044-2313.

Buiel, E., and J. R. Dahn, (1999), Li-insertion in hard carbon anode materials for Li-ion batteries: *Electrochimica Acta*, Vol. 45, No. 1-2, pp. 121-130, 0013-4686.

Campbell, S. A., et al., (1990), The Electrochemical-Behavior of Tetrahydrofuran and Propylene Carbonate without Added Electrolyte: *Journal of Electroanalytical Chemistry*, Vol. 284, No. 1, pp. 195-204, 0022-0728.

Cernosek, Z., et al., (1997), Raman scattering in GeS_2 glass and its crystalline polymorphs compared: *Journal of Molecular Structure*, Vol. 435, No. 2, pp. 193-198, 0022-2860.

Chang, C. C., et al., (2000), Synthesis and electrochemical characterization of divalent cation-incorporated lithium nickel oxide: *Journal of the Electrochemical Society*, Vol. 147, No. 5, pp. 1722-1729, 0013-4651.

Chhowalla, M., et al., (2001), Growth process conditions of vertically aligned carbon nanotubes using plasma enhanced chemical vapor deposition: *Journal of Applied Physics*, Vol. 90, No. 10, pp. 5308-5317, 0021-8979.

Chiang, Y. M., et al., (1998), Synthesis of $LiCoO_2$ by decomposition and intercalation of hydroxides: *Journal of the Electrochemical Society*, Vol. 145, No. 3, pp. 887-891, 0013-4651.

Cho, K., et al., (2007), Fabrication of Li_2O-B_2O_3-P_2O_5 solid electrolyte by aerosol flame deposition for thin film batteries: *Solid State Ionics*, Vol. 178, No. 1-2, pp. 119-123, 0167-2738.

Chowdari, B. V. R., et al., (2001), Cathodic behavior of (Co, Ti, Mg)-doped $LiNiO_2$: *Solid State Ionics*, Vol. 140, No. 1-2, pp. 55-62, 0167-2738.

Chrisey, D. B., and G. K. Hubler, (1994), Pulsed laser deposition of thin films: New York, J. Wiley, 613 p., 0471592188.

Croce, F., et al., (2001), A High-Rate, Long-Life, Lithium Nanocomposite Polymer Electrolyte Battery: *Electrochemical and Solid-State Letters*, Vol. 4, No. 8, pp. A121-A123,

Crowther, O., and A. C. West, (2008), Effect of electrolyte composition on lithium dendrite growth: *Journal of the Electrochemical Society*, Vol. 155, No. 11, pp. A806-A811, 0013-4651.

Cui, P., et al., (2011), Preparation and characteristics of Sb-doped $LiNiO_2$ cathode materials for Li-ion batteries: *Journal of Physics and Chemistry of Solids*, Vol. 72, No. 7, pp. 899-903, 0022-3697.

Cunningham, P. T., et al., (1972), Phase Equilibria in Lithium-Chalcogen Systems: *Journal of The Electrochemical Society*, Vol. 119, No. 11, pp. 1448-1450,

Dahn, J. R., et al., (1991), Rechargeable $LiNiO_2$ Carbon Cells: *Journal of the Electrochemical Society*, Vol. 138, No. 8, pp. 2207-2211, 0013-4651.

David, W. I. F., et al., (1987), Structure Refinement of the Spinel-Related Phases $Li_2Mn_2O_4$ and $Li_{0.2}Mn_2O_4$: *Journal of Solid State Chemistry*, Vol. 67, No. 2, pp. 316-323, 0022-4596.

Davidse, P. D., (1967), Theory and practice of RF sputtering: *Vacuum*, Vol. 17, No. 3, pp. 139-145, 0042-207X.

Dickens, P. G., et al., (1979), Phase-Relationships in the Ambient-Temperature $Li_xV_2O_5$ System (0.1-Less-Than-X-Less-Than-1.0): *Materials Research Bulletin*, Vol. 14, No. 10, pp. 1295-1299, 0025-5408.

Dudney, N. J., (2000), Addition of a thin-film inorganic solid electrolyte (LiPON) as a protective film in lithium batteries with a liquid electrolyte: *Journal of Power Sources*, Vol. 89, No. 2, pp. 176-179, 0378-7753.

Dudney, N. J., (2005), Solid-state thin-film rechargeable batteries: *Materials Science and Engineering B-Solid State Materials for Advanced Technology*, Vol. 116, No. 3, pp. 245-249, 0921-5107.

Dudney, N. J., et al., (1999), Nanocrystalline $Li_xMn_{2-y}O_4$ cathodes for solid-state thin-film rechargeable lithium batteries: *Journal of the Electrochemical Society*, Vol. 146, No. 7, pp. 2455-2464, 0013-4651.

Elmoudane, M., et al., (2000), Glass-forming region in the system Li_3PO_4-$Pb_3(PO_4)_2$-$BiPO_4$ (Li_2O-PbO-Bi_2O_3-P_2O_5) and its ionic conductivity: *Materials Research Bulletin*, Vol. 35, No. 2, pp. 279-287, 0025-5408.

Fauteux, D., et al., (1995), Lithium polymer electrolyte rechargeable battery: *Electrochimica Acta*, Vol. 40, No. 13-14, pp. 2185-2190, 0013-4686.

Fenton, D. E., et al., (1973), Complexes of Alkali-Metal Ions with Poly(Ethylene Oxide): *Polymer*, Vol. 14, No. 11, pp. 589-589, 0032-3861.

Foix, D., et al., (2002), Electronic structure of thiogermanate and thioarseniate glasses: experimental (XPS) and theoretical (ab initio) characterizations: *Solid State Ionics*, Vol. 154, pp. 161-173, 0167-2738.

Foix, D., et al., (2001), The structure of ionically conductive chalcogenide glasses: a combined NMR, XPS and ab initio calculation study: *Solid State Sciences*, Vol. 3, No. 1-2, pp. 235-243, 1293-2558.

Frumarova, B., et al., (1996), Synthesis and physical properties of the system $(GeS_2)(80-x)$-$(Ga_2S_3)(20)$:xPr glasses: *Optical Materials*, Vol. 6, No. 3, pp. 217-223, 0925-3467.

Gao, Y. A., et al., (1998), Novel $LiNi_{1-x}Ti_{x/2}Mg_{x/2}O_2$ compounds as cathode materials for safer lithium-ion batteries: *Electrochemical and Solid State Letters*, Vol. 1, No. 3, pp. 117-119, 1099-0062.

Gonbeau, D., et al., (2005), Photoinduced changes in the valence band states of $Ge_xAs_{40-x}S_{60}$ thin films: *Journal of Optoelectronics and Advanced Materials*, Vol. 7, No. 1, pp. 341-344, 1454-4164.

Gummow, R. J., et al., (1994), Improved Capacity Retention in Rechargeable 4V Lithium Lithium Manganese Oxide (Spinel) Cells: *Solid State Ionics*, Vol. 69, No. 1, pp. 59-67, 0167-2738.

Gummow, R. J., et al., (1993), Lithium Extraction from Orthorhombic Lithium Manganese Oxide and the Phase-Transformation to Spinel: *Materials Research Bulletin*, Vol. 28, No. 12, pp. 1249-1256, 0025-5408.

Gummow, R. J., and M. M. Thackeray, (1994), An Investigation of Spinel-Related and Orthorhombic Limno2 Cathodes for Rechargeable Lithium Batteries: *Journal of the Electrochemical Society*, Vol. 141, No. 5, pp. 1178-1182, 0013-4651.

Guyomard, D., and J. M. Tarascon, (1995), High-voltage stable liquid electrolytes for $Li_{1+x}Mn_2O_4$ carbon rocking-chair lithium batteries: *Journal of Power Sources*, Vol. 54, No. 1, pp. 92-98, 0378-7753.

Haizheng, T., et al., (2004), Raman spectroscopic study on the microstructure of GeS_2-Ga_2S_3-KCl glasses: *Journal of Molecular Structure*, Vol. 697, No. 1-3, pp. 23-27, 0022-2860.

Hayashi, A., et al., (1996), Preparation of $Li_6Si_2S_7$-$Li_6B_4X_9$ (X=S, O) glasses by rapid quenching and their lithium ion conductivities: *Solid State Ionics*, Vol. 86-8, pp. 539-542, 0167-2738.

Hayashi, A., et al., (2005), All-solid-state lithium secondary batteries with SnS-P_2S_5 negative electrodes and Li_2S-P_2S_5 solid electrolytes: *Journal of Power Sources*, Vol. 146, No. 1-2, pp. 496-500, 0378-7753.

Hayashi, A., et al., (2010), Development of sulfide glass-ceramic electrolytes for all-solid-state lithium rechargeable batteries: *Journal of Solid State Electrochemistry*, Vol. 14, No. 10, pp. 1761-1767, 1432-8488.

Hayashi, A., et al., (2009), All-solid-state lithium secondary batteries using nanocomposites of NiS electrode/Li_2S-P_2S_5 electrolyte prepared via mechanochemical reaction: *Journal of Power Sources*, Vol. 189, No. 1, pp. 629-632, 0378-7753.

Hayashi, A., et al., (2003), All-solid-state Li/S batteries with highly conductive glass-ceramic electrolytes: *Electrochemistry Communications*, Vol. 5, No. 8, pp. 701-705, 1388-2481.

Hayashi, A., et al., (2002), Characterization of Li_2S-SiS_2-Li_xMO_y (M=Si, P, Ge) amorphous solid electrolytes prepared by melt-quenching and mechanical milling: *Solid State Ionics*, Vol. 148, No. 3-4, pp. 381-389, 0167-2738.

Hintenlang, D. E., and P. J. Bray, (1985), NMR studies of B_2S_3-based glasses: *Journal of Non-Crystalline Solids*, Vol. 69, No. 2-3, pp. 243-248, 0022-3093.

Hirai, K., et al., (1995), Thermal and electrical properties of rapidly quenched glasses in the systems Li_2S-SiS_2-Li_xMO_y (Li_xMO_y = Li_4SiO_4, Li_2SO_4): *Solid State Ionics*, Vol. 78, No. 3-4, pp. 269-273, 0167-2738.

Hirano, A., et al., (1995), Relationship between Nonstoichiometry and Physical-Properties in $LiNiO_2$: *Solid State Ionics*, Vol. 78, No. 1-2, pp. 123-131, 0167-2738.

Huang, H., et al., (2001), Approaching theoretical capacity of $LiFePO_4$ at room temperature at high rates: *Electrochemical and Solid State Letters*, Vol. 4, No. 10, pp. A170-A172, 1099-0062.

Huang, H. T., et al., (1999), $LiAl_xCo_{1-x}O_2$ as 4 V cathodes for lithium ion batteries: *Journal of Power Sources*, Vol. 82, pp. 690-695, 0378-7753.

Iio, K., et al., (2002), Mechanochemical synthesis of high lithium ion conducting materials in the system Li_3N-SiS_2: *Chemistry of Materials*, Vol. 14, No. 6, pp. 2444-2449, 0897-4756.

Iriyama, Y., et al., (2005), Charge transfer reaction at the lithium phosphorus oxynitride glass electrolyte/lithium cobalt oxide thin film interface: *Solid State Ionics*, Vol. 176, No. 31-34, pp. 2371-2376, 0167-2738.

Ito, Y., et al., (1983), Ionic conductivity of Li_2O-B_2O_3 thin films: *Journal of Non-Crystalline Solids*, Vol. 57, No. 3, pp. 389-400, 0022-3093.

Itoh, K., et al., (2006), Structural observation of Li_2S-GeS_2 superionic glasses: *Physica B: Condensed Matter*, Vol. 385-386, No. Part 1, pp. 520-522, 0921-4526.

Jak, M. J. G., et al., (1999a), Defect Structure of Li-Doped BPO_4: A Nanostructured Ceramic Electrolyte for Li-Ion Batteries: *Journal of Solid State Chemistry*, Vol. 142, No. 1, pp. 74-79, 0022-4596.

Jak, M. J. G., et al., (1999b), Dynamically compacted all-ceramic lithium-ion batteries: *Journal of Power Sources*, Vol. 80, No. 1-2, pp. 83-89, 0378-7753.

Jamal, M., et al., (1999), Sodium ion conducting glasses with mixed glass formers NaI-Na_2O-V_2O_5-B_2O_3: application to solid state battery: *Materials Letters*, Vol. 39, No. 1, pp. 28-32, 0167-577X.

Jeon, E. J., et al., (2001), Characterization of all-solid-state thin-film batteries with V_2O_5 thin-film cathodes using ex situ and in situ processes: *Journal of the Electrochemical Society*, Vol. 148, No. 4, pp. A318-A322, 0013-4651.

Jin, B. J., et al., (2000), Violet and UV luminescence emitted from ZnO thin films grown on sapphire by pulsed laser deposition: *Thin Solid Films*, Vol. 366, No. 1-2, pp. 107-110, 0040-6090.

Jones, S. D., and J. R. Akridge, (1992), A thin film solid state microbattery: *Solid State Ionics*, Vol. 53-56, No. Part 1, pp. 628-634, 0167-2738.

Jones, S. D., and J. R. Akridge, (1996), A microfabricated solid-state secondary Li battery: *Solid State Ionics*, Vol. 86-8, pp. 1291-1294, 0167-2738.

Jourdaine, L., et al., (1988), Lithium solid state glass-based microgenerators: *Solid State Ionics*, Vol. 28-30, No. Part 2, pp. 1490-1494, 0167-2738.

Kaczmarek, S., (1996), Pulsed laser deposition - Today and tomorrow: *Laser Technology V: Applications in Materials Sciences and Engineering*, Vol. 3187, pp. 129-134

Kanehori, K., et al., (1986), Titanium Disulfide Films Fabricated by Plasma Cvd: *Solid State Ionics*, Vol. 18-9, pp. 818-822, 0167-2738.

Kanert, O., et al., (1994), Significant differences between nuclear-spin relaxation and conductivity relaxation in low-conductivity glasses: *Physical Review B*, Vol. 49, No. 1, pp. 76,

Kang, S. H., and J. B. Goodenough, (2000), $Li[Li_yMn_{2-y}]O_4$ spinel cathode material prepared by a solution method: *Electrochemical and Solid State Letters*, Vol. 3, No. 12, pp. 536-539, 1099-0062.

Kawamoto, Y., and M. Nishida, (1976), Ionic conduction in As_2S_3-Ag_2S, GeS_2-GeS-Ag_2S and P_2S_5-Ag_2S glasses: *Journal of Non-Crystalline Solids*, Vol. 20, No. 3, pp. 393-404, 0022-3093.

Kenji, F., et al., Foliated natural graphite as the anode material for rechargeable lithium-ion cells: *Journal of Power Sources*, Vol. 69, No. 1-2, pp. 165-168, 0378-7753.

Kennedy, J. H., (1989), Ionically conductive glasses based on SiS_2: *Materials Chemistry and Physics*, Vol. 23, No. 1-2, pp. 29-50, 0254-0584.

Kim, C., et al., (2000), Microstructure and electrochemical properties of boron-doped mesocarbon microbeads: *Journal of the Electrochemical Society*, Vol. 147, No. 4, pp. 1257-1264, 0013-4651.

Kim, Y., et al., (2005), Glass formation in and structural investigation of $Li_2S + GeS_2 + GeO_2$ composition using Raman and IR spectroscopy: *Journal of Non-Crystalline Solids*, Vol. 351, No. 49-51, pp. 3716-3724, 0022-3093.

Kim, Y., et al., (2006), Anomalous ionic conductivity increase in $Li_2S + GeS_2 + GeO_2$ glasses: *Journal of Physical Chemistry B*, Vol. 110, No. 33, pp. 16318-16325, 1089-5647.

Komaba, S., et al., (2000), Preparation of Li-Mn-O thin films by r.f.-sputtering method and its application to rechargeable batteries: *Journal of Applied Electrochemistry*, Vol. 30, No. 10, pp. 1179-1182, 0021-891X.

Komiya, R., et al., (2001), Solid state lithium secondary batteries using an amorphous solid electrolyte in the system $(100\text{-}x)(0.6Li_2S+0.4SiS_2)+xLi_4SiO_4$ obtained by mechanochemical synthesis: *Solid State Ionics*, Vol. 140, No. 1-2, pp. 83-87, 0167-2738.

Kondo, S., et al., (1992), New lithium ion conductors based on $Li_2S\text{-}SiS_2$ system: *Solid State Ionics*, Vol. 53-56, No. Part 2, pp. 1183-1186, 0167-2738.

Kong, Y. C., et al., (2001), Ultraviolet-emitting ZnO nanowires synthesized by a physical vapor deposition approach: *Applied Physics Letters*, Vol. 78, No. 4, pp. 407-409, 0003-6951.

Kubota, Y., et al., (2008), Electronic structure of LiSi: *Journal of Alloys and Compounds*, Vol. 458, No. 1-2, pp. 151-157, 0925-8388.

Kumta, P. N., et al., (1998), Synthesis of $LiCoO_2$ powders for lithium-ion batteries from precursors derived by rotary evaporation: *Journal of Power Sources*, Vol. 72, No. 1, pp. 91-98, 0378-7753.

Lapp, T., et al., (1983), Ionic conductivity of pure and doped Li_3N: *Solid State Ionics*, Vol. 11, No. 2, pp. 97-103, 0167-2738.

Larson, R. W., and D. E. Day, (1986), Preparation and characterization of lithium phosphorus oxynitride glass: *Journal of Non-Crystalline Solids*, Vol. 88, No. 1, pp. 97-113, 0022-3093.

Lee, C. H., et al., (2002), Characterizations of a new lithium ion conducting $Li_2O\text{-}SeO_2\text{-}B_2O_3$ glass electrolyte: *Solid State Ionics*, Vol. 149, No. 1-2, pp. 59-65, 0167-2738.

Lee, S.-J., et al., (2001), Stress effect on cycle properties of the silicon thin-film anode: *Journal of Power Sources*, Vol. 97-98, pp. 191-193, 0378-7753.

Lee, S., et al., (2007), Modification of network structure induced by glass former composition and its correlation to the conductivity in lithium borophosphate glass for solid state electrolyte: *Solid State Ionics*, Vol. 178, No. 5-6, pp. 375-379, 0167-2738.

Lee, S. W., et al., (2004), Electrochemical characteristics of Al_2O_3-coated lithium manganese spinel as a cathode material for a lithium secondary battery: *Journal of Power Sources*, Vol. 126, No. 1-2, pp. 150-155, 0378-7753.

Li, W., et al., (1992), Crystal-Structure of Lixni2-Xo2 and a Lattice-Gas Model for the Order-Disorder Transition: *Physical Review B*, Vol. 46, No. 6, pp. 3236-3246, 0163-1829.

Li, X. X., et al., (2007), Electrochemical lithium intercalation/deintercalation of single-crystalline V_2O_5 nanowires: *Journal of the Electrochemical Society*, Vol. 154, No. 1, pp. A39-A42, 0013-4651.

Liu, R. S., and C. H. Shen, (2003), Structural and electrochemical study of cobalt doped $LiMn_2O_4$ spinels: *Solid State Ionics*, Vol. 157, No. 1-4, pp. 95-100, 0167-2738.

Machida, N., et al., (2002), All-solid-state lithium battery with $LiCo_{0.3}Ni_{0.7}O_2$ fine powder as cathode materials with an amorphous sulfide electrolyte: *Journal of the Electrochemical Society*, Vol. 149, No. 6, pp. A688-A693, 0013-4651.

Machida, N., et al., (2004), A new amorphous lithium-ion conductor in the system $Li_2S\text{-}P_2S_3$: *Chemistry Letters*, Vol. 33, No. 1, pp. 30-31, 0366-7022.

Malta, M., and R. M. Torresi, (2005), Electrochemical and kinetic studies of lithium intercalation in composite nanofibers of vanadium oxide/polyaniline: *Electrochimica Acta*, Vol. 50, No. 25-26, pp. 5009-5014, 0013-4686.

Marchand, R., et al., (1988), Characterization of Nitrogen Containing Phosphate-Glasses by X-Ray Photoelectron-Spectroscopy: *Journal of Non-Crystalline Solids*, Vol. 103, No. 1, pp. 35-44, 0022-3093.

Martin, S. W., (1991), Ionic-conduction in phosphate-glasses: *Journal of the American Ceramic Society*, Vol. 74, No. 8, pp. 1767-1784, 0002-7820.

Martin, S. W., and C. A. Angell, (1984), Conductivity Maximum in Sodium Aluminoborate Glass: *Journal of the American Ceramic Society*, Vol. 67, No. 7, pp. C148-C150, 0002-7820.

Mercier, R., et al., (1981a), Superionic conduction in Li_2S-P_2S_5-LiI - glasses: *Solid State Ionics*, Vol. 5, pp. 663-666, 0167-2738.

Mercier, R., et al., (1981b), Superionic Conduction in Li_2S-P_2S_5-LiI-Glasses: *Solid State Ionics*, Vol. 5, No. Oct, pp. 663-666, 0167-2738.

Minami, K., et al., (2011), Electrical and electrochemical properties of glass-ceramic electrolytes in the systems Li_2S-P_2S_5-P_2S_3 and Li_2S-P_2S_5-P_2O_5: *Solid State Ionics*, Vol. 192, No. 1, pp. 122-125, 0167-2738.

Minami, K., et al., (2008), Structure and properties of the $70Li_2S$ + $(30\text{-}x)P_2S_5$ + xP_2O_5 oxysulfide glasses and glass-ceramics: *Journal of Non-Crystalline Solids*, Vol. 354, No. 2-9, pp. 370-373, 0022-3093.

Minami, T., et al., (2006), Recent progress of glass and glass-ceramics as solid electrolytes for lithium secondary batteries: *Solid State Ionics*, Vol. 177, No. 26-32, pp. 2715-2720, 0167-2738.

Mitkova, M., et al., (2004), Thermal and photodiffusion of Ag in S-rich Ge-S amorphous films: *Thin Solid Films*, Vol. 449, No. 1-2, pp. 248-253, 0040-6090.

Miyauchi, K., et al., (1983), Lithium Ion Conductive Thin-Films and Their Electrical-Properties: *Denki Kagaku*, Vol. 51, No. 1, pp. 211-212, 0366-9297.

Mizuno, F., et al., (2005), New, highly ion-conductive crystals precipitated from Li_2S-P_2S_5 glasses: *Advanced Materials*, Vol. 17, No. 7, pp. 918-+, 0935-9648.

Mizushima, K., et al., (1980), Li_xCoO_2 "(Oless-Thanxless-Than-or-Equal-To1) - a New Cathode Material for Batteries of High-Energy Density: *Materials Research Bulletin*, Vol. 15, No. 6, pp. 783-789, 0025-5408.

Mount, E., (2003), Principles of Chemical Vapor Deposition (Book): *Sci-Tech News*, Vol. 57, No. 4, pp. 80, 00368059.

Murayama, M., et al., (2002), Structure of the thio-LISICON, Li_4GeS_4: *Solid State Ionics*, Vol. 154-155, pp. 789-794, 0167-2738.

Murayama, M., et al., (2004), Material design of new lithium ionic conductor, thio-LISICON, in the Li_2S-P_2S_5 system: *Solid State Ionics*, Vol. 170, No. 3-4, pp. 173-180, 0167-2738.

Myung, S.-T., et al., (2000), Capacity fading of $LiMn_2O_4$ electrode synthesized by the emulsion drying method: *Journal of Power Sources*, Vol. 90, No. 1, pp. 103-108, 0378-7753.

Nakayama, M., et al., (2003), Grain size control of $LiMn_2O_4$ cathode material using microwave synthesis method: *Solid State Ionics*, Vol. 164, No. 1-2, pp. 35-42, 0167-2738.

Nalwa, H. S., (2002), Handbook of thin film materials: San Diego, Academic Press, 0125129084 (set acid-free paper)

Narayan, J., et al., (1992), Epitaxial growth of Tin Films on (100) silicon substrates by laser physical vapor deposition: *Applied Physics Letters*, Vol. 61, No. 11, pp. 1290-1292, 0003-6951.

Neudecker, B. J., et al., (2000), "Lithium-Free" Thin-Film Battery with In Situ Plated Li Anode: *Journal of The Electrochemical Society*, Vol. 147, No. 2, pp. 517-523,

Nishida, Y., et al., (1997), Synthesis and properties of gallium-doped $LiNiO_2$ as the cathode material for lithium secondary batteries: *Journal of Power Sources*, Vol. 68, No. 2, pp. 561-564, 0378-7753.

Nogami, M., and Y. Moriya, (1982), Glass formation of the SiO_2-B_2O_3 system by the gel process from metal alkoxides: *Journal of Non-Crystalline Solids*, Vol. 48, No. 2-3, pp. 359-366, 0022-3093.

Ohtomo, T., et al., (2005), Electrical and electrochemical properties of Li_2S-P_2S_5-P_2O_5 glass-ceramic electrolytes: *Journal of Power Sources*, Vol. 146, No. 1-2, pp. 715-718, 0378-7753.

Ohtsuka, H., et al., (1990), Solid-State Battery with Li_2O-V_2O_5-SiO_2 Solid Electrolyte Thin-Film: *Solid State Ionics*, Vol. 40-1, pp. 964-966, 0167-2738.

Ohtsuka, H., and J. Yamaki, (1989), Electrical Characteristics of Li_2O-V_2O_5-SiO_2 Thin-Films: *Solid State Ionics*, Vol. 35, No. 3-4, pp. 201-206, 0167-2738.

Ohzuku, T., et al., (1991), Electrochemistry of Manganese-Dioxide in Lithium Nonaqueous Cells .4. Jahn-Teller Deformation of MnO_6-Octahedron in Li_xMnO_2: *Journal of the Electrochemical Society*, Vol. 138, No. 9, pp. 2556-2560, 0013-4651.

Padhi, A. K., et al., (1997), Phospho-olivines as positive-electrode materials for rechargeable lithium batteries: *Journal of the Electrochemical Society*, Vol. 144, No. 4, pp. 1188-1194, 0013-4651.

Pang, S. P., et al., (2005), Synthesis of polyaniline-vanadium oxide nanocomposite nanosheets: *Macromolecular Rapid Communications*, Vol. 26, No. 15, pp. 1262-1265, 1022-1336.

Park, H. Y., et al., (2007), $LiCoO_2$ thin tilm cathode fabrication by rapid thermal annealing for micro power sources: *Electrochimica Acta*, Vol. 52, No. 5, pp. 2062-2067, 0013-4686.

Park, K. S., et al., (2005), The effects of sulfur doping on the performance of O.3Li-0.7$[Li_{1/12}Ni_{1/12}Mn_{5/6}]O_2$ powder: *Korean Journal of Chemical Engineering*, Vol. 22, No. 4, pp. 560-565, 0256-1115.

Park, S. H., and Y. K. Sun, (2003), Synthesis and electrochemical properties of layered $Li[Li_{0.15}Ni(0.275_{x/2})Al_xMn(0.575_{x/2})]O_2$ materials: *Journal of Power Sources*, Vol. 119, pp. 161-165, 0378-7753.

Park, Y. S., et al., (1999), All-solid-state lithium thin-film rechargeable battery with lithium manganese oxide: *Electrochemical and Solid State Letters*, Vol. 2, No. 2, pp. 58-59, 1099-0062.

Patel, H K., and S. W. Martin, (1992), Fast ionic conduction in Na_2S+B_2S_3 glasses: Compositional contributions to nonexponentiality in conductivity relaxation in the extreme low-alkali-metal limit: *Physical Review B*, Vol. 45, No. 18, pp. 10292,

Patil, A., et al., (2008), Issue and challenges facing rechargeable thin film lithium batteries: *Materials Research Bulletin*, Vol. 43, No. 8-9, pp. 1913-1942, 0025-5408.

Peled, E., et al., (1996), Improved Graphite Anode for Lithium-Ion Batteries Chemically: *Journal of the Electrochemical Society*, Vol. 143, No. 1, pp. L4-L7,

Peng, Z. S., et al., (1998), Synthesis by sol-gel process and characterization of $LiCoO_2$ cathode materials: *Journal of Power Sources*, Vol. 72, No. 2, pp. 215-220, 0378-7753.

Pouillerie, C., et al., (2000), Synthesis and characterization of new $LiNi_{1-y}Mg_yO_2$ positive electrode materials for lithium-ion batteries: *Journal of the Electrochemical Society*, Vol. 147, No. 6, pp. 2061-2069, 0013-4651.

Pradel, A., et al., (1985), Effect of rapid quenching on electrical properties of lithium conductive glasses: *Solid State Ionics*, Vol. 17, No. 2, pp. 147-154, 0167-2738.

Prosini, P. P., et al., (2001), Improved electrochemical performance of a $LiFePO_4$-based composite cathode: *Electrochimica Acta*, Vol. 46, No. 23, pp. 3517-3523, 0013-4686.

Rabenau, A., (1982), Lithium Nitride and Related Materials - Case-Study of the Use of Modern Solid-State Research Techniques: *Solid State Ionics*, Vol. 6, No. 4, pp. 277-293, 0167-2738.

Rajendran, S., and T. Uma, (2000), Conductivity studies on PVC/PMMA polymer blend electrolyte: *Materials Letters*, Vol. 44, No. 3-4, pp. 242-247, 0167-577X.

Reimers, J. N., et al., (1993), Structure and Electrochemistry of $Li_xFe_yNi_{1-y}O_2$: *Solid State Ionics*, Vol. 61, No. 4, pp. 335-344, 0167-2738.

Sakamoto, R., et al., (1999), Preparation of fast lithium ion conducting glasses in the system Li_2S-SiS_2-Li_3N: *Journal of Physical Chemistry B*, Vol. 103, No. 20, pp. 4029-4031, 1089-5647.

Sakka, S., (1986), Oxynitride Glasses: *Annual Review of Materials Science*, Vol. 16, pp. 29-46, 0084-6600.

Saunier, J., et al., (2004), Plasticized microporous poly(vinylidene fluoride) separators for lithium-ion batteries. II. Poly(vinylidene fluoride) dense membrane swelling behavior in a liquid electrolyte - Characterization of the swelling kinetics: *Journal of Polymer Science Part B-Polymer Physics*, Vol. 42, No. 3, pp. 544-552, 0887-6266.

Schalkwijk, W. A. v., and B. Scrosati, (2002), Advances in lithium-ion batteries: New York, NY, Kluwer Academic/Plenum Publishers, x, 513 p. pp., 0306473569.

Scholz, F., and B. Meyer, (1994), ELECTROCHEMICAL SOLID-STATE ANALYSIS - STATE-OF-THE-ART: *Chemical Society Reviews*, Vol. 23, No. 5, pp. 341-347, 0306-0012.

Seo, I., and S. W. Martin, (2011a), Fast lithium ion conducting solid state thin-film electrolytes based on lithium thio-germanate materials: *Acta Materialia*, Vol. 59, No. 4, pp. 1839-1846, 1359-6454.

Seo, I., and S. W. Martin, (2011b), Preparation and Characterization of Fast Ion Conducting Lithium Thio-Germanate Thin Films Grown by RF Magnetron Sputtering: *Journal of the Electrochemical Society*, Vol. 158, No. 5, pp. A465-A470, 0013-4651.

Seo, I., and S. W. Martin, (2011c), Structural Properties of Lithium Thio-Germanate Thin Film Electrolytes Grown by Radio Frequency Sputtering: *Inorganic Chemistry*, Vol. 50, No. 6, pp. 2143-2150, 0020-1669.

Sharma, R. A., and R. N. Seefurth, (1976), Thermodynamic Properties of the Lithium-Silicon System: *Journal of the Electrochemical Society*, Vol. 123, No. 12, pp. 1763-1768,

Shi, S. Q., et al., (2003), Enhancement of electronic conductivity of $LiFePO_4$ by Cr doping and its identification by first-principles calculations: *Physical Review B*, Vol. 68, No. 19, pp. -, 1098-0121.

Shokoohi, F. K., et al., (1991), Fabrication of Thin-Film $LiMn_2O_4$ Cathodes for Rechargeable Microbatteries: *Applied Physics Letters*, Vol. 59, No. 10, pp. 1260-1262, 0003-6951.

Sigala, C., et al., (1995), Positive Electrode Materials with High Operating Voltage for Lithium Batteries - $LiCr_yMn_{2-y}O_4$ (0-Less-Than-or-Equal-to-Y-Less-Than-or-Equal-to-1): *Solid State Ionics*, Vol. 81, No. 3-4, pp. 167-170, 0167-2738.

Song, J. Y., et al., (1999), Review of gel-type polymer electrolytes for lithium-ion batteries: *Journal of Power Sources*, Vol. 77, No. 2, pp. 183-197, 0378-7753.

Souquet, J. L., and M. Duclot, (2002), Thin film lithium batteries: *Solid State Ionics*, Vol. 148, No. 3-4, pp. 375-379, 0167-2738.

Tabata, H., et al., (1994), Formation of artificial $BaTiO_3$-$SrTiO_3$ superlattices using pulsed laser deposition and their dielectric properties: *Applied Physics Letters*, Vol. 65, No. 15, pp. 1970-1972, 0003-6951.

Takada, K., et al., (1995), Electrochemical behavior of Li_xMO_2 (M = Co, Ni) in all solid state cells using a glass electrolyte: *Solid State Ionics*, Vol. 79, pp. 284-287, 0167-2738.

Takada, K., et al., (1996), Solid state lithium battery with oxysulfide glass: *Solid State Ionics*, Vol. 86-8, pp. 877-882, 0167-2738.

Tarascon, J. M., and M. Armand, (2001), Issues and challenges facing rechargeable lithium batteries: *Nature*, Vol. 414, No. 6861, pp. 359-367, 0028-0836.

Unuma, H., and S. Sakka, (1987), Electrical-Conductivity in Na-Si-O-N Oxynitride Glasses: *Journal of Materials Science Letters*, Vol. 6, No. 9, pp. 996-998, 0261-8028.

Volynsky, A. B., et al., (2001), Low-temperature transformations of sodium sulfate and sodium selenite in the presence of pre-reduced palladium modifier in graphite furnaces for electrothermal atomic absorption spectrometry: *Spectrochimica Acta Part B-Atomic Spectroscopy*, Vol. 56, No. 8, pp. 1387-1396, 0584-8547.

Wada, H., et al., (1983), Preparation and ionic conductivity of new B_2S_3-Li_2S-LiI glasses: *Materials Research Bulletin*, Vol. 18, No. 2, pp. 189-193, 0025-5408.

Wang, B., et al., (1995a), Ionic conductivities of lithium phosphorus oxynitride glasses, polycrystals and thin films: *Solid State Ionics Iv*, Vol. 369, pp. 445-456, 0272-9172.

Wang, B., et al., (1995b), Synthesis, Crystal-Structure, and Ionic-Conductivity of a Polycrystalline Lithium Phosphorus Oxynitride with the Gamma-Li3po4 Structure: *Journal of Solid State Chemistry*, Vol. 115, No. 2, pp. 313-323, 0022-4596.

Wang, B., et al., (1995c), Ionic Conductivities and Structure of Lithium Phosphorus Oxynitride Glasses: *Journal of Non-Crystalline Solids*, Vol. 183, No. 3, pp. 297-306, 0022-3093.

Wang, H. F., et al., (1999), TEM study of electrochemical cycling-induced damage and disorder in LiCoO2 cathodes for rechargeable lithium batteries: *Journal of the Electrochemical Society*, Vol. 146, No. 2, pp. 473-480, 0013-4651.

Wang, Y., and G. Z. Cao, (2008), Developments in nanostructured cathode materials for high-performance lithium-ion batteries: *Advanced Materials*, Vol. 20, No. 12, pp. 2251-2269, 0935-9648.

Wang, Z., et al., (2002), Structural and electrochemical characterizations of surface-modified LiCoO2 cathode materials for Li-ion batteries: *Solid State Ionics*, Vol. 148, No. 3-4, pp. 335-342, 0167-2738.

West, W. C., et al., (2004), Chemical stability enhancement of lithium conducting solid electrolyte plates using sputtered LiPON thin films: *Journal of Power Sources*, Vol. 126, No. 1-2, pp. 134-138, 0378-7753.

Whittingham, M. S., (1976), Role of Ternary Phases in Cathode Reactions: *Journal of the Electrochemical Society*, Vol. 123, No. 3, pp. 315-320, 0013-4651.

Whittingham, M. S., (2004), Lithium batteries and cathode materials: *Chemical Reviews*, Vol. 104, No. 10, pp. 4271-4301, 0009-2665.

Willmott, P. R., and J. R. Huber, (2000), Pulsed laser vaporization and deposition: *Reviews of Modern Physics*, Vol. 72, No. 1, pp. 315,

Wu, H. Z., et al., (2000), Low-temperature epitaxy of ZnO films on Si(0 0 1) and silica by reactive e-beam evaporation: *Journal of Crystal Growth*, Vol. 217, No. 1-2, pp. 131-137, 0022-0248.

Wu, Y. P., et al., (2003), Carbon anode materials for lithium ion batteries: *Journal of Power Sources*, Vol. 114, No. 2, pp. 228-236, 0378-7753.

Xia, H., et al., (2009), Thin film Li electrolytes for all-solid-state micro-batteries: *International Journal of Surface Science and Engineering*, Vol. 3, No. 1-2, pp. 23-43, 1749-785X.

Xu, K., et al., (2002), Lithium Bis(oxalato)borate Stabilizes Graphite Anode in Propylene Carbonate: *Electrochemical and Solid-State Letters*, Vol. 5, No. 11, pp. A259-A262,

Yamada, A., et al., (2001), Optimized $LiFePO_4$ for lithium battery cathodes: *Journal of the Electrochemical Society*, Vol. 148, No. 3, pp. A224-A229, 0013-4651.

Yamaki, J., et al., (1996), Rechargeable lithium thin film cells with inorganic electrolytes: *Solid State Ionics*, Vol. 86-8, pp. 1279-1284, 0167-2738.

Yamashita, M., et al., (1999), Studies on magnetron sputtering assisted by inductively coupled RF plasma for enhanced metal ionization: *Japanese Journal of Applied Physics Part 1-Regular Papers Short Notes & Review Papers*, Vol. 38, No. 7B, pp. 4291-4295, 0021-4922.

Yamashita, M., and H. Yamanaka, (2003), Formation and ionic conductivity of Li_2S-GeS_2-Ga_2S_3 glasses and thin films: *Solid State Ionics*, Vol. 158, No. 1-2, pp. 151-156, 0167-2738.

Yamashita, M., et al., (1996a), Thin-film preparation of the Li_2S-GeS_2-Ga_2S_3 glass system by sputtering: *Solid State Ionics*, Vol. 89, No. 3-4, pp. 299-304, 0167-2738.

Yamashita, M., et al., (1996b), Thin-film preparation of the Li_2S-GeS_2-Ga_2S_3 glass system by sputtering: *Solid State Ionics*, Vol. 89, No. 3-4, pp. 299-304, 0167-2738.

Yonco, R. M., et al., (1975), Solubility of Nitrogen in Liquid Lithium and Thermal-Decomposition of Solid Li3n: *Journal of Nuclear Materials*, Vol. 57, No. 3, pp. 317-324, 0022-3115.

Yonezawa, S., et al., (1998), Effect of LiF addition at preparation of $LiCoO_2$ on its properties as an active material of lithium secondary battery: *Journal of Fluorine Chemistry*, Vol. 87, No. 2, pp. 141-143, 0022-1139.

Yoshitake, T., et al., (2001), Microstructure of β-$FeSi_2$ thin films prepared by pulsed laser deposition: *Thin Solid Films*, Vol. 381, No. 2, pp. 236-243, 0040-6090.

Yu, M. X., et al., (2001), Investigation on structure and conductivity of microporous pan electrolyte: *Acta Polymerica Sinica*, No. 5, pp. 665-669, 1000-3304.

Yu, X., et al., (1997), A Stable Thin-Film Lithium Electrolyte: Lithium Phosphorus Oxynitride: *Journal of The Electrochemical Society*, Vol. 144, No. 2, pp. 524-532,

Zhang, B., et al., (2004), Study on the structure and dielectric properties of BaO-SiO_2-B_2O_3 glass-doped (Ba,Sr)TiO_3 ceramics: *Ceramics International*, Vol. 30, No. 7, pp. 1767-1771, 0272-8842.

Zhang, Z., and J. H. Kennedy, (1990), Synthesis and characterization of the B_2S_3-Li_2S, the P_2S_5-Li_2S and the B_2S_3-P_2S_5-Li_2S glass systems: *Solid State Ionics*, Vol. 38, No. 3-4, pp. 217-224, 0167-2738.

Zhecheva, E., and R. Stoyanova, (1993), Stabilization of the Layered Crystal-Structure of $LiNiO_2$ by Co-Substitution: *Solid State Ionics*, Vol. 66, No. 1-2, pp. 143-149, 0167-2738.

Zhou, Z. H., et al., (1999), Structure of Na_2S-GeS_2 sulphide glasses - comparison with Na_2O-GeO_2 oxide glasses: *Physics and Chemistry of Glasses*, Vol. 40, No. 3, pp. 146-152, 0031-9090.

6

Electrolyte and Solid-Electrolyte Interphase Layer in Lithium-Ion Batteries

Alexandre Chagnes[1] and Jolanta Swiatowska[2]
[1]LECIME, CNRS (UMR 7575),
[2]LPCS, CNRS (UMR 7045),
Ecole Nationale Supérieure de Chimie de Paris (Chimie ParisTech), Paris, France

1. Introduction

The supply and the management of the energy are particularly at the centre of our daily concerns and represent a socio-economic priority. Indeed, while cars use fossil fuel as the main source of energy for over a century, the depletion of the oil reserves and the necessity to reduce the carbon dioxide emissions, stimulate the development of electric vehicles. Therefore, one of the main challenges for the coming decades is the development of new technologies for the storage of electrochemical energy.

Lithium-ion battery (LIB) seems to be the best choice for electric vehicles, and perhaps for the storage of electricity from wind turbines or photovoltaic cells. Even if the lithium-ion technology has known remarkable improvements over the last two decades by doubling the energy density, a technological breakthrough seems to be necessary to further increase the energy density, the charge rate, the safety and the longevity. The performances of the LIB can be improved either by optimizing the electrolyte, or by developing electrode materials more efficient in terms of energy density and cycling ability.

The internal resistance in a LIB should be maintained as low as possible throughout its life especially if the battery is dedicated to applications needing a high charge rate such as the electric vehicles. The internal resistance of a battery (R_b) can be expressed as follows:

$$R_b = R_{el} + R_{in}(N) + R_{in}(P) + R_c(N) + R_c(P) \quad (1)$$

Where R_{el}, $R_{in}(P)$, $R_{in}(N)$, $R_c(P)$ and $R_c(N)$ denote the electrolyte resistance, the interfacial resistance at the positive (P) and the negative (N) electrodes, and the resistance of the current collector at the positive (P) and the negative electrodes (N), respectively.

The resistance of the collector depends mainly on the conductivity of the material used as current collector, i.e. the conductivity of the material should be as high as possible. The electrolyte resistance depends on the distance between the positive and the negative electrodes (L), the geometric area of the electrodes (A) and the ionic conductivity of the electrolyte (κ):

$$R_{el} = L / (\kappa A) \quad (2)$$

Equation (2) shows that the electrolyte resistance can be reduced by decreasing the distance between the electrodes and by increasing the ionic conductivity (and decreasing the viscosity as conductivity and viscosity are related to each other) of the electrolyte and the geometric area of the electrodes. Nevertheless, the area of the electrodes should not be too large as the interfacial resistance (R_{in}) is proportional to A/A_{sp} where A_{sp} denotes the interfacial area at the electrode/electrolyte interface (which can be designated to the specific surface of the electrode). On the other side, the interfacial resistance can be lowered by increasing the specific surface of the electrode, i.e. by using porous electrodes with nanoparticles providing a good electrical contact between the nanoparticles.

Then, the internal resistance of a battery can be reduced by optimizing the geometry of the battery, by using porous electrodes and by increasing the ionic conductivity of the electrolyte. Nevertheless, the performances of LIBs do not depend only on the internal resistance of the battery. For instance, the longevity and the charge rate of a battery are governed by the nature of the electrode materials (diffusion coefficient of lithium ions into the host material, resistance of the material against large volume variation, etc.) and the electrode/electrolyte interface that results from the reactivity of the electrode material towards the electrolyte and especially from the reduction or the oxidation of the electrolyte on the surface of the negative or positive electrodes, respectively. These interfacial reactions are key issues for LIBs, playing a major role in the chemical and physical stability of the electrodes, the cycling stability, the lifetime and the reversible capacity of the battery. Despite numerous studies on the SEI layer, there are still lots of doubts concerning the composition of the SEI layer and its mechanism of formation.

The first part of this chapter is focused on the physicochemical properties of the electrolytes such as wettability and transport properties including the main models that were developed to predict the ionic conductivity and the viscosity of the electrolytes used in the lithium batteries. The different families of solvents and lithium salts used in LIBs are discussed in the second part of this chapter. As the electrode/electrolyte interface plays an important role in the operation of a battery, the mechanisms of formation of the SEI, the composition of the SEI on different types of negative electrodes like carbonaceous-, conversion-, and alloying-type materials as well as positive electrodes analyzed by the most appropriate techniques are also presented.

2. Physicochemical properties of electrolytes

Physicochemical properties of the electrolytes such as viscosity, ionic conductivity, thermal stability and wettability are governed by the composition of the electrolyte, i.e. the salt and the solvent. For instance, ionic conductivity and fluidity of the electrolyte should be as high as possible in order to minimize the resistance of the battery whereas the temperature range at which the electrolyte remains liquid should be as high as possible with respect to technical and safety considerations.

2.1 Transport properties

Ionic conductivity and viscosity are related to each other by the well-known Walden's product which states that the product of the limiting molar ionic conductivity (Λ_0) of an electrolyte and the viscosity (η) are constant providing that the radius of the solvated ions remains constant [Bockris, 1970]. Thus, a decrease of the viscosity of the electrolyte results in an increase of the ionic conductivity. Therefore, the development of models for evaluating the viscosity and the ionic conductivity of an electrolyte is of great importance for the optimization of the electrolyte formulation.

2.1.1 Viscosity

Viscosity is a foremost electrolyte property that has a prominent influence on key transport properties. There are several factors that can influence electrolyte viscosity such as temperature, salt concentration, and the nature of the interactions between solvents and ions. The addition of lithium salts in dipolar aprotic organic solvents is responsible for an increase of the viscosity due to the appearance of new interactions in solution (ion-solvent and ion-ion interactions) and the solvent structuration [Chagnes, 2010]. The variation of the viscosity *vs* the salt concentration can be described by the Jones-Dole equation [Jones & Dole, 1929]:

$$\eta_r = \eta / \eta_0 = 1 + A\,C^{1/2} + BC + DC^2 \qquad (3)$$

In this equation, η and η_0, are the viscosities of the solution and the pure solvent respectively and A, B and D are coefficients. This first term, in $C^{1/2}$, on the right-hand side of Eq.(3), is linked to the interaction of a reference ion with its ionic environment and may be calculated by using the Falkenhagen theory but, usually, this term is vanished in organic solvents when the salt concentration is above $C \approx 0.05M$.

The BC term is predominant at $C > 0.05M$ and has been attributed to ion-solvent interactions as well as to volume effects. These interactions can induce an increase of the viscosity due to the solvation of the ions or the effect of the electric field generated by the ions on the solvent molecules. These interactions can be also responsible for a decrease of the viscosity of structured media such as water or alcohols due to the destructuration of the solvent in the presence of big ions such as K^+ or Cs^+. In the case of the electrolytes for LIB's, the ion-solvent interactions are always responsible for an increase of the viscosity and the value of B is then always positive.

The third term in C^2, appears at the highest concentrations in salt (0.5 to 2M or more), i.e. when the mean interionic distance decreases and becomes of the order of magnitude of a few solvent molecules diameters. It is mainly related to ion-ion and/or ion-solvent interactions as it does not appear in the case of weak electrolytes. The D value is always positive in LIB electrolytes. The validity of Eq. (3) was confirmed in γ-butyrolactone, oxazolidinone, dimethyl carbonate and propylene carbonate in the presence of lithium salts [Chagnes, 2001a; Chagnes, 2002; Gzara, 2006].

Usually, the electrolytes for the LIB are mixtures of two or three solvents (ethylene carbonate-dimethylcarbonate or propylene carbonate-ethylene carbonate-dimethylcarbonate) and lithium salts. Modelling of the viscosity dependence on the concentration can be carried out by using semi-empirical equations such as the Jones-Dole equation (Eq. (3)). The physicochemical properties of the mixtures of dipolar aprotic solvents cannot be described by regular laws due to the deviation from non-ideality. In dimethylcarbonate (DMC)-ethylene carbonate (EC) and γ-butyrolactone (BL)-EC mixtures, η^E exhibits negative values over the whole range of mole fraction and the minimum of the curve is located at x_{DMC}=0.3 and η^E=-0.37 mPa·s. [Mialkowski, 2002]. The asymmetrical shape of the curve $\eta^E = f(x_{DMC})$ indicates that the addition of a small amount of DMC to BL involves a more important effect of breaking the structure than occurs with the addition of BL to DMC. In the absence of strong specific interactions leading to complex formation, negative deviations generally occur when dispersion forces are primarily responsible for interaction, or even when they interact more strongly via dipole-dipole interactions. This means also that the interaction between pairs of like molecules is stronger than between pairs of unlike molecules like cyclic lactones and open chain carbonates [Mialkowski, 2002].

The investigation of the excess volume, the excess dielectric constant and the excess Gibbs energy of activation of flow shows that the interaction between BL and DMC is such that the basic networks of intermolecular association in the pure solvents are disrupted and that the individual BL and DMC molecules are loosely bound together to give rise to a less structured solution. The loss in association of molecules gives rise to a better molecular packing as the molecular volumes are different (V_m^E is negative) and a slight negative deviation in the variation of viscosity (η^E is negative) is observed [Mialkowski, 2002].

The study of other excess thermodynamics functions [Mialkowski, 2002] gives evidence of weak dipole-dipole interactions in DMC rich mixtures and stronger interaction in BL rich mixtures. The dipole-dipole interactions are more important in (PC+DMC) or (EC+DMC) mixtures which exhibit large values of the Kirkwood parameter (g_K). The Kirkwood parameter, which provides the correlation between dipoles, is in fair agreement with XRD structure of BL at low temperature [Papoular, 2005], dielectric fraction model and excess Gibbs energy for the activation flow.

Then, the non-ideality of the viscosity should be taken into account to model the viscosity of solvent mixtures. Gering [Gering, 2006] developed such a model (tested for aqueous systems and organic electrolytes such as EC-DMC+$LiPF_6$, EC-EMC+$LiPF_6$, EC-PC+$LiClO_4$ and EC-DEC+$LiClO_4$) based on the exponentially modified associative mean spherical approximation [Barthel, 1998; Anderson & Chandler, 1970; Chandler & Anderson, 1971; Anderson & Chandler, 1971] that gives an accurate description of ion-ion interactions even at very high salt concentration. In this model, input parameters include solvated ion sizes, solution densities, permittivity, temperature, ionic number densities, and governing equations covering ion association (e.g. ion pair formation) under equilibrium conditions. Effects from ion solvation are explicitly considered in terms of solvent residence times and average ion-solvent ligand distances, both of which influence the effective solvated ionic radii.

In this model, Eq. (4) governs the viscosity:

$$\eta = \eta_{mix}{}^{\circ}\,(1 + f_{pos,h} - f_{neg,h} + f_{coul,h})f_{DS} \tag{4}$$

The salt-free viscosity ($\eta_{mix}{}^{\circ}$) of each solvent in the mixture at the system temperature T towards a reference temperature T_r=298.15 K can be expressed as follows:

$$\eta_{mix}{}^{\circ} = \exp\,[y_1 \ln(\eta_1{}^{\circ}) + y_2 \ln(\eta_2{}^{\circ}) + y_1 y_2 a(1+2y_1 b)(1+2y2c)(T_r / T)^5] \tag{5}$$

Where in $\eta_i{}^{\circ}$ is the pure component viscosity of each solvent at the system temperature, y_i the mole fractions, a–c are adjustable mixing parameters. Subscripts 1 and 2 denote first and second solvents. For simple mixtures, a=1 and a-c=0 whereas for non ideal mixing behaviour a≠1 and a-c≠0 and depends on the nature of the interactions between the molecules.

Eq. (4) is divided into (i) a positive contribution ($f_{pos,\eta}$) due to association as contact ion pairs, solvent-shared ion pairs and triple ion species, (ii) a negative contribution ($f_{neg,\eta}$) due to the solvent structure breaking in the presence of ions and (iii) a positive or negative contribution ($f_{coul,\eta}$) due to a net attraction or repulsion between ions that increases or decreases the viscosity depending on the sign of this term. For more information about this model and the exact mathematical expression of these terms, the reader could read the paper written by Gering [Gering, 2006].

Viscosity depends also drastically on the temperature. The variation of the viscosity as a function of the temperature follows the Eyring theory as viscosity is an activated process:

$$\eta = hN_a / V_m \exp(\Delta S^{\neq} / R).\exp\,[\Delta H^{\neq} / RT)] \quad (6)$$

In Eq.(4), h is the Planck's constant, V_m the molar volume of solvent, $\Delta S^{\neq}$ the activation entropy and $\Delta H^{\neq}$ the activation enthalpy, generally identified to the activation energy of the viscous flow $E_{a,\eta}$. Non-associated solvents and non glass-forming ionic and molecular liquids, usually confirm this equation.

A linear relation between the activation energy for the viscous flow $E_{a,\eta}$ and the salt concentration C has been proposed [Chagnes, 2000]:

$$E_{a,\eta} = E_{a,\eta}{}^{0} + V_m E_{a,\eta}{}^{salt} C \quad (7)$$

Where $E_{a,\eta}{}^{0}$ and $E_{a,\eta}{}^{salt}$ are respectively, the energy of activation for the pure solvent and the contribution of the salt (per mole of the solute) to the activation energy for the transport process.

As $E_{a,\eta}$ is always positive, the increase of the salt concentration in the solvent is responsible for an increase of the sensibility of the viscosity towards a variation of the temperature. Therefore, a good electrolyte for LIB is an electrolyte with a low value of $E_{a,\eta}{}^{salt}$ in order to avoid a high increase of viscosity when the temperature decreases.

2.1.2 Ionic conductivity

The Debye-Hückel-Onsager theory can be used to calculate quantitatively the dependence of the molar ionic conductivity ($\Lambda=\kappa/C$ with κ the specific conductivity) on concentration [Hamman et al., 2007]. For a completely dissociated 1:1 electrolyte:

$$\Lambda = \Lambda^{0} - kC^{1/2} \quad (8)$$

Where C is the electrolyte concentration and k is a constant which can be calculated by using the Debye-Hückel theory.

Nevertheless this relationship can only be used in very diluted electrolytic solutions ($C<0.001$ M) due to the limitation of application of the Debye-Hückel theory for evaluation of the non-ideal behaviour in electrolytes. The Debye-Hückel theory was modified by adding empirical terms for the calculation of the activity coefficients at higher concentration or other theories were developed such as the Specific Interaction Theory (SIT) or the Pitzer theory but too few data are available in the literature concerning lithium salts dissolved in the dipolar aprotic solvents used in lithium ion batteries.

Another approach was provided by the pseudo-lattice theory adapted for the conductivity. This theory assumes that, the ions are placed in the nodes of a pseudolattice, and that even at moderate to high concentrations, the classical Debye–Hückel random picture of ionic solutions can be preserved if Debye's length is replaced by the average distance between ions of opposite charge. The conductance of electrolyte solutions thus follows a linear $C^{1/3}$ law instead of the Debye-Hückel-Onsager $C^{1/2}$ one, reflecting the underlying pseudo-lattice structure. This model was tested successfully by Chagnes et al. [Chagnes, 2001; Chagnes (2002); Gzara, 2006] in electrolytic solutions containing a lithium salt dissolved in a dipolar aprotic solvent such as $LiPF_6$, $LiAsF_6$, $LiBF_4$, $LiClO_4$ or LiTFSI (Lithium Bis(Trifluoromethanesulfonyl)Imide) dissolved in γ–butyrolactone, carbonate propylene or 3-methyl-2-oxazolidinone. When the concentration in salt is raised, the number of charge carriers increases but, at the same time, the viscosity increases and the competition between the increase in number of charges and the decrease of their mobilities lead to a maximum in the conductivity-concentration curve [Lemordant, 2002].

Conductance also depends strongly on the temperature as described by the Eyring theory that leads to the following relationship:

$$\Lambda = A\exp(E_{a,\Lambda}) / RT \tag{9}$$

Where $E_{a,\Lambda}$, A, R and T represent the activation energy for the conductivity, a constant, the ideal gas constant and T the absolute temperature, respectively.

Chagnes et al. used the quasi-lattice theory to write a new relationship between the activation energy for the conductivity ($E_{a,\Lambda}$) and the salt concentration [Chagnes, 2003b; Lemordant, 2005]:

$$E_{a,\Lambda} = E_{a,\Lambda}^{0} + \frac{2N_{av}^{4/3}e^{2}MV_{m}}{4\pi\varepsilon_{0}\varepsilon_{r}Z^{1/3}}C^{4/3} + E_{id} \tag{10}$$

Where $E_{a,\Lambda}{}^{0}$ is the activation energy for the conductivity at infinitesimal dilution and is closed to the activation energy for the viscosity of the pure solvent and E_{id} is the ion-dipole energy. M is a Madelung-like constant (the value M=1.74, corresponding to fcc lattice, is often used), N_{av} the Avogadro number, e=1.6 10^{-19} C, ε_r the static dielectric constant of the solvent, ε_0=8.82 10^{-12} S.I, and Z is the number of ions in a unit cell of the anion or cation sub-lattice (Z=4 in a fcc lattice). In Eq. (4), only ion-ion interactions were hitherto taken into account. The experimental $E_{a,\Lambda}$ values follow effectively a $C^{4/3}$ dependency on the salt concentration

($E_{a,\Lambda}=E_{a,\Lambda}{}^{0} + b\ C^{4/3}$), but the slope value (b) has been found to differ significantly from the calculated values. This has been attributed to the fact that the interaction energy between ions and solvent dipoles has been neglected as confirmed by a mathematical algorithm that permits to calculate the value of E_{id} in the quasi-lattice framework. By accounting the ion-dipole interaction, the quasi-lattice model successfully describes the variation of the activation energy for the conductivity *vs* the salt concentration [Chagnes, 2003b].

Later, Varela et al. [Varela, 1997; Varela, 2010] used a statistical mechanical framework based on the quasi-lattice theory to model satisfactorily experimental values of ionic conductivity *vs* salt concentration in conventional aqueous electrolytes. In this approach, the ion distribution is treated in a mean-field Bragg–Williams-like fashion, and the ionic motion is assumed to take place through hops between cells of two different types separated by non-random-energy barriers of different heights depending on the cell type. Assuming non-correlated ion transport, this model permits to observe the maximum of conductivity of $\kappa=f(C)$. This model could be likely extended to the electrolytes used in LIB and provide another theoretical background that confirm the validity of the quasi-lattice theory.

2.1.3 Thermal behaviour

The lithium-ion technology is efficient in term of specific energy and energy density compared to other technologies but efforts should be made to enhance the safety at high temperature and the performances of the LIB at low temperature. The charge capacity of LIB's decreases at low temperature due to the increase of the resistance of the battery (the ionic conductivity decreases and the viscosity increases) and the decrease of the lithium diffusion coefficient in the electrode materials. At high temperature, the decomposition of the electrolyte onto the positive electrode can lead to safety problems such as thermal runaway causing explosion, overpressure due to the generation of gas, etc. Therefore, the formulation of an electrolyte should take into account these aspects.

At low temperature, the wettability of the electrolyte towards the separator and the electrodes, the ionic conductivity of the electrolyte and the crystallisation point of the electrolyte mainly limit the operation of a battery. Basically, the electrolyte would have a low crystallisation point providing that the crystallization point of the solvents mixture used in the composition of the electrolyte is low. Phase diagrams of mixtures of dipolar aprotic solvents used in LIB give useful information about the low temperature behaviour of electrolytes.

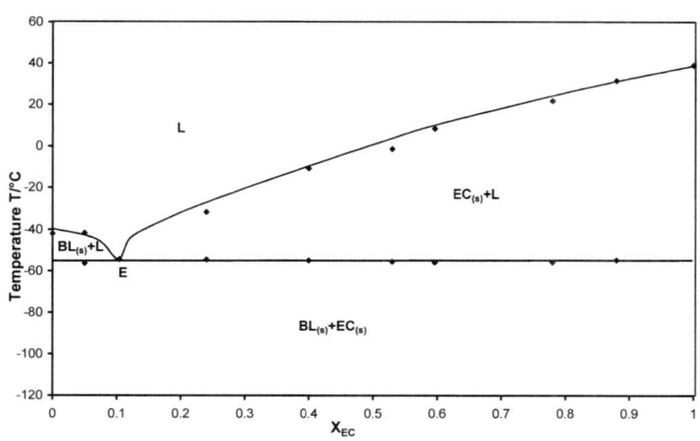

Fig. 1. Phase diagram of BL-EC mixture [Chagnes, 2003].

For instance, the phase diagrams of mixtures of cyclic ethylene carbonate (EC) and linear carbonates as dimethylcarbonate (DMC) or ethylmethylcarbonate (EMC) have been studied by Ding [Ding, 2000; Ding, 2001]. The obtained phase diagrams are simple and characterised by an eutectic point which is close to the compound having the lowest melting point. Besides, there is no miscibility gap in the liquid state whereas in solid state there is no mutual solubility. For a sake of illustration, Figure 1 shows a phase relation between γ-butyrolactone and ethylene carbonate [Chagnes, 2003a] determined by using differential scanning calorimetry (DSC) coupled with X-ray diffraction (XRD) at low temperatures.

This phase diagram shows the presence of an eutectic point for x_{EC} =0.1 and T=-56.3 °C as it was already observed for EC-DMC (eutectic point at x_{DMC}=0.12 and T=-57.5 °C) [Chagnes, 2001]. No significant influence of the carbonate co-solvent on the composition of the eutectic point was observed in BL-carbonate mixtures. Then, from a thermal viewpoint, it seems that the most interesting composition for this kind of solvent mixtures corresponds to a mole fraction of carbonate close to 0.1.

At high temperature due to outside warm or thermal runaway under abuse conditions, the electrolyte can react with the positive electrode and feed up the runaway process. Lithium hexafluorophosphate is the most used salt in lithium battery though its poor thermal stability.

Indeed, this salt decomposes to LiF and PF_5 and the latter readily hydrolyzes to form HF and PF_3O [Ravdel, 2003; Yang, 2006]. Ping et al. [Ping, 2010] confirmed that the addition of lithium salt reduces drastically the thermal stability of the solvent due to the strong Lewis acidity of PF_5 in the case of $LiPF_6$ and BF_3 in the case of $LiBF_4$ even if BF_3 is a weaker Lewis acid than PF_5. The following mechanism of degradation was proposed according to different works reported in the literature [Ping, 2010; Sloop, 2003; Wang, 2005 ; Gnanaraj, 2003a ; Gnanaraj, 2003b] :

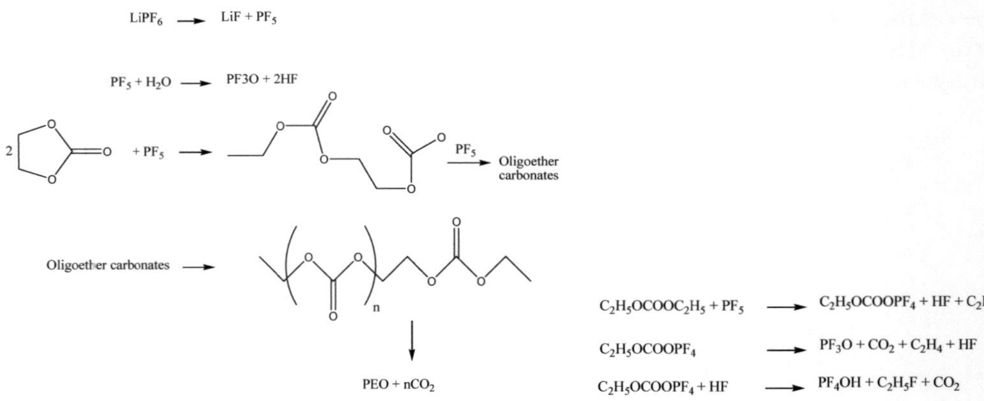

The results of the thermal stability studies show that the thermal stabilities of lithium salt in inert atmosphere can be ranked as LiTFSI (lithium bis(trifluoromethylsulfonyl)imide) $<LiPF_6<$LiBOB (lithium bis(oxalate)borate)$<LiBF_4$ and the thermal stabilities of EC electrolytes follows this order: 1 M $LiPF_6$/EC + DEC<1 M $LiBF_4$/EC + DEC<1 M LiTFSI/EC + DEC< 0.8 M LiBOB/EC + DEC. Nevertheless, it may be pointed out that electrocatalytic reactions onto the positive electrode associated with high reactivity of the electrolyte at high temperature can significantly reduce the thermal stability compared to the thermal stability of the electrolyte without any contact with a positive electrode.

2.1.4 Wettability

The wettability of the electrolytes towards the separator or the electrodes in a LIB is of great importance for the performances of the battery, especially at low temperature. Lowering the viscosity and increasing the surface tension improve the wettability. For instance, it was observed that increasing the amount of EC and/or lithium salts decreases the wettability of $LiCoO_2$ electrodes due to poorer electrolyte spreading and penetration [Wu, 2004].

The cycling ability of BL-EC+$LiBF_4$ at -20 °C in a full-cell (graphite as negative electrode and $LiCoO_2$ as positive electrode) was improved by adding a surfactant (tetraethylammonium perfluorooctanesulfonate) [Chagnes, 2003c]. This additive at 0.014 M does not perturb the quality of the solid electrolyte interphase (SEI) at the negative electrode and increases the wettability on the electrodes and the Celgard® separator by lowering the surface tension of the electrolyte.

Another approach for improving the wettability of an electrolyte on a separator consists to modify its surface by plasma treatment or by grafting acrylic acid and diethyleneglycol-dimethacrylate onto the surface of the separator [Gineste & Pourcelly, 1995].

3. Electrolyte formulation

The electrolytes for LIB are composed of a lithium salt dissolved in a dipolar aprotic solvent or a mixture of dipolar aprotic solvents. This electrolyte must meet the following specifications:

- high conductivity even at low temperature (-20 °C for electric vehicle),
- low viscosity,
- good wettability towards the separator and the electrodes,
- low melting point ($T<<-20$ °C) and high boiling point ($T>180$ °C),
- high flash point,
- large electrochemical window,
- environmentally friendly,
- low cost.

Usually, all of these criteria cannot be gathered by one solvent and the formulation of the LIB's electrolytes involves a mixture of two or three solvents and one lithium salt.

3.1 Solvents

3.1.1 Dipolar aprotic solvents

Solvents compatible with LIB are dipolar aprotic solvents because dipolar molecules can dissolve inorganic salts such as lithium salts due to the existence of ion-dipole interactions. Furthermore, aprotic solvents do not react violently with lithium as there is no proton in solution.

The main relevant solvents properties for LIB are the dipolar moment, the permittivity, the melting point, the boiling point, the flash point and the viscosity. The dipolar moment should be high to get easy the solubilisation of the lithium salt by complexing lithium ions. Usually, solvents for LIB contain electronegative atoms such as oxygen, nitrogen or sulphur to favour the complexation of lithium ions. The permittivity should be high to dissociate the lithium salt and limit the formation of ion pairs in the electrolyte as ion pairs do not participate to the ionic conductivity (neutral species). The flash point should be as high as possible for safety considerations and the viscosity should be as low as possible to facilitate the mobility of the ions in solution and ensure a good conductivity. These properties vary significantly from one category of solvents to another. The main categories of solvents used or studied for LIB's electrolytes are alkycarbonates, ethers, lactones, sulfones and nitriles. The physiscochemical properties of these solvents are collected in Table 1. Alkylcarbonates are usually used as electrolyte in lithium batteries. Propylene carbonate (PC) and ethylene carbonate (EC) exhibit a high permittivity due to the high polarity of these solvents but they are very viscous due to strong intermolecular interactions. On the other side, dimethyl carbonate (DMC) and diethyl carbonate (DEC) have a low permittivity and a low viscosity due to their molecular structure (linear carbonate) that permits to increase the degree of freedom of the molecule (rotation of alkyl groups). Besides, alkylcarbonate solvents such as ethylene carbonate form a stable passivating film (solid electrolyte interphase, *vide infra*) required for reversible intercalation at the graphite electrode. Electrolytes for LIB are mostly constituted of mixture of solvents with high permittivity (such as EC) and low viscosity (DMC, DEC) in order to promote simultaneously ionic dissociation and ion mobility. Asymmetric alkylcarbonates such as methylpropylcarbonate or ethylpropylcarbonate are promising solvents to replace conventional alkylcarbonates such as EC, DMC or PC because

Category of solvent	Solvent	T_m (°C)	T_b (°C)	ε_r	η (cP)	μ_D (D)
AlkylCarbonates	Ethylene carbonate	39-40	248.0	89.6[b]	1.860[b]	4.80
	Propylene carbonate	-49.2	241.7	64.4	2.530	5.21
	Dimethylcarbonate	3	90	3.12	0.585	0.76
	Diethylcarbonate	-43	127	3.83	0.750	0.96
	Methylpropylcarbonate	-49	130	5	1.08	4.84
	Ethylpropylcarbonate	-81	148	5	1.13	5.25
Ethers	Diethylether	-116.2	34.6	4.3	0.224	1.18
	1,2-dimethoxyethane	-58.0	84.7	7.2	0.455	1.07
	Tetrahydrofurane	-108.5	65.0	7.25[a]	0.460[a]	1.71
	Diglyme	-64.0	162.0	6.25	0.98	-
	2-methyltetrahydrofurane	-137	80.0	6.2	0.457	1.6
Lactones	γ-butyrolactone	-42.0	206.0	39.1	1.751	4.12
	γ-valerolactone	-31	208	34	2	4.29
Sulfones	Sulfonomethane	28.9	287.3	42.5[a]	9.870[a]	4.70
Nitriles	Acetonitrile	-45,7	81,8	38,0	0,345	3,94
	Adiponitrile	2	295	30	6	-

Table 1. Physicochemical properties of the main families of dipolar aprotic solvents for LIB's electrolytes. T_m : melting point, T_b: boiling point, μ_D: dipolar moment, η: absolute viscosity (25 °C), ε: permittivity (25 °C). [a] : à 30°C ; [b] : à 40°C ; [c] : à 20°C. [Hayashi et al., 1999; Hayashi et al., 1999; Smart et al., 1999; Wakihara, M., 1998; Geoffroy et al., 2000 ; Xu (2004); Abu-Lebdeh & Davidson, 2009].

these solvents suffer from their poor low temperature behaviour and exhibit low flash point (18 and 31°C, respectively) [Geoffroy et al., 2002]. In spite of their relatively low permittivity (ε_r=3-8), asymmetric alkyl carbonate solvents present promising properties such as low melting points, relatively high boiling points (>100°C), low viscosities and large electrochemical windows.

Ethers was studied as solvent for LIB to replace PC because they exhibit a low viscosity (<1 cP at 25 °C) and a low melting point. This category of solvent seems to be less and less interesting as their oxidation potentials are lower than 4V, especially on traditional positive electrodes for LIB. Electrolytes containing γ-butyrolactone (BL) or γ-valerolactone (VL) are very promising because these solvents have a large electrochemical window, a high flame point, a high boiling point, a low vapour pressure and a high conductivity at low temperatures in spite of a moderate permittivity and an absolute viscosity of 1.75 cP at 25°C. Sulfones such as ethylmethylsulfone (EMS), methoxy-methylsulfone (MEMS) or tetramethylsulfone (TMS) are good candidates for the high voltage electrolytes (electric vehicles) as their electrochemical stability in the presence of $LiPF_6$ remains good up to 5 V *vs* Li/Li^+ on platinum electrode [Watanabe et al., 2008; Sun & Angel, 2009; Abouimrane et al., 2009]. Unfortunately, these solvents cannot be used with graphite electrodes as they do not form a stable and protective SEI onto graphite. Nitriles solvents have a low viscosity and a good anodic stability (about 6.3 V *vs* Li/Li^+). Nagahama et al. [Nagahama, 2010] showed that the electrochemical window of dinitrile based electrolytes such as sebaconitrile mixed with EC and DMC in the presence of $LiBF_4$ can reach 6V at a vitreous carbon electrode and promising result were obtained with $LiFePO_4$ electrode.

Room temperature ionic liquids (RTIL) belong to another category of "solvent" which are more and more studied for various applications including the LIB's electrolytes. RTIL's are liquid at room temperature and contain big organic cations associated with small inorganic or organic anions by strong electrostatic interactions. They are popular because they have a low vapour pressure, a large electrochemical window and a high conductivity in spite of a high viscosity [Galinski, 2006]. Furthermore, they can be mixed with numerous organic solvents. For instance, the electrochemical and thermal behaviours of 1-butyl-3-methylimidazolium tetrafluoroborate or 1-butyl-3-methylimidazolium hexafluorophosphate mixed with BL were studied by Chagnes et al. [Chagnes, 2005a; Chagnes et al., 2005b]. They showed that these mixtures exhibit a very good thermal stability (>350 °C) and remains liquid at very low temperature (<-110°C) providing that the molar fraction of BL was greater than 0.3. Unfortunately, these ionic liquids mixed with BL and lithium salts undergo an easy reduction at a graphite electrode near 1V that leads to the formation of a blocking film which prevents any further cycling. However, the titanate oxide electrode can be cycled with a high capacity without any significant fading but cycling at the positive cobalt oxide electrode was unsuccessfully owing to an oxidation reaction at the electrode surface that prevents the intercalation or de-intercalation of Li ions in and from the host material. Then, less reactive positive material than cobalt oxide must be employed with this kind of RTIL's. Furthermore, the high viscosity and poor wettability of RTIL's seems to get impossible to apply them directly as electrolyte for LIB if they are not mixed with an organic solvent.

3.1.2 Additives

Numerous additives can be added to the electrolyte to improve the performance of LIB:

Category of additives	Molecules
SEI forming improver	Vinylene carbonate, vinyl ethylene carbonate, allyl ethyl carbonate, 2-vinyl pyridine, maleic anhydride, etc.
SEI forming improver and poisoning electrocatalytic effect	SO_2, CS_2, polysulfide, ethylene sulfite, propylene sulfite aryl sulfites.
SEI stabilizer (cycle life improvement)	B_2O_3, organic borates, trimethoxyboroxine, trimethylboroxin, Lithium bis(oxalato) borate.
Cathode protection agent	Butylamine, N,N' -dicyclohexylcarbodiimide, N,N-diethylamino trimethyl- silane, Lithium bis(oxalato) borate.
$LiPF_6$ salt stabilizer	tris(2,2,2-trifluoroethyl) phosphite, amides, carbamates, and fluorocarbamates pyrrolidinone, hexamethyl-phosphoramide.
Fire retardant	Trimethyl phosphate, triethyphosphate
Overcharge protection	Tetracyanoethylene,tetramethylphenylenedi-amine, dihydrophenazine, Ferrocene.

Table 2. Electrolyte additives. [Zang, 2006]

However, the use of such additives can be responsible for negative effects if they are used at high concentration or if they interfere with other compounds. Table 2 gathers the main additives reported in the literature. The reader could have more information in the paper of Zang [Zang, 2006].

3.2 Salts

Lithium salts for LIB must be soluble in dipolar aprotic solvents at a concentration close to 1 M. Such lithium salts should usually have a large anion to ensure a good dissociation in the solvents and limit the formation of ion pairs. Furthermore, these salts should be safe, environmentally friendly and they must exhibit a high oxidation potential especially for high energy applications such as the electric vehicle while they should participate in the formation of a good passivative layer at the negative electrode if necessary. In the literature, the most studied salts are lithium perchlorate ($LiClO_4$), lithium hexafluorophosphate ($LiAsF_6$), lithium tetrafluoroborate ($LiBF_4$), lithium Bis(Trifluoromethanesulfonyl)Imide (LiTFSI), lithium triflate (LiTf) and lithium hexafluorophosphate ($LiPF_6$), the most commercialized salt.

It is well known that $LiAsF_6$ and $LiClO_4$ cannot be used in LIB because these salts are not safe as $LiAsF_6$ can release toxic gases and $LiClO_4$ is an explosive material at high temperature. LiTFSI exhibits an interesting ionic conductivity in dipolar aprotic solvent but it can be responsible for the corrosion of the collectors whereas $LiBF_4$ has a low ionic conductivity in dipolar aprotic solvents. $LiPF_6$ is the most used lithium salts in LIB because it forms a good SEI when dissolved in EC-DMC or PC-EC-DMC electrolytes and the ionic conductivity of such electrolytes is high enough to minimize the internal resistance of the battery. Indeed, the success of $LiPF_6$ is mainly due to a combination of well-balanced properties such as ion mobility, ion pair dissociation, solubility, chemical inertness, surface chemistry (SEI) and collector passivation. Nevertheless, this salt is expensive and it reacts with trace of water in the solvent to form HF, a corrosive product that can degrade the SEI. A major drawback of $LiPF_6$ solutions is its poor stability at elevated temperatures. The decomposition reaction may be written as:

$$LiPF_6 = PF_5 + LiF \text{ and } PF_5 + H_2O = 2\,HF + POF_3$$

which may be summed up as:

$$LiPF_6 + H_2O = LiF + POF_3 + 2\,HF$$

The properties of $LiPF_6$ are compared to those of other lithium salts and are classified from best to worse in table 3 [Marcus, 2005].

Property	From best		to		worse
Ion mobility	$LiBF_4$	$LiClO_4$	$LiPF_6$	$LiAsF_6$	LiTf	LiTFSI
Ion pair dissociation	LiTFSI	$LiAsF_6$	$LiPF_6$	$LiClO_4$	$LiBF_4$	LiTf
Solubility	LiTFSI	$LiPF_6$	$LiAsF_6$	$LiBF_4$	LiTf	
Thermal stability	LiTFSI	LiTf	$LiAsF_6$	$LiBF_4$	$LiPF_6$	
Chem. Inertness	LiTf	LiTFSI	$LiAsF_6$	$LiBF_4$	$LiPF_6$	
SEI formation	$LiPF_6$	$LiAsF_6$	LiTFSI	$LiBF_4$		
Al corrosion	$LiAsF_6$	$LiPF_6$	$LiBF_4$	$LiClO_4$	LiTf	LiTFSI

Table 3. Classification of lithium salts.

The properties of LiBOB were enhanced by adding fluorine atoms to LiBOB. The corresponding salt, lithium difluoromono(oxalato)borate (LiDFOB), shows excellent Al-corrosion-protection properties, excellent cycling behavior of lithiated carbon anodes and $LiNi_{1/3}Co_{1/3}Mn_{1/3}O_2$ positive electrode, no HF formation by hydrolysis, and a good solubility in dipolar aprotic solvent [Zugmann et al., 2011].

4. Solid Electrolyte Interphase Layer (SEI)

This part of the chapter is designed to present of the passivation layer formed on the surface of electrode materials due to electrolyte decomposition (solvent and salt) in Li-ion battery systems. The understanding of the electrodes surface chemistry (Fig. 2) *i.e.* the electrochemical behaviour of the electrolyte (salt and solvent) including the formation of the passivating layer is prerequisite for a good battery design and functioning. Electrochemical processes taking place at the surface of a noble electrode (in this case Au) in a typical Li-salt/alkyl carbonate solution ($LiClO_4$/PC) are schematically presented in Fig.2.

The schema was plotted on the basis of the voltammetric studies and EQCM (Electrochemical Quartz Crystal Microbalance) presented by Aurbach et al. [Aurbach et al., 2001]. The numerous cathodic (reduction) and anodic (oxidation) processes undergoing on the electrode surface can have a big influence on the LIBs performance. In a case of some undesirable water and oxygen traces in the electrolyte, the reduction of both (oxygen at around 2V and water at around 1.5 V *vs* Li/Li^+) can take place which is then followed by reduction of electrolyte (solvent and salt) starting at potential inferior to 1 V *vs* Li/Li^+. Another important irreversible process which has a significant influence on the LIB system stability and the formation of a passive layer on surface of positive electrode is electrolyte oxidation at potentials superior to 3.5 V *vs* Li/Li^+. Other reversible processes, such as gold oxidation or lithium underpotential deposition (Li UPD), which also appear in the scheme, are irrelevant to our discussion.

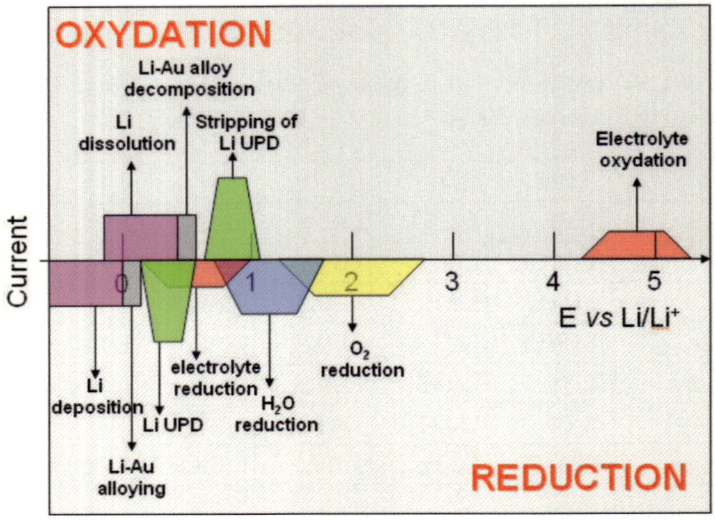

Fig. 2. Schematic presentation of reactions taking place on a noble metal electrode (Au) in alkyl carbonate/Li salt solution depending on the applied potential.

The passivating layer formed on the electrode surface due to reductive decomposition of electrolyte was named Solid Electrolyte Interphase (SEI) layer by Peled [Peled 1979]. The properties of the SEI layer affect the LIBs cyclability, life time, power and rate capability, and even their safety. Therefore, the comprehension of the formation of the SEI layer and the possibility of tuning the quality of the SEI layer on the surface of the negative as well as on the positive electrodes are essential for optimising the LIB systems.

To understand the mechanisms of the SEI formation and to control the quality of the SEI layer, it is necessary to correctly analyse the chemical composition of this layer as well as its morphology. The most appropriate techniques for the SEI analysis are the surface analytical techniques. When studying the bibliography, one can notice that the most popular analytical techniques used for the SEI analysis are the FTIR and XPS but other new techniques complementary to XPS and FTIR can be used, like in situ microscopic techniques and spectroscopic techniques (time-of-flight secondary ion mass spectrometry - ToF-SIMS).

4.1. SEI formation on negative electrodes

4.1.1 Role of the SEI

The SEI layer is formed onto the electrode surface due to electrolyte decomposition (solvent and salt) at potential below 1 V *vs* Li/Li^+ (typically 0.7 V *vs* Li/Li^+ for most electrolytes). For instance, in the case of carbonaceous electrodes, the formation of the SEI layer proceeds from the deposition of organic and inorganic compounds during the first five charge-discharge cycles. The role of the SEI layer is to prevent further decomposition of the electrolyte in the successive cycles in order to ensure a good cycling ability of the electrode, low irreversible capacity (constant and high reversible capacity) all along the charge/discharge cycles. Therefore, the SEI layer should be well adhered onto the electrode material, be insoluble in the electrolyte even at high temperatures, and also be a good electronic insulator and a good ionic conductor for lithium ions, while removing the solvation shell around the lithium ions to avoid solvent co-intercalation which is associated to exfoliation of active material [Besenhard et al., 1995]. The potential of formation, the composition and the stability of the SEI depend on numbers of factors like the formulation of the electrolyte (solvent, additives and salt) [Ein-Eli et al. 1994; Liebenow et al. 1995], the charge and discharge rate, *etc*. Few strategies can be implemented to improve the cycling ability of LIB systems. For instance, it is possible to modify the electrolyte formulation (see above) or the electrode surface in order to decrease the reactivity of the material towards the electrolyte. Recently, some studies have shown an improvement of the electrochemical performances of the graphite electrodes after modification of the surface structure by mild oxidation [Buka et al., 2001], deposition of metals oxides [Lee et al., 2000], polymer coatings [Wang et al., 2002] and coatings with other kinds of carbons [Fu et al., 2006]. Another approach for improving the electrical performances of negative material for lithium-ion batteries could be a deposition of a "synthetic" passivating layer on the graphite electrode as it was investigated by Swiatowska et al., where a thin ceria layer was deposited onto graphite electrode by electroprecipitation [Swiatowska et al., 2011].

4.1.2 SEI formation on lithium, carbonaceous electrodes, conversion and alloying-type electrodes

The composition of the SEI depends on the nature of the electrolyte and the type of active material on which the SEI is formed. The proportion of the electrolyte decomposition products can vary depending on the electrode material. For instance, in the case of carbonaceous electrode materials, the type of carbon affects the composition and the quality of the SEI [P. Verma et al., 2010]. In addition of surface specific area, particle morphology, crystallographic structure of graphite which influence the formation the SEI layer, edges and surface imperfections like defects, crevices, and active sites act as catalytic sites for electrolyte reduction [Peled at al. 2004]. The formation of the SEI layer can be also dependent on the surface finishing of the carbonaceous electrodes. Thus different kind of pre-treatments can be applied in order to modify the morphology and the chemical composition of the electrode like chemical reduction [Scott et al. 1998] or oxidation [Ein-Eli et al. 1997], electrochemical [Liu et al. 1999] or thermal treatment [Ohzuku et al., 1993].

Various spectroscopic techniques allowed for identification of the SEI layer components formed on electrode materials of LIBs. Numerous spectroscopic studies have been performed on the SEI layer in classical organic electrolytes like ethylene carbonate based electrolytes in the presence of $LiPF_6$ which resulted in proposition of different mechanisms of SEI formation. The very recent study performed by Swiatowska et al. [Swiatowska et al., 2011] the graphite-type electrodes by means of XPS showed that the SEI layer is composed of polymeric compounds such as poly(ethylene oxide) PEO ($-CH_2-CH_2-O-)_n$), ROLi (e.g.

$LiOCH_3$), R-CH_2OCO_2Li and R-CH_2OLi, and fluorinated carbons -CF_2- in agreement with previous studies [Fong et al., 1990; Aurbach et al., 1995; Laruelle et al., 2004; Andersson et al., 2001; Peled et al., 2001; Lee et al., 2000]. The formation of some SEI products like the fluorinated carbon results from reaction between the PVDF binder of the graphite and the electrolyte [Stjerndahl et al., 2007]. The decomposition of the lithium salt ($LiPF_6$) can lead to the formation of LiF, but also some other products like $Li_xPF_yO_z$ [Andersson et al., 2001; Herstedt et al. 2004] and polymer compounds *i.e.* organic-fluorinated and/or organofluoro-phosphorous (*i.e.* $C_2H_4OF_2P$) as already reported by Aurbach et al. can be formed [Aurbach et al. 2002].

Much less information can be found in the literature about the formation of the SEI layer on negative electrodes that undergo conversion/deconversion reactions. These types of electrodes (for example Sn-, Si-based and several transition metal oxide materials like Fe, Cr, Ni, Co, Cu, etc.) have attracted intense scientific attention as LIB negative electrode materials due to much higher capacities than commercially used graphite materials and no problem with so-called co-intercalation present in graphite electrodes [Bosenhard et al. 1997]. The conversion-type materials exhibit an outstanding initial irreversible capacity in comparison to the intercalation-type materials (i.e. graphite) which is always higher than it can be expected [Binotto et al., 2007]. The irreversible capacity is related to the reductive decomposition of the electrolyte, which leads to the formation of the SEI layer. Besides, the first deconversion reaction is incomplete which contributes to a large initial irreversible capacity, and the SEI layer is unstable on transition oxide materials due to volume change effects induced by the conversion/deconversion reaction [Hu et al., 2006].

The formation of the SEI layer on thin film of Cr_2O_3 in PC-1M $LiClO_4$ and its evolution during the cycling was studied by Li et al. [Li et al. 2009]. The cycling voltammetry data showed that the reductive decomposition of the electrolyte resulting from the formation of the SEI layer on the surface of the Cr_2O_3 film starts at 1.1 V *vs* Li/Li^+ and gives a cathodic peak at 0.59 *vs* Li/Li^+ in the first scan. The initial irreversible capacity due to the electrolyte reduction and the incomplete deconversion process during the first cycle is 70% of the first discharge capacity. A stable charge/discharge capacity of 460 $mAhg^{-1}$ was obtained between the 3rd to 10th cycles. XPS and FM-IRRAS reveal that the main composition of the SEI layer grown on the Cr_2O_3 film by reductive decomposition of PC is Li_2CO_3. The XPS data show that the chemical composition of the SEI layer is stable and the thickness and/or density changes in a function of the conversion/deconversion process. The formation of the Li_2CO_3 can be also confirmed from the analysis of the Li1s binding energy peak which can be found at around 55.3 ± 0.2 eV. The significant presence of Li_2CO_3 compound on the conversion-type electrodes cycled in PC-$LiClO_4$ can be justified by a reductive decomposition of PC. From the analysis of the O1s peak observed at around 533.2±0.2 eV it can be deduced that some other SEI layer components like Li-alkoxides (R-CH_2OLi) are present on the electrode surface. The XPS analysis can also give an approximate estimation of the SEI layer thickness. In the case of the thin Cr_2O_3 film the SEI homogenous layer is already well formed after one cycle of conversion/deconversion process and the thickness exceeds the 5 nm as the Cr2p core level signal (coming from the electrode surface) is completely attenuated.

The exact mechanism of the SEI layer formation on the negative electrode materials made of transition metal oxides is not well known. Some recent studies performed by Li et al. [Li et al. 2009] using ToF-SIMS analyses show possible reconstruction of the SEI layer on the electrode surface during the following cycles of conversion/deconversion process. The analysis of the SEI layer and bulk composition of the thin Cr_2O_3 film during lithiation/delithiation process performed by ToF-SIMS using negative and/or positive ion depth profiles evidences volume expansion and contraction of the electrode material. The volume expansion on the lithiated sample presumably generates cracks in the SEI layer that are filled by the immediate adsorption of the electrolyte products decomposition at low potential, thus increasing the surface content in Li_2CO_3. The volume shrink of the delithiated oxide, also evidenced by ToF-SIMS after deconversion, is thought to generate the loss of fragments of the SEI layer due to compressive stress.

a) b)

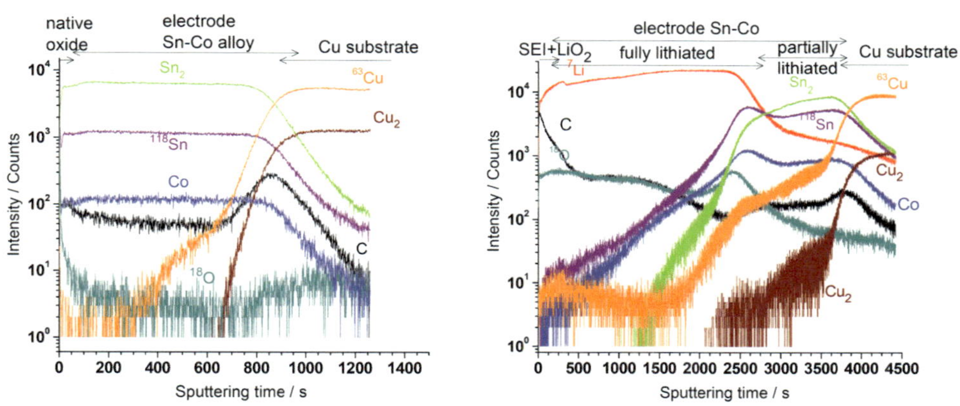

Fig. 3. ToF-SIMS negative ion depth profiles of (a) pristine and (b) lithiated Sn–Co alloy electrode at 0.02 V *vs* Li/Li^+ in 1M $LiClO_4$/PC.

Alloying-type materials are considered to be very promising alternative negative electrodes for LIBs due to their high capacity and good cycling ability [Winter et al., 1999; Huggins, 1999; Mao et al., 1999]. Despite the progress of electrochemical performance of alloying materials such as Sn-Co electrodes, the mechanisms of interfacial reactions, especially the formation, the stability, the variation and the composition of the SEI layer are not yet completely known. The formation of the SEI layer on the electrode surface and bulk modification of the alloys Sn-based electrodes deposited on a metallic copper substrate by electroplating was studied by Li et al. [Li et al. 2010; Li et al. 2011] using XPS and ToF-SIMS. A significant chemical surface modification can be observed by the changes of the C1s and O1s core level peaks which indicate that the SEI layer formed on the Sn-Co electrode is composed of Li_2CO_3, lithium alkyl carbonate $ROCO_2Li$ and alcoholates (ROLi). These findings are supported also by XPS analysis of other research groups [Ehinon et al. 2008; Dedryvere et al. 2006; Naille et al. 2006; Leroy et al. 2007]. As evidenced by the XPS data [Li et al. 2010], the quantity of Li_2CO_3 in the SEI layer increases with increasing the number of cycling. Similarly to conversion-type electrodes, the SEI layer formed on the alloying-type electrodes is thicker than the detection limit of XPS in depth, which means that it exceeds 5 nm. Tof-SIMS analysis performed on the Sn-Co electrode sample reveals some complementary information about the composition and distribution of the SEI layer on this type of electrode. Comparing the ToF-SIMS negative ion profiles performed on Sn-Co sample before and after lithiation (Fig. 3a and 3b), significant differences are observed concerning the surface and the bulk of the Sn-Co thin electrode.

Apart significant volume expansion of the electrode material (thin Sn-Co layer) after lithiation evidenced by the increase of sputtering time, significant changes can be observed in the first seconds of sputtering. The Tof-SIMS results confirm the formation of the SEI layer on the extreme surface of the alloy-type electrode and then the presence of a second layer composed of Li_2O. The products of the electrolyte decomposition are also present in the bulk and in the cracks of the alloyed (Li_xSn) electrode.

4.1.3 Influence of the SEI on the electrochemical performances of LIBs

It is well known that graphite electrode cannot be cycled in propylene carbonate electrolyte (PC+$LiPF_6$ or $LiClO_4$) because the SEI layer is not able to prevent the co-intercalation of solvent molecules in the graphite leading to the formation of gas responsible for an increase of the pressure in graphene layers and then electrode destruction by exfoliation. The exfoliation phenomenon is governed by low quality SEI layer which depends on its composition and morphology. The use of ethylene carbonate (EC) as co-solvent with PC increases significantly the cycling ability. Indeed, the composition of the SEI influences the electronic insulating properties of the SEI layer and its chemical stability. The morphology of the SEI layer, its porosity and thickness, will govern the conduction of lithium ions through the SEI layer. Therefore, the charge capacity and the irreversible capacity will depend strongly on the quality of the SEI layer. For instance, Chagnes et al. studied the influence of the composition and the morphology of the SEI layer on the cycling ability in a function of salt nature ($LiPF_6$ or $LiBF_4$) in γ-butyrolactone (BL)-ethylene carbonate (EC) [Chagnes et al., 2003c].

This study shows that the graphite electrode can be successfully cycled in the electrolyte of $LiBF_4$ used as a salt in a mixture of BL-EC solvent. Indeed, the galvanostatic charge/discharge tests show that the $LiPF_6$ used as a salt is responsible for an important ohmic resistance that prevents lithium insertion/deinsertion. This behaviour has been explained by a different SEI layer morphology and its chemical composition formed in presence of $LiPF_6$ or $LiBF_4$. SEM micrographs show that the SEI layer formed onto graphite in the presence of $LiBF_4$ is more dense and thicker than the SEI layer obtained in electrolyte containing $LiBF_4$. The XPS analyses of these SEI layers show the presence of Li_xPF_y and Li_xPO_y when $LiPF_6$ is used as salt in BL-EC and Li_xBF_y and Li_xBO_y when the electrode is cycled in $LiBF_4$+BL-EC electrolyte. Besides, in both cases LiF is present in the whole width of the SEI layer. Therefore, Li_xPF_y and Li_xPO_y might be responsible for the ionic insulating properties of the SEI.

4.2 SEI formation on positive electrodes

Numerous research efforts are still focused on the formation of SEI layer on negative electrode material. Even the SEI layer formation on the surface of negative electrode is so extensively studied since decades it should be noted that especially its formation mechanism and chemical composition are still controversial. Contrary to negative electrode materials, the positive electrode materials were not thoroughly studied in the meaning of interfacial electrode/electrolyte processes due to difficult experimental implementation resulting from very thin SEI layer on positive electrodes. The most appropriate techniques can be the *in situ* measurements like Fourier transform infrared (FTIR), Raman spectroscopy chemical analysis or Atomic Force Microscopy (AFM). However, these techniques present numerous disadvantages due to difficulties in interpretation of origin of molecules having, for example, similar FTIR signature etc. One of the most appropriate and powerful tool for analysis of thin layers is the XPS technique by combination of core and valence band spectra as it has been already proposed by Dedryvère et al. [Dedryvère et al., 2005].

Formation of the surface layer on the positive electrode material has different properties and origins. Thus, it has been proposed that this layer can be called Solid Permeable Interface (SPI) rather than a Solid-Electrolyte Interphase (SEI) [Balbuena & Wuang, 2004]. This layer formed on the positive electrode material has some limits to passivate the electrode surface as is principally composed of organic compounds rather than inorganic compounds [Levi et al., 1999].

4.2.1 Oxidation potentials of the electrolytes

The formation of the SEI layer on the positive electrode surface depends on the anodic stability of the electrolyte. The oxidation potentials increase with a following order: alkyl carbonates > esters > ethers. To study the anodic stability of organic solvents one should take into account that the solvent reactivity depends on the nature of the electrode and even solvents with relatively high anodic stability, like alkyl carbonates, undergo anodic reactions on noble metal (Au, Pt) at potentials below 4V versus Li/Li^+ [Moshkovich et al., 2001], whereas these solvents are stable on the classical cathodes ($LiNiO_2$, $LiCoO_2$, $LiMn_2O_4$, etc.). Nevertheless, the stability of the electrolytes with 4V positive electrodes is assured due to formation of a film at the surface of the positive electrode [Aurbach, 2000]. In-situ FTIR and EQCM measurements do not show the formation of passive layer on the surface of metallic electrode, hence it can be concluded that the oxidation products of these systems are formed in the solution phase and do not precipitate as surface films. Various possible reaction mechanisms of oxidation of DMC and EC were proposed where one the final reaction product can be $CH_3OCO_2CH_2CH_2OCO_2CH_3$ as observed by the nuclear magnetic resonance (NMR) analysis. Gas chromatographic mass spectroscopy (GCMS) analysis of an oxidized EC-DMC solution shows the formation of an oxidation product issue of EC-DMC combination which is $CH_3OCO_2CH_2CH_2OCO_2CH_2CH_2\text{-}OCO_2CH_3$. The following both products have been detected by NMR, GCMS and FTIR analysis $HOCH_2CH_2OCO_2CH_2CH{=}O$, and $O{=}CH\text{-}CH_2OCO_2CH_2CH_2OCH{=}O$ [Moshkovich et al., 2001]. However, the information about oxidized alkyl carbonate solutions that can be obtained from these three experimental techniques is not fully conclusive, thus it is impossible to propose the exact reaction patterns and identify all the reaction products. The identification of the products is difficult due to impossible separation of the products of reaction and the compounds originating from the mother solution. Moreover, the solution storage can lead to formation of polymer compounds.

4.2.2 SEI formation on lithiated metal transition oxides, phosphates

As already stated above, the major reason of film formation on the positive electrode is the oxidation of electrolyte. The process could be driven by the reduction of unstable metal ions in the active electrode material (in a case of $LiMn_2O_4$ electrode it can be Mn^{4+}). The corrosion of electrode materials provoked by the electrolyte depends on the pH and it was reported that the dissolution of spinel-type electrode decreases in the order: $LiCF_3SO_3$ > $LiPF_6$ > $LiClO_4$ > $LiAsF_6$ > LiBF [Jang et al., 1997]. The XPS spectra performed on the $LiMn_2O_4$ electrode showed that the passive layer is constituted of both organic and inorganic materials like: LiF, and Li_xPF_y-type compounds, poly(oxyethylene), $ROCO_2Li$ and Li_2CO_3 [Balbuena & Wuang, 2004]. The XPS profile performed on this electrode evidenced that the passive film has a layered structure and that the film is not dense enough to serve as a barrier between the electrolyte and the oxidizing environment which is in contrast with the properties of the SEI layer formed on the negative electrode material.

Similar compounds are found to be formed on the surface of LiNi0.8Co0.2O2, which are organic polycarbonates, polymeric hydrocarbons and salt-based products [Balbuena & Wuang, 2004].

The XPS analysis of the V_2O_5 and MoO_3 used as positive electrodes show the irreversible build-up of a surface layer including lithium carbonates and Li-alkyl carbonates ($ROCO_2Li$) and likely Li-alkoxides ($R\text{-}CH_2OLi$) [Swiatowska-Mrowiecka et al., 2007; Swiatowska-Mrowiecka et al., 2008]. The layer formed on the positive electrode transition metal oxides is very thin and does not exceed few nanometers in comparison to the SEI layer formed on the negative electrode materials. After multiple cycles, the composition of the passive layer formed on the V_2O_5 is modified and the lithium carbonates disappears, whereas Li-alkyl carbonates and/or Li-alkoxides remain on the surface, indicating the dissolution and/or conversion of the SEI layer [Swiatowska-Mrowiecka et al. 2007].

The formation of the passive layer on the phosphate-type electrodes is significantly different than on the spinel-type electrodes. The most typical products of electrolyte decomposition (i.e., polycarbonates, semicarbonates and Li_2CO_3) could not be detected on the carbon-coated $LiFePO_4$ surface that indicates that the phosphate group does not react with the solvents. The film formed on the $LiFePO_4$ surface consist of products issued the salt decomposition like : LiF, $LiPF_6$, Li_xF_y- and $Li_x PO_yF_z$-type compounds.

5. Conclusions

The formulation of electrolytes for lithium batteries (LIB's) is difficult as this task should take into account numerous constraints depending on the application (high energy density, temperature, etc.). Furthermore, the electrolyte must be compatible with each battery component, i.e. the electrode materials, the current collector, and the separator. The following specification must be taking into account:

- high conductivity even at low temperature (-20 °C for electric vehicle),
- low viscosity,
- good wettability towards the separator and the electrodes,
- low melting point (T<<-20 °C) and high boiling point (T>180 °C),
- high flash point,
- large electrochemical window,
- environmentally friendly,
- low cost.

These specifications cannot be reached with only one dipolar aprotic solvent in which a lithium salt is dissolved. Usually, the formulation of electrolytes for LIB implies two or three solvents and one lithium salt. Most of the electrolytes contain ethylene carbonate because this solvent permits to enhance the cycling ability of graphite-types electrodes by improving the quality of the Solid Electrolyte Interface (SEI) formed on the electrode surface before reaction of lithium insertion. The morphology of the SEI is governed by the formulation of the electrolyte, and then by the composition of the SEI. The morphology of the SEI is of great importance as a thick SEI with a poor porosity can be responsible for insulating properties towards lithium ions and poor cycling ability of the electrode. Therefore, the electrode-electrolyte interface plays a crucial role in lithium batteries at negative electrode and at positive electrode as well. Although an electrolyte exhibits good electrochemical window on inert electrodes such as glassy carbon electrode, an oxidation can occur on the surface of positive electrode due to catalytic effects of electrode materials (i.e. lithiated transition metal oxides). The main challenges for the next generation of LIB's for the high energy density applications such as the electric vehicle will be to find material for positive electrode and the electrolyte operating at high voltages. At present, sebaconitrile seems to be a promising solvent for this application but the formulation should be adapted to be used with classical negative electrode materials.

6. References

Abouimrane, A; Belharouak, I & Amine, K. (2009). Sulfone-based electrolytes for high-voltage Li-ion batteries. *Electrochemistry Communications,* Vol. 11, 5, pp. 1073-1076, ISSN: 1388-2481.

Abu-Lebdeh, Y. & Davidson, I. (2009). Storage High-Voltage Electrolytes Based on Adiponitrile for Li-Ion Batteries. *The Journal of the Electrochemical Society,* Vol. 156, 1, (2009), pp. A60-A65, ISSN: 0013-4651.

Andersson, A.M. & Edström, K. (2001). Chemical Composition and Morphology of the Elevated Temperature SEI on Graphite. The *Journal of the Electrochemical Society,* Vol. 148, (2001), pp. AA1100-1109, ISSN: 0013-4651.

Anderson, H.C. & Chandler, D. (1970). Mode expansion in equilibrium statistical mechanics. I. General theory and application to the classical electron gas. *Journal of Chemical Physics,* Vol.53, (1970), pp. 547, ISSN 0021-9606.

Anderson, H.C. & Chandler, D. (1971). Mode expansion in equilibrium statistical mechanics. III. Optimized convergence and application to ionic solution theory. *Journal of Chemical Physics,* Vol .55, (1971), pp. 1497-1504, ISSN 0021-9606.

Aurbach, D. ; Levi, M.D. ; Levi, E. ; Teller, H. ; Markovsky, B. & Salitra, G. (1998). Common Electroanalytical Behavior of Li Intercalation Processes into Graphite and Transition Metal Oxides. The *Journal of the Electrochemical Soc*iety, Vol. 145, (1998), pp. 3024-3034, ISSN: 0013-4651.

Aurbach, D.; Moshkovich, M. & Gofer, Y. (2001). Investigation of the Electrochemical Windows of Aprotic Alkali Metal (Li, Na, K) Salt Solutions. The *Journal of the Electrochemical Soc*iety, Vol. 148, (2001), pp. E155-E167, ISSN: 0013-4651.

Aurbach, D., Markovsky; B., Rodkin; A., Cojocaru, M.; Levi, E. & Kim, H.-J. (2002). An analysis of rechargeable lithium-ion batteries after prolonged cycling. *Electrochimica Acta,* Vol. 47 (12), (2002), pp. 1899-1911, ISSN: 0013-4686.

Aurbach, D. (2000). Review of selected electrode-solution interactions which determine the performance of Li and Li ion batteries. *Journal of Power Sources,* Vol. 89, (2000), pp. 206-218, ISSN: 0378-7753.

Aurbach, D; Talyosef, Y.; Markovsky, B.; Markevich, E.; Zinigrad, E.; Asraf, L.; Gnanaraj, J.S. & Kim H.-J. (2004). Design of electrolyte solutions for Li and Li-ion batteries: a review. *Electrochimica Acta,* Vol.50, (2004), pp. 247–254, ISSN: 0013-4686.

Aurbach, D.; Ein-Eli, Y.; Markovsky, B.; Zaban, A.; Luski, S.; Carmeli, Y. & Yamin, H. (1995). The Study of Electrolyte Solutions Based on Ethylene and Diethyl Carbonates for Rechargeable Li Batteries. *The Journal of the Electrochemical Society,* Vol. 142, (1995) pp. 2882-2890, ISSN: 0013-4651.

Balbuena, B.P. & Wuang & X. (2004). *Lithium ion batteries – Solid electrolyte interphase,* Imperial College Press, London, ISBN: 1-86094-362-4.

Barthel, J.M.G. ; Krienke, H. & Kunz, W. (1998). *Physical Chemistry of Electrolyte Solutions, Modern Aspects,* Steinkopff-Verlag Darmstadt, ISBN: 78-3798510760, London.

Besenhard, J.O.; Winter, M.; Yang, J. & Biberacher, W. (1995). Filming mechanism of lithium-carbon anodes in organic and inorganic electrolytes. *Journal of Power Sources,* Vol. 54, pp. 228-231, ISSN: 0378-7753.

Binotto, G.; Larcher, D.; Prakash, A.S.; Urbina, R.H.; Hegde, M.S. & Tarascon, J.M. (2007). Synthesis, Characterization, and Li-Electrochemical Performance of Highly Porous Co_3O_4 Powders. *Chemistry of Materials,* Vol. 19 (2007) 3032-3040, ISSN: 1520-5002.

Bockris, R. (1970). *Modern Electrochemistry,* Plenum Press, Vol.1, pp.386, ISBN: 0-306-37036-0, New-York.

Besenhard, J.O.; Yang, J. & Winter, M. (1997). Will advanced lithium-alloy anodes have a chance in lithium-ion batteries? *Journal of Power Sources,* Vol. 68, (1997), pp. 87–90, ISSN: 0378-7753.

Buka, H.; Golob, P.; Winter, M. & Bensenhard, J.O. (2001). Modified carbons for improved anodes in lithium ion cells. *Journal of Power Sources,* Vol. 97-98, p. 122-125, ISSN: 0378-7753.

Chagnes, A.; Carré, B.; Lemordant, D. & Willmann, P. (2001a). Ion transport theory of nonaqueous electrolytes. LiClO4 in γ-butyrolactone: The quasi lattice approach. *Electrochimica Acta,* Vol.46, (2001), pp. 1783-1791, ISSN: 0013-4686.

Chagnes, A.; Mialkowski, C. ; Carré, B. ; Lemordant, D. ; Agafonov, V. & Willmann, P. (2001b). Phase diagram of Lactone - Carbonate mixture. *Journal de Physique IV,* Vol.11, (2001), pp. 10-27, ISBN 2-86883-723-9.

Chagnes, A.; Carré, B.; Lemordant, D. & Willmann, P. (2002). Modeling viscosity and conductivity of lithium salts in γ-butyrolactone. Application of the Quasi-Lattice theory. *Journal of Power Sources,* Vol.109, (2002), pp. 203-213, ISSN: 0378-7753.

Chagnes, A. ; Allouchi, H. ; Carré, B. ; Oudou, G. ; Willmann, P. & Lemordant, D. (2003a). γ-Butyrolactone-Ethylene carbonate based electrolytes for lithium batteries. *Journal of Applied Electrochemistry,* Vol.33, (2003), pp. 589-595, ISSN: 0021-891X.

Chagnes, A. ; Nicolis, S. ; Carré, B. ; Willmann, P. & Lemordant, D. (2003b). Ion-dipole interaction in concentrated organic electrolytes. *ChemPhysChem,* Vol.4, (2003), pp. 559-566, 1439-4235.

Chagnes A; Carre, B ; Willmann, P ; Dedryvere, R ; Gonbeau, D & Lemordant, D. (2003c). Cycling Ability of γ-Butyrolactone-Ethylene Carbonate Based Electrolytes. *Journal of the Electrochemical Society,* Vol. 150 (9), (2003), pp. A1255-A1261, ISSN: 0013-4651.

Chagnes, A.; Allouchi, H.; Carré, B. & Lemordant, D. (2005a). Phase diagram of imidazolium-butyrolactone binary mixtures. *Solid State Ionics,* Vol.176, (2005), pp. 1419-1427, ISSN: 0167-2738.

Chagnes, A.; Diaw, M.; Carré, B.; Willmann, P. & Lemordant, D. (2005b). Imidazolium-Organic Solvent Mixtures as Electrolytes for Lithium Batteries. *Journal of Power Sources,* Vol.145, (2005), pp. 82-88, ISSN: 0378-7753.

Chagnes, A. (2010). *Les batteries lithium-ions-Formulation de l'électrolyte.* Edition Universitaires Européennes, Reha GmbH, Saarbrucken, (2010), ISBN: 978-613-1-51035-9.

Chandler, D. & Anderson, H.C. (1971). Mode expansion in equilibrium statistical mechanics. II. A rapidly converging theory of ionic solutions. *Journal of Chemical Physics,* Vol.54, (1971) pp. 26, ISSN: 0021-9606.

Chen ,R.J.; He, Z.Y. & Wu, F. (2011). Lithium Organic Borate Salt and Sulfite Functional Electrolytes. *Progress in Chemistry,* Vol.23, 2-3, (2001), pp. 382-389, ISSN: 1005-281X.

Dedryvère, R.; Leroy, S. ; Martinez, H. ; Blanchard, F. ; Lemordant, D. & Gonbeau, D., 2006. The Journal of Physical Chemistry B, Vol. 110, (2006), pp. 12986–12992, ISSN: 1520-5207.

Dedryvère, R.; Laruelle, S.; Grugeon, S.; Gireaud, L.; Tarascon, J.-M. & Gonbeau, D. (2005). XPS Identification of the Organic and Inorganic Components of the Electrode/Electrolyte Interface Formed on a Metallic Cathode. *The Journal of the Electrochemical Society,* Vol. 152, (2005), pp. A689-A696, ISSN: 0013-4651.

Ding, M.S.; Xu, K. & Jow, T.R. (2000). Liquid-Solid Phase Diagrams of Binary Carbonates for Lithium Batteries. *Journal of the Electrochemical Society,* Vol.147, (2000), pp. 1688-1694, ISSN: 0013-4651.

Ding, M.S.;Xu, K., Zhang, S. & Jow, T.R. (2001). Liquid/Solid Phase Diagrams of Binary Carbonates for Lithium Batteries-Part II. The *Journal of the Electrochemical Society,* Vol.148, 4, (2001), pp.A299-A304, ISSN: 0013-4651.

Ehinon, K.K.D.; Naille, S.; Dedryvere, R.; Lippens, P.E.; Jumas, J.C. & Gonbeau, D. (2008). *Chemistry of Material,* Vol. 20, (2008), pp. 5388–5398, ISSN: 1520-5002.

Ein-Eli, Y.; Markovsky, B.; Aurbach, D.; Carmeli, Y.; Yamin, H. & Luski, S. (1994). The dependence of the performance of li-c intercalation anodes for li-ion secondary batteries on the electrolyte solution composition. *Electrochimica Acta,* Vol. 39, (1994), pp. 2559-2569, ISSN: 0013-4686.

Ein-Eli Y. & Koch, V.R., 1997. Chemical Oxidation: A Route to Enhanced Capacity in Li-Ion Graphite Anodes. *The Journal of the Electrochemical Society,* Vol. 144 (1997), pp. 2968-2973, ISSN: 0013-4651.

Fong, R., Von Sacken; U. & Dahn; J.R., 1990. Studies of Lithium Intercalation into Carbons Using Nonaqueous Electrochemical Cells. *The Journal of the Electrochemical Society,* Vol. 137, (1990), pp. 2009-2013, ISSN: 0013-4651.

Fu, L.J.; Liu, H.; Li, C.; Wu, Y.P.; Rahm, E.; Holze, R. & Wu, H.Q. (2006). Surface modifications of electrode materials for lithium ion batteries . *Solid State Sciences,* Vol. 8, (2006), pp. 113-128, ISSN: 1293-2558.

Galinski, M. ;Lewandowski, A. & Stepniak, I. (2006). Ionic liquids as electrolytes. *Electrochimica Acta,* Vol. 51, (2006), pp. 5567-5580, ISSN: 0013-4686.

Geoffroy, I.; Chagnes, A.; Carré, B.; Lemordant, D.; Biensan, P. & Herreyre, S. (2002). Electrolytic characteristics of asymetric alkyl carbonates solvents for lithium batteries. *Journal of Power Sources,* Vol. 112, 1, (2002), pp. 191-198, ISSN: 0378-7753.

Gisneste, J.L. & Pourcelly, G. (1995). Polypropylene separator grafted with hydrophilic monomers for lithium batteries. *Journal of Membrane Science,* Vol.197, 1-2, (1995), pp. 155-164, ISSN: 0376-7388.

Guering, K.L. (2006). Prediction of electrolyte viscosity for aqueous and non aqueous systems: Results from a molecular model based on ion solvation and a chemical physic framework. *Electrochimica Acta,* Vol. 51, (2006), pp. 3125-3138, ISSN: 0013-4686.

Gnanaraj, J.S. ; Zinigrad, E. ; Asraf, L. ; Gottlieb, H.E. ; Sprecher, M. ; Schmidt, M. ; Geissler, W. & Aurbach, D. (2003a). A Detailed Investigation of the Thermal Reactions of $LiPF_6$ Solution in Organic Carbonates Using ARC and DSC. *Journal of the Electrochemical Society,* Vol. 150, (2003), pp. A1533-1537, ISSN: 0013-4651.

Gnanaraj, J.S. ; Zinigrad, E. ; Asraf, L. ; Gottlieb, H.E. ; Sprecher, M. ; Geissler, W. ; Aurbach, D. & Schmidt, M. (2003b). The use of accelerating rate calorimetry (ARC) for the study of the thermal reactions of Li-ion battery electrolyte solutions. *Journal of Power Sources,* Vol. 119–121, (2003), pp.794-798, ISSN: 0378-7753.

Gzara, L.; Chagnes, A.; Carré, B.; Dhahbi, M. & Lemordant, D. (2006). Is 3-Methyl-2-Oxazolidinone a suitable solvent for lithium-ion batteries? *Journal of Power Sources,* Vol.156, (2006), pp. 634-644, ISSN: 0378-7753.

Hamann, C.H.; Hamnett, A., & Vielstich, W. (2007). *Electrochemistry, second, completely revised and updated,* Wiley-VCH, Verlag GmbH & Co KGaA, ISBN 978-3-527-31069-2, Weinheim.

Hayashi, K; Nemoto, Y.; Tobishima, S.L. & Yamaki, J.I. (1999). Mixed solvent electrolyte for high voltage lithium metal secondary cells. *Electrochimica Acta,* Vol.44, (1999), pp. 2337-2344, ISSN: 0013-4686.

Herstedt, M.; Abraham, D.P.; Kerr, J.B. & Edström, K. (2004). X-ray photoelectron spectroscopy of negative electrodes from high-power lithium-ion cells showing various levels of power fade. *Electrochimica Acta,* Vol. 49, 28, (2004), pp. 5097-5110. ISSN: 0013-4686.

Hu, J.; Li, H.; Huang, X.J. & Chen, L.Q. (2006). *Solid State Ionic*s, Vol. 177, 26-32, (2006), pp. 2791-2699, ISSN: 0167-2738.

Huggins, R.A. (1999). Lithium alloy negative electrodes. *Journal of Power Sources,* Vol. 81–82, (1999), pp. 13–19 , ISSN: 0378-7753.

Jang, D. H. & Oh, S. M. (1997). Electrolyte effects on spinel dissolution and cathodic capacity losses in 4 V Li/Li{sub x}Mn{sub 2}O{sub 4} rechargeable cells. *The Journal of the Electrochemical Society,* Vol. 144, (1997), pp. 3342. ISSN: 0013-4651.

Jones, G. & Dole, M (1929). The viscosity of aqueous solutions of strong electrolytes with special reference to barium chloride. *Journal of the American Chemical Society,* Vol.51, pp. 2950-2964, ISSN 0002-7863.

Laruelle, S., Pilard, S., Guenot, P., Grugeon, S. & Tarascon, J.-M. (2004). Identification of Li-Based Electrolyte Degradation Products Through DEI and ESI High-Resolution Mass Spectrometry. The *Journal of the Electrochemical Society,* Vol. 151, (2004), pp. 1202-1209. ISSN: 0013-4651.

Lee, J.-C. & Litt, M. (2000). Ring-Opening Polymerization of Ethylene Carbonate and Depolymerization of Poly(ethylene oxide-co-ethylene carbonate). *Macromolecules,* Vol. 33, (2000), pp. 1618-1627, ISSN: 0024-9297.

Lemordant, D. ; Montigny, B. ; Chagens, A. ; Caillon-Caravanier, M. ; Blanchard, F. ; Bosser, G. ; Carré, B. & Willmann, P. (2002). *Viscosity-conductivity relationships in concentrated lithium salt-organic solvent électrolytes. Materials chemistry in lithium batteries,* Kumagai, N. ; Komaba, S. (Ed.), Research Signpost, Trivandrum, pp. 343-367, ISBN 81-7736-115-5.

Lemordant, D. ; Blanchard, F. ; Bosser, G. ; Caillon-Caravanier, M. ; Carré, B. ; Chagnes, A. ; Montigny, B. & Naejus, R. (2005). *Physicochemical properties of fluorine-containing electrolytes for lithium batteries.* Fluorine Materials for energy conversion, Chap. 7, Nakajima, T. ; Groult, H. (Ed.), Elsevier, London, pp.137-172, ISBN: 008-044472-5.

Leroy, S.; Martinez, H.; Dedryvere, R.; Lemordant, D. & Gonbeau, D. (2007). Influence of the lithium salt nature over the surface film formation on a graphite electrode in Li-ion batteries: An XPS study. *Applied Surface Sci*ence,Vol. 253, (2007), pp. 4895–4905, ISSN: 0169-4332.

Lee, J.; Zhang, R. & Liu, Z. (2000). Dispersion of Sn and SnO on carbon anodes. *Journal of Power Sources,* Vol. 90, p. 70-75, ISSN: 0378-7753.

Levi, M.D.; Salitra, G.; Markovsky, B.; Teller, H.; Aurbach; D., Heider, U. & Heider, L. (1999). Solid-State Electrochemical Kinetics of Li-Ion Intercalation into Li1–xCoO2: Simultaneous Application of Electroanalytical Techniques SSCV, PITT, and EIS. *The Journal of the Electrochemical Society,* Vol. 146, (1999), pp. 1279-1289. ISSN: 0013-4651.

Li, J.-T.; Maurice, V.; Swiatowska-Mrowiecka, J.; Seyeux, A.; Zanna, S.; Klein, L.; Sun, S.-G. & Marcus, P. (2009). XPS, Time-of-Flight-SIMS and Polarization Modulation IRRAS study of Cr_2O_3 thin film materials as anode for lithium ion battery, *Electrochimica Acta,* Vol. 54, (2009), pp. 3700-3707. ISSN: 0013-4686.

Li, J.–T.; Swiatowska, J.; Seyeux, A.; Huang, L.; Maurice, V.; Sun, S.–G. & Marcus, P. (2010). XPS and ToF-SIMS study of Sn-Co alloy thin films as anode for lithium ion battery, *Journal of Power Sources,* (2010), Vol. 195, pp. 8251–8257, ISSN: 0378-7753.

Li, J.-T.; Światowska, J.; Maurice, V.; Seyeux, A.; Huang, L.; Sun, S.-G. & Marcus, P., 2011, XPS and ToF-SIMS Study of Electrode Processes on Sn-Ni Alloy Anodes for Li-ion Batteries, *Journal of Physical Chemistry C,* Vol. 115, (2011), pp. 7012-7018, ISSN: 1932-7447.

Liebenow, C.; Wagner, M.W.; Lühder, K.; Lobitz, P. & Besenhard, J.O. (1995). Electrochemical-behavior of coated lithium-carbon electrodes. *Journal of Power Sources,* Vol.54, (1995), pp. 369-372 , ISSN: 0378-7753.

Liu, Z.; Yu, A. & Lee, J.Y. (1999). Modifications of synthetic graphite for secondary lithium-ion battery applications. *Journal of Power Sources,* Vol. 81–82, (1999), pp. 187, ISSN: 0378-7753.

Mao, O., Turner, R.L.; Courtney, I.A.; Fredericksen, B.D.; Buckett, M.I.; Krause & L.J.,Dahn, J.R., 1999. *Electrochemical and Solid State Letters,* Vol. 2, (1999), pp. 3–5, 1099-0062.

Marcus, Y (1985). Ion Solvation, John Wiley and Sons (Ed.), 1985, pp. 135.

Mialkowski, C. ; Chagnes, A. ; Carré, B. ; Willmann, P. & Lemordant, D. (2002). Excess thermodynamic properties of binary liquid mixtures containing dimethyl carbonate and γ-butyrolactone. *The Journal of Chemical Thermodynamics,* Vol.34, 11, pp. 1845-1854, ISSN: 0021-9614.

Moshkovich, M.; Cojocaru, M.; Gottlieb, H.E. & Aurbach, D. (2001). The study of the anodic stability of alkyl carbonate solutions by in situ FTIR spectroscopy, EQCM, NMR and MS. *Journal of Electroanalytical Chemistry,* Vol. 497, (2001), pp. 84- 96, ISSN: 0022-0728.

Nagahama, M.; Hasegawa, N. & Okada, S. (2010). High Voltage Performances of Li_2NiPO_4F Cathode with Dinitrile-Based Electrolytes. The *Journal of the Electrochemical Society,* Vol.157, 6, (2010), pp. A748-A757, ISSN: 0013-4651.

Naille, S.; Dedryvere, R.; Martinez, H.; Leroy, S.; Lippens, P.E; Jumas, J.C. & Gonbeau, D. 2007. XPS study of electrode/electrolyte interfaces of η-Cu6Sn5 electrodes in Li-ion batteries. Journal of Power Sources, Vol. 174, (2007), pp. 1086–1090, ISSN: 0378-7753.

Ohzuku, T.; Iwakoshi, Y. & Sawai, K., 1993. Formation of Lithium-Graphite Intercalation Compounds in Nonaqueous Electrolytes and Their Application as a Negative Electrode for a Lithium Ion (Shuttlecock) Cell. *The Journal of the Electrochemical Society*, Vol. 140, (1993), pp. 2490-2498, ISSN: 0013-4651.

Papoular, R.J.; Allouchi, H.; Chagnes, A.;Dzyabchenko, A.; Carré, B.; Lemordant, D. & Agafonov, (2005). X-ray Powder Diffraction Structure Determination of γ-butyrolactone at 180 K. Phase problem solution from the lattice energy minimization with two independent molecules. *Acta Cristallography Section B*, Vol. B61, (2005), pp. 312-320, ISSN: 0108-7673.

Peled, E.; Golodnitsky, D.; Menachem, C. & Bar-Tow, D. (1998). An Advanced Tool for the Selection of Electrolyte Components for Rechargeable Lithium Batteries. *The Journal of the Electrochemical Society*, Vol. 145, (1998), pp. 3482-3486 , ISSN: 0013-4651.

Peled, E.; Bar Tow, D.; Merson, A.; Gladkich, A.; Burstein L. & Golodnitsky, D. (2001). Composition, depth profiles and lateral distribution of materials in the SEI built on HOPG-TOF SIMS and XPS studies. *Journal of Power Sources*, Vol. 97-98, (2001), pp. 52-57, ISSN: 0378-7753.

Peled, E.; Golodnitsky, D.; Ulus, A. & Yufit, V. (2004). Effect of carbon substrate on SEI composition and morphology. *Electrochimica Acta*, Vol. 50, (2004), pp. 391-395, ISSN: 0013-4686.

Ping, P.; Qingsong, W; Jinhua, S. ; Hongfa X. & Chen, C.H. (2010). Thermal Stabilities of Some Lithium Salts and Their Electrolyte Solutions With and Without Contact to a $LiFePO_4$ Electrode. *Journal of the Electrochemical Society*, Vol. 157, 11, pp. A1170-A1176, ISSN: 0013-4651.

Ravdel, B.; Abraham, K.M.; Gitzendanner, R. ; DiCarlo, J. ; Lucht, B. & Campion, C. (2003). Thermal Stability of Li-ion battery Electrolytes. *Journal of Power Sources*, Vol.805, pp. 119-121, ISSN: 0378-7753.

Scott, M.G.; Whitehead, A.H. & Owen, J.R. (1998). Chemical Formation of a Solid Electrolyte Interface on the Carbon Electrode of a Li-Ion Cell. *The Journal of the Electrochemical Society*, Vol. 145, (1998), pp. 1506-1510, ISSN: 0013-4651.

Sloop, S.E. ; Kerr, J.B. & Kinoshita, K., (2003). The role of Li-ion battery electrolyte reactivity in performance decline and self-discharge. *Journal of Power Sources*, Vol. (119–121), pp. 330-337, ISSN: 0378-7753.

Stjerndahl, M.; Bryngelsson, H.; Gustafsson, T.; Vaughey, J.T.; Thackeray; M.M., Edström & K. (2007). Surface chemistry of intermetallic AlSb-anodes for Li-ion batteries. *Electrochimica Acta*, Vol. 52, (2007), pp. 4947-4955. ISSN: 0013-4686.

Swiatowska, J.; Lair, V.; Pereira-Nabais, C.; Cote, G.; Marcus, P.; Chagnes, A., (2011). *Applied Surface Science*, Vol. 257, (2011), pp. 9110– 9119. ISSN: 0169-4332.

Swiatowska-Mrowiecka; J., Maurice; V., Zanna, S.; Klein, L. & Marcus, P. (2007). XPS study of Li ion intercalation in V2O5 thin films prepared by thermal oxidation of vanadium metal. *Electrochimica Acta*, Vol. 52, pp. 5644–5653. ISSN: 0013-4686.

Swiatowska-Mrowiecka, J.; de Diesbach, S.; Maurice, V.; Zanna, S.; Klein, L.; Briand, E.; Vickridge, I. & Marcus, P. (2008). Li-ion intercalation in thermal oxide thin films of

MoO3 as studied by XPS, RBS and NRA. *Journal of Physical Chemistry C,* Vol. 112, (2008), pp. 11050–11058, ISSN: 1932-7447.

Swiatowska-Mrowiecka, J.; Maurice, V.; Zanna, S.; Klein, L.; Briand, E.; Vickridge, I. & Marcus, P. (2007). Ageing of V2O5 thin films induced by Li intercalation multi-cycling. *Journal of Power Sources,* Vol. 170, (2005), pp. 160–172. ISSN: 0378-7753.

Smart, M.C.; Ratnakumar, B.V. & Surampudi, S. (1999). Electrolytes for low-temperature lithium batteries based on ternary mixtures of aliphatic carbonates. *The Journal of the Electrochemical Society,* Vol.146, 2, (1999), pp. 486-492, ISSN: 0013-4651.

Sun, X.G & Angell, C.A (2009). Doped sulfone electrolytes for high voltage Li-ion cell applications. *Electrochemistry Communications,* Vol.11, 7, (2009), pp. 1418-1421, ISSN: 1388-2481.

Varela, L.M.; Garcia, M.; Sarmiento, F.; Attwood, D. & Mosquera, V. (1997) Pseudolattice theory of strong electrolyte solutions. *Journal of Chemical Physics,* 107, 16, (1997), pp. 6415- 6419, ISSN 0021- 9606.

Varela, L.M.; Carretea, J. ; Garcí M. ; Gallegoa, L.J. ; Turmine, M. ; Riloc, E. & Cabezac, O (2010). Pseudolattice theory of charge transport in ionic solutions: Corresponding states law for the electric conductivity. *Fluid Phase Equilibria,* Vol.298, pp. 280–286,

Verma, P.; Maire, P. & Novák, P. (2010). A review of the features and analyses of the solid electrolyte interphase in Li-ion batteries. *Electrochimica Acta,* Vol.55, (2010), pp. 6332–6341. ISSN: 0013-4686.

Wakihara, M. (1998). Lithium Ion Batteries-Fundamentals and Performance, O. Yamamoto (Ed), Wiley-VCH, Berlin (1998).

Wang, Q.S. ; Sun, J.H. ; Yao, H.L. & Chen, C.H. (2005). Thermal stability of $LiPF_6$/EC + DEC electrolyte with charged electrodes for lithium ion batteries. *Thermochimica Acta* Vol.437, (2005), pp. 12-15 ISSN: 0040-6031.

Wang, H.; Yoshio, M.; Abe, T. & Ogumi, Z. (2002). Characterization of Carbon-Coated Natural Graphite as a Lithium-Ion Battery Anode Material. *The Journal of the Electrochemical Society,* 149, (2002), pp. A499, ISSN: 0013-4651.

Watanabe, Y.; Kinoshita, S.I., Wada S.; Hoshino K.; Morimoto H. & Tobishima, S.I. (2008). Electrochemical properties and lithium ion solvation behavior of sulfone-ester mixed electrolytes for high-voltage rechargeable lithium cells. *Journal of Power Sources,* Vol.179, 2, (2002), pp. 770-779, ISSN: 0378-7753.

Winter, M. & Besenhard, J.O. (1999). Electrochemical lithiation of tin and tin-based intermetallics and composites. *Electrochimica. Acta,* Vol. 45, (1999),pp. 31–50, ISSN: 0013-4686.

Wu, M.S; T.L. Liao, Y.Y. Wang & C.C. Wan (2004). Assessment of the wettability of porous electrodes for lithium-ion batteries. *Journal of Applied Electrochemistry,* Vol. 34, 8, (2004), pp. 797-805, SSN: 0021-891X.

Wu, Y.P.; Rahm, E. & Holze, R. (2003). Carbon anode materials for lithium ion batteries. *J. Power Sources,* Vol. 114, (2003), pp. 228-236, ISSN: 0378-7753.

Xu, K. (2004). Nonaqueous liquid electrolytes for lithium bases rechargeable batteries. *Chemical Review,* Vol. 104, 4303-4417, ISSN: 1520-6890.

Yang, H; Zhuang, G.V. & Ross, P.N. (2006). Thermal Stability of $LiPF_6$ Salt and Li-Ion Battery Electrolytes Containing $LiPF_6$. *Journal of Power Sources,* Vol.161 (1), (2006), pp. 573-579, ISSN: 0378-7753.

Zang, S.H. (2006). A review on electrolyte additives for lithium-ion batteries. *Journal of Power Sources*, Vol.162, (2006), pp. 1379-1394, ISSN: 0378-7753.

Zugmann, S; Moosbauer, D.; Amereller, M.; Schreiner, C.; Wudy, F.; Schmitz, R., Schmitz, R.; Isken, P.; Dippel, C.; Müller, R.; Kunze, M.; Lex-Balducci, A.; Winter, M. & Gores, H.J. (2011). Electrochemical characterization of electrolytes for lithium-ion batteries based on lithium difluoromono(oxalato)borate. *Journal of Power Sources*, Vol.196, (2011), pp. 1417-1424, ISSN: 0378-7753.

7

Redox Shuttle Additives for Lithium-Ion Battery

Lu Zhang, Zhengcheng Zhang and Khalil Amine
Chemical Sciences and Engineering Division, Argonne National Laboratory, Argonne, USA

1. Introduction

Overcharge of lithium-ion batteries can be dangerous. Overcharge generally occurs when a current is forced through a cell, and the charge delivered exceeds its charge-storing capability.[1-3] Overcharge of lithium-ion batteries can lead to the chemical and electrochemical reaction of batteries components, rapid temperature elevation, self-accelerating reactions, and even explosion.

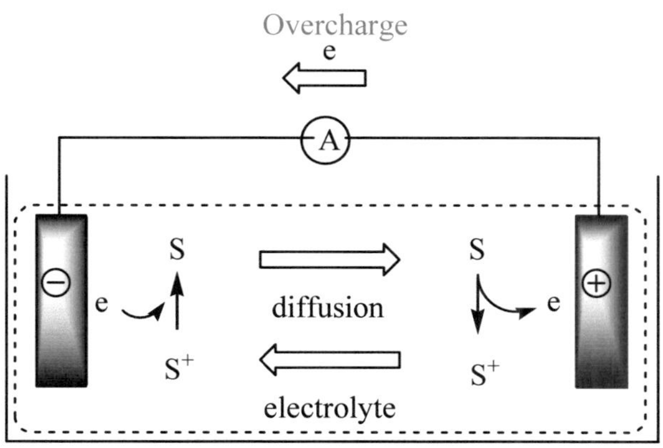

Fig. 1. Mechanism of redox shuttle overcharge protection

Redox shuttle additives have been proposed for overcharge protection of secondary lithium-ion batteries for decades.[4-8] Generally, the redox shuttle molecule can be reversibly oxidized and reduced at a defined potential slightly higher than the end-of-charge potential of the cathode. This mechanism can protect the cell from overcharge by locking the potential of the cathode at the oxidation potential of the shuttle molecules. The detailed mechanism is shown in Figure 1. On the overcharged cathode surface, the redox shuttle molecule (S) is oxidized to its (radical) cation form (S^+), which, via diffusion across the cell electrolyte, would be reduced back to its original or reduced state on the surface of the anode. The reduced form would then diffuse back to the cathode and oxidize again. The "oxidation-diffusion-reduction-diffusion" cycle can be repeated continuously due to the reversible nature of the redox shuttle to shunt the overcharge current. The redox shuttling mechanism at overcharge can be regarded as a controlled internal short, and the net result of the shuttling is to convert the overcharge electricity power into heat, which avoids the reactions that occur between the electrodes and electrolyte at high voltage. Redox shuttles can also be used for automatic capacity balancing during battery manufacturing and repair.

2. Brief history

The research on redox shuttles for overcharge protection of lithium-ion batteries can be traced back to the 1980s, when Behl et al. reported that I^-/I_2 (**1** in Figure 2) has its first oxidation potential at about 3.25 V vs. Li^+/Li and, hence, is suitable for overcharge protection of 3-V class lithium batteries.[9,10] Behl et al. did not fully evaluate the overcharge protection performance due to the lack of a proper cell system and high-voltage limit test procedures; nonetheless, I^-/I_2 represents the first example that showed promising overcharge protection by using an electrochemical quasi-reversible system in lithium cells. Soon thereafter, Behl evaluated Br/Br_2 (**2** in Figure 2) as a redox shuttle additive and obtained similar results, except for the higher oxidation potential deriving from the more compact electron configuration of the Br atom.[11]

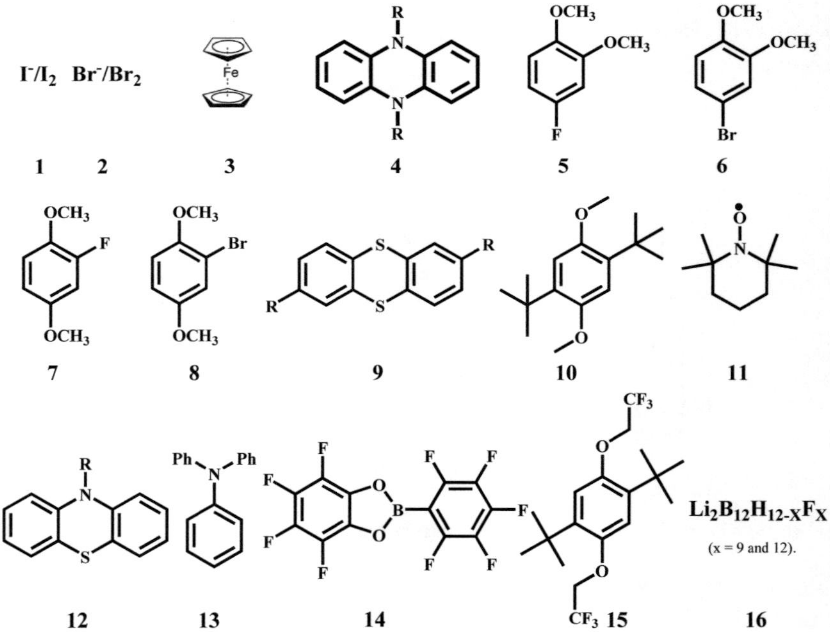

Fig. 2. Representative examples of redox shuttle additives

Later, organometallic ferrocene derivatives were extensively studied as redox shuttles[6,12] (**3** in Figure 2). Ferrocene can undergo reversible one-electron oxidation to form a stable cation, called "ferrocenium," and, therefore, is electrochemically reversible.[13] Some ferrocene derivatives, such as methoxymethylferrocene, carbomethoxyferrocene, carbamoylferrocene, dimethylaminomethylferrocene, and 1,l'-dimethylferrocene,[6,12] have been investigated to explore their overcharge protection performance. Most of them could provide long overcharge protection to lithium cells, and by tuning the substituents, upper voltage limits from 3.0 to 3.5 V vs. Li/Li^+ can be achieved.

Owing to the success with ferrocenes, due to the versatile design and feasible modification, organic systems have drawn more and more attention to the development of novel redox shuttles. In 1999, Delabouglise and co-workers first examined the dihydrophenazine system (**4** in Figure 2) as a redox shuttle.[14] The introduction of alkyl groups into the N atoms played a key role in enhancing the overcharge performance of the dihydrophenazine system.[14] Around the same time, Adachi and co-workers conducted a screening study, investigating a series of metallocene and dimethoxybenzene (**5-8** in Figure 2) derivatives.[15] Metallocene compounds were observed to be electrochemical reversible systems with potentials around 4 V, but the low solubility and large molecular size hindered their overcharge performance in electrolytes. Dimethoxybenzene derivatives, on the other hand, did show promising results, especially for those compounds with methoxy groups at the *ortho* and *para* positions. The substitution of halogen atoms could lift the redox potential to around 4 V. This was the first time that the dimethoxybenzene system was investigated as a redox shuttle candidate.

In 2002, Lee and co-workers discovered a series of thiantlurene derivatives (**9** in Figure 2) that showed potential as redox shuttles for lithium-ion batteries.[16] By introducing acetyl, alkyl, and halogen groups, they developed novel compounds that provide reversible redox potential higher than 4 V, making them applicable to overcharge protection for 4-V cathode materials.

In 2005, Dahn and co-workers screened numerous molecules with various substituents on the aromatic rings.[17,18] Among them 2,5-di-*tert*-butyl-1,4-dimethoxybenzene (later termed DDB) (**10** in Figure 2) stood out as having excellent overcharge protection performance. DDB is electrochemically reversible at 3.9 V vs. Li/Li^+ and can provide over 300 cycles of 100% overcharge per cycle for lithium-ion cells with $LiFePO_4$ as the cathode. The DDB was evaluated to confirm its superior performance in terms of high-rate overcharge protection[19] and overdischarge protection,[17] and a theoretical and mechanistic analysis for its high stability was also conducted.[18,20,21] Dahn's group later discovered another two stable redox shuttle systems, which are phenothiazine derivatives (**12** in Figure 2) and 2,2,6,6-tetramethylpiperinyl-oxide (TEMPO) (**11** in Figure 2) derivatives. Both systems work well in terms of long-time overcharge protection; for instance, use of 10-methylphenothiazine (MPT) in a cell system provided 153 cycles of 100% overcharge at the C/10 rate, and TEMPO survived 124 cycles. However, the relatively low potentials of these two systems (3.5~3.7 V vs. Li/Li^+) limit their applications in real batteries. Triphenylamine derivatives (**13** in Figure 2) were later investigated as potential redox shuttle candidates.[22] Even though those compounds exhibited reversible cyclic voltammetry signals and tunable redox potentials using various electronic substituents, they are not comparable to DDB in terms of either stability or redox potentials. Because of the lack of a high-potential electrochemical reversible system, the redox potential became a major limitation of the redox shuttles for lithium-ion battery technology. This obstacle remained unresolved until 2007, when Chen and co-workers developed a novel redox shuttle, 2-(pentafluorophenyl)-tetrafluoro-1,3,2-benzodioxaborole (PFPTFBB) (**14** in Figure 2), which has a high redox potential of 4.43 V vs. Li/Li^+.[23] Under various charging rates and aggressive conditions in cell testing with a $LiNi_{0.8}Co_{0.15}Al_{0.05}O_2$ cathode, PFPTFBB was able to survive after more than 160 cycles with 100% overcharge. PFPTFBB was the first stable redox shuttle that could be used for 4-V class cathodes. In addition, the incorporated boron center of this additive is a strong Lewis acid and can act as an anion receptor to dissolve LiF generated during the operation of lithium-ion batteries, making it a bifunctional electrolyte additive.

Another recent example is 1,4-di-*tert*-butyl-2,5-bis(2,2,2-trifluoroethoxy)benzene (**15** in Figure 2), developed by Dahn's group in 2009.[24] This shuttle is a high potential version of DDB and keeps the same redox center structure. The fluoride groups increase the potential to 4.2 V vs. Li/Li^+, high enough for the $LiCoO_2$ cathode. However, in half-cell tests with $LiCoO_2$, this molecule was only able to provide 46 cycles of 100% overcharge at C/10, and no graphite anode full-cell study was reported.

In 2010, Chen and Amine reported another high-potential redox shuttle system, lithium borate cluster salt, $Li_2B_{12}H_{12-x}F_x$ (x = 9 and 12) (**16** in Figure 2).[25] The redox potential of this molecule can be tuned by the degree of fluorination; therefore, it is possible to design the redox shuttle for various cathode materials. In addition, the form of this lithium salt makes it a bi-functional solute of the electrolyte for long-life and safe lithium-ion batteries.

3. Characteristics of ideal redox shuttles

From this brief history of the development of redox shuttles, some general ideas on what makes an idea redox shuttle have emerged.

First, redox shuttle molecules have to be electrochemically reversible, which is the most important requirement. From the redox shuttle mechanism shown in Figure 1, the oxidized form of the redox shuttle, i.e., the radical cation for the neutral compound or cation for the TEMPO-type compound has to be stable enough to survive the diffusion circle, which is also a determining factor of an electrochemically reversible system.

Second, the redox potential of the system must be slightly higher than the end of charge potential of the cathode materials, which determines whether or not the redox shuttle could be used for certain cathode materials. According to a previous study,[7,26] the redox potential of the shuttle additive should be about 0.3–0.4 V above the normal maximum operating potential of the cathode so that the cell can be normally charged before the shuttle molecule begins to function, thereby minimizing the self-discharge effect. Also, the potential should not exceed the electrochemical window of state-of-art electrolytes, i.e., 4.5 V *vs.* Li/Li^+;[27] otherwise, the electrolytes can be oxidized and lead to safety issues. Therefore, for state-of-art cathode materials, the typical potentials of redox shuttles should be tunable from 3.8 V (for $LiFePO_4$ olivine) to 4.5 V [for $LiMnO_4$ or $Li_{1.1}(Co_{1/3}Mn_{1/3}Ni_{1/3})_{0.9}O_2$] *vs.* Li/Li^+. However, with the high-voltage cathode materials and electrolytes recently emerging, redox shuttles with higher potentials (4.4 V to 4.9 V *vs.* Li/Li^+) are also greatly needed

Third, the electrochemical stability is of vital importance in determining the longevity of overcharge protection resulting from redox shuttles.[20,21,23,28] This stability actually depends upon the stability of the oxidative species (S^+ in Figure 1) of the redox shuttles generated during overcharge protection. However, evaluation of redox shuttle stability does not have a clear standard. Overcharge protection time and cycle numbers are mostly used to depict the overcharge longevity or stability. But considering most research uses various overcharge abuse test procedures and experimental setups to evaluate overcharge protection performance, it is difficult to compare redox shuttle performance from different research groups.[29] The variations of experimental details, such as cell types, amounts of electrolytes, formulation of electrolytes, concentrations of redox shuttles, charge rate, overcharge percentage, electrode loadings, electrode match-ups, use of other additives, and test temperatures, could all affect the evaluation results. To compare different redox shuttles, parallel tests under identical conditions have to be undertaken.

Finally, some other factors, such as a good solubility[19,30] and high diffusion coefficient of the redox shuttles in non-aqueous electrolytes, are also highly desired to maximize the shuttle molecule's mobility through the cell, therefore delivering high-current overcharge protection.

These aforementioned characteristics allow targeting the development of redox shuttle additives with superior overcharge protection; however, besides overcharge, the normal cell performance resulting from the addition of redox shuttles has rarely been investigated and evaluated. Actually, the normal cell performance is of equal, if not greater, importance compared with the overcharge performance. After all, it is the merits of the normal cell performance that enables application of lithium-ion technology to various power sources.

The effect of redox shuttle additives on normal cell performance, or the so-called "compatibility" to the cell system, should be carefully evaluated and established as standards for idea redox shuttles.[28] The compatibility can be determined for many different properties relative to different cell components. For instance, for the compatibility of the redox shuttle to the electrolyte solutions, the effects of redox shuttle solubility, conductivity, co-reaction, etc., have to be investigated and considered. For compatibility to electrode materials, the effects of the solid-electrolyte interface, interface resistance, co-reaction, and thermal property should be taken into account. As for the compatibility to other cell components, such as separators, current collectors, and binders, more studies may be needed to explore and set up standards for better redox shuttles. The characteristics regarding an ideal redox shuttle are summarized in Table 1 with respect to both overcharge protection and cell compatibility.

Overcharge performance factors	*Cell compatibility factors*
Electrochemical reversibility	Solubility in electrolytes
Redox potential	Effect on self-discharge
Electrochemical stability	Effect on conductivity
Molecular weight	Effect on impedance
Diffusion coefficient	Reactivity with cell components

Table 1. Characteristics for an ideal redox shuttle

4. Theoretical studies of redox shuttles

Development of redox shuttles with superior performance and excellent compatibility is of great importance to the lithium-ion battery technology. Various theoretical studies have been conducted to establish an efficient screening procedure for redox shuttles, including calculation of oxidation potentials[31] and estimation of electrochemical stability.[21] However, most theoretical results have been used to support experimental results and provide design guidance for development of novel redox shuttles. With few exceptions, the development of redox shuttle additives has been based on trial and error or experience. Reports on redox shuttles rarely state the design thinking or strategy that could help to build redox shuttles with tunable properties, such as potentials, electrochemical stabilities. A possible reason is that the redox shuttle design requires multi-discipline knowledge, including not only electrochemistry and lithium-ion battery technology, but also physical organic chemistry and organic synthesis. For example, researchers with electrochemical knowledge and battery experience may understand the requirements of redox shuttles, but owing to a lack of knowledge about organic chemistry, would encounter serious impediments to practical design of a new system.

As one of the pioneering groups, Dahn and co-workers did some systematic work to calculate the relation between chemical structure and redox potential.[31] The oxidation potentials of seventeen molecules used as candidate shuttle additives in Li-ion cells were calculated by density functional theory, and the results compared with experiment. Based on the correlation between the highest occupied molecular orbital (HOMO) energy and the oxidation potential, a good empirical relation was established (shown in equation 1), which can be used as a quick estimate of the oxidation potential for novel molecules:

$$E_{est}(\varepsilon) = -(\varepsilon / e) - 1.46 \text{ V} \quad (1)$$

where ε is the orbital energy in solution, in electron volts, and represents the energies of the molecule's HOMO; e is the electron charge; and $E_{est}(\varepsilon)$ is the estimated oxidation potential, in volts, relative to a Li/Li^+ reference electrode.

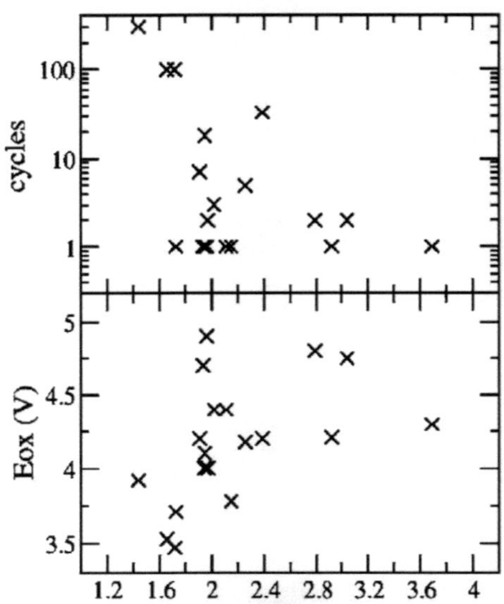

Fig. 3. Number of shuttle protected overcharge cycles sustained and the oxidation potentials of 19 shuttle molecules as a function of their binding energies with C_2H_5. (Reprinted from Journal of the Electrochemical Society, 2006, 153, A1922, Copyright (2006), with permission from the Electrochemical Society.)

Later, Dahn's group conducted another computational study to estimate the stability of redox shuttle additives,[21] which is of even more importance for development of redox shuttles because compared with redox potentials, the stability of redox shuttles is even harder to tune and improve. In their report, Dahn et al. employed the binding energy, E_b, between oxidized shuttle molecules and an ethyl radical, ER, to estimate the relative reactivity of oxidized shuttle molecules, E_{ox}. As shown in Figure 3, the smaller the value of

$E_b(ER)$, the more stable the redox shuttle. The calculated binding energies of 19 selected oxidized shuttle molecules were found to agree approximately with the experimentally measured stability. Even though this method still has some limitations and inconsistencies, for the first time, it provided a quick evaluation to predict redox shuttle stability.

The computation prediction method for the redox potential and stability of redox shuttles was applied to *tert*-butyl- and methoxy-substituted benzene derivatives,[32] which were chosen because of the earlier success of DDB. Both the oxidation potential and the stability of *tert*-butyl- and methoxy-substituted benzene molecules agreed well with the calculated results. Of the 43 molecules evaluated, DDB proved to be the most stable shuttle molecule suitable for $LiFePO_4$-based cells.

Chen and co-workers used molecular orbital theory to explain the stability of redox shuttles.[20] Their theoretical calculations clearly demonstrated that the π-π interaction between the aromatic ring and the substitution groups is critical to maximize the stability of radical cations. They also proposed that intermolecular polymerization is the major decomposing pathway, and that this source of instability can be minimized by full substitution on the aromatic ring.

Aromatic theory was later applied to explain the stability of the dimethoxybenzene system.[1] Aromaticity is a chemical property in which a conjugated ring of unsaturated bonds, lone pairs, or empty orbitals exhibits stabilization stronger than would be expected by the stabilization of the conjugation alone. It can also be considered a manifestation of cyclic delocalization and resonance.[33] A cyclic ring molecule follows Hückel's rule when the number of its π-electrons equals $4n+2$; where n is zero or any positive integer, the molecule can be regarded as an aromatic system. In the case of dimethoxybenzene, the lone pairs of the two oxygen atoms can participate in the conjugated benzene π-electrons, making 10 delocalized electrons and thus forming a large stable conjugated system. This process partially explains the excellent stability of DDB. The *tert*-butyl groups in DDB also play an important role in stabilizing the radical cation.[17] The steric hindrance effects prevent the active radical cation from intermolecular annihilation, leading to enhanced longevity of overcharge protection.

5. Rational design of redox shuttles with tunable properties

5.1 Overcharge protection performance

For an ideal redox shuttle, the desirable additive needs to be tunable with regard to many properties (Table 1), to accommodate different cell chemistries, and to resolve issues that not only could occur during overcharge protection, but also during normal operation. For instance, to minimize the self-discharge effect for different cathode materials, tunable redox potentials are desired and, ideally, should be 0.3~0.4 V higher than the end-of-charge potential. The electrochemical stability is also important when designing a redox shuttle; therefore, factors like electron donating effect or steric hindrance should be considered.

In 2005 Caudia et al. developed DDB (**10** in Figure 2) on the basis of a screening selection,[34] and this molecule stood out as possessing many unique and desirable features. It is built on the benzene ring with two methoxy groups helping to stabilize the radical cation that could be produced during the overcharge process, and two tert-butyl groups were attached to the benzene ring to protect the radical cation from intermolecular annihilation.[20,28] With this design, DDB displays perfect electrochemical reversibility at 3.9 V *vs.* Li/Li+ and provides over 200 cycles of overcharge protection to the lithium-ion cells using $LiFePO_4$ cathodes. DDB meets most of the aforementioned characteristics in terms of overcharge performance and is very close to practical application. DDB has a simple chemical structure, leaving ample room for organic modification. In addition, the benzene ring is known as the smallest hydrocarbon conjugated system with the least electron density so that, compared with other heterocyclic aromatic systems, it has the most chance to achieve high oxidation potential, a favorable feature for the design of novel redox shuttles. Therefore, DDB provides an excellent starting point from which to tune properties, such as redox potential, and cell compatibility.

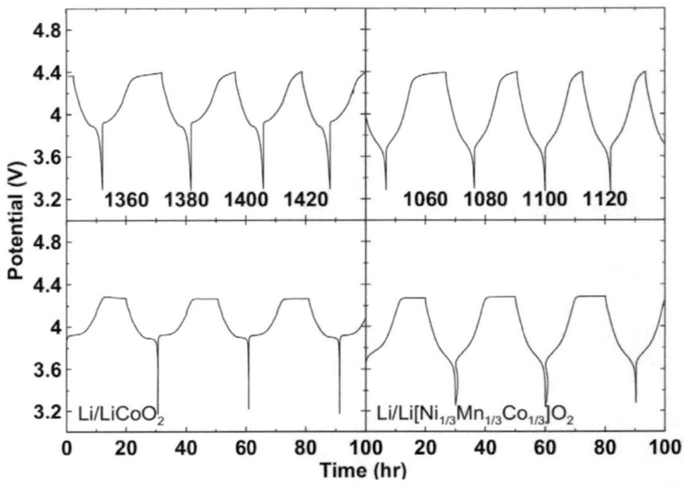

Fig. 4. Coin cell data for $Li/LiCoO_2$ cells (left) and $Li/Li(Ni_{1/3}Mn_{1/3}Co_{1/3})O_2$ cells (right) containing 0.1 M 1,4-di-*tert*-butyl-2,5-bis(2,2,2-trifluoroethoxy)benzene in an electrolyte composed of 0.5 M $LiPF_6$ in propylene carbonate (PC): dimethyl carbonate (DMC): ethylene carbonate (EC): diethyl carbonate (DEC) in a 1:2:1:2 ratio by volume. Cells were charged at C/10 rate for 20 h and then discharged at C/10 rate. (Reprinted from Journal of the Electrochemical Society, 2009, 156, A309., Copyright (2009), with permission from the Electrochemical Society.)

DDB is electrochemically reversible at 3.9 V, which is high enough for the $LiFePO_4$ cathode but not for 4-V class cathode materials. Therefore, increasing the redox shuttle of this molecule is of great importance. According to molecular orbital theory, increasing the oxidation potential means lowering the energy or electron density of the HOMO orbital.[35] A recent example was reported in 2009,[24] when the potential of 1,4-di-*tert*-butyl-2,5-bis(2,2,2-trifluoroethoxy)benzene (**15** in Figure 2) was increased to 4.25 V (*vs.* Li/Li+) by introducing electron-withdrawing trifluoroethyl groups. As shown in Figure 4, this new redox shuttle provided overcharge protection for $Li/LiCoO_2$ and $Li/Li(Ni_{1/3}Mn_{1/3}Co_{1/3})O_2$ cells for around 40 cycles. This report signified a promising direction for development of novel redox shuttles: that is, use of a DDB platform to retain the excellent electrochemical properties and introduction of electron withdrawing groups to raise the potential. However, because the incorporated trifluoroethyl groups are not directly attached to the conjugated ring, the resultant potential is still not very high, especially for the new high-energy 5-V materials.[36,37]

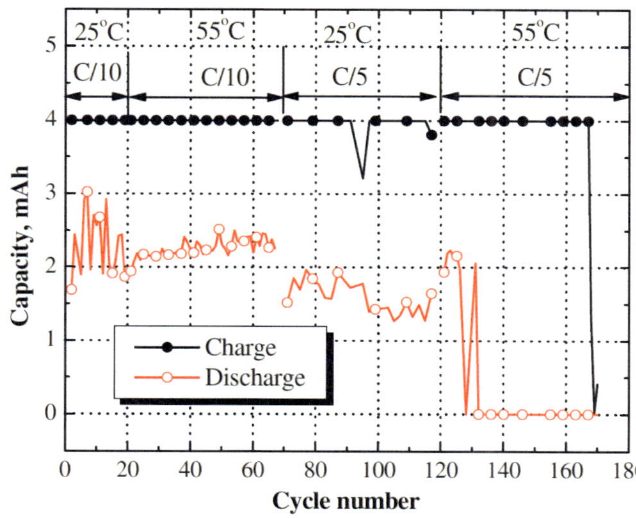

Fig. 5. Charge and discharge capacity of graphite/$LiNi_{0.8}Co_{0.15}Al_{0.05}O_2$ cell during the whole course of overcharge test. The electrolyte used contained 5 wt% PFPTFBB. (Reprinted from Electrochem. Commun., 9: 703–707, Copyright (2007), with permission from Elsevier).

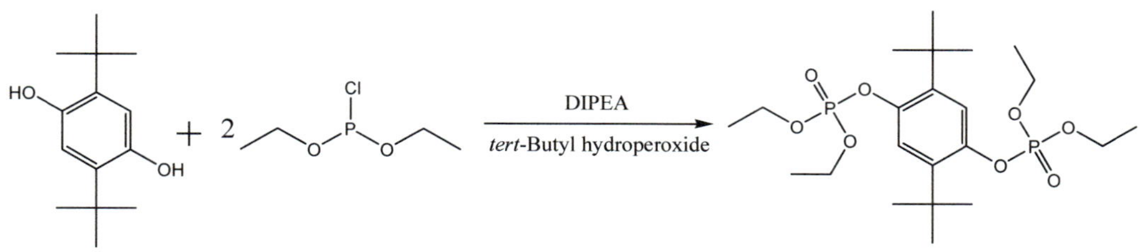

Fig. 6. Synthesis route of tetraethyl-2,5-di-*tert*-butyl-1,4-phenylene diphosphate (TEDBPDP).

Another promising molecule using fluorine groups to increase the redox potential is 2-(pentafluorophenyl)-tetrafluoro-1,3,2-benzodioxaborole (PFPTFBB) (**14** in Figure 2).[23] This shuttle is also based on the dimethoxybenzene platform, but the two methoxy groups are in the ortho positions. As shown in Figure 2, PFPTFBB was fully substituted by fluorine atoms. The fluorine atoms were directly attached to the conjugated ring and increased the reversible potential to 4.46 V vs. Li/Li^+, high enough for most 4-V cathode materials. In addition, the fully substituted structure made this molecule stable against possible polymerization during the overcharge protection process; therefore, PFPTFBB can provide long-time overcharge protection. For instance, as shown in Figure 5, 5 wt% PFPTFBB in the electrolyte can provide more than 150 overcharge cycles, even under elevated temperature and at a high charge rate. However, this molecule is extremely difficult to synthesize and is not chemically stable against moisture.[38]

Even though 1,4-di-*tert*-butyl-2,5-bis(2,2,2-trifluoroethoxy)benzene and PFPTFBB achieved increased oxidation potentials, the potentials of these shuttle molecules are still not high enough to be used for cathode materials with potentials higher than 4.4 V. With numerous high-voltage and high-energy cathode materials closer to application as a large-scale power source,[36,37,39,40] the potential of the corresponding redox shuttle additives must be increased.

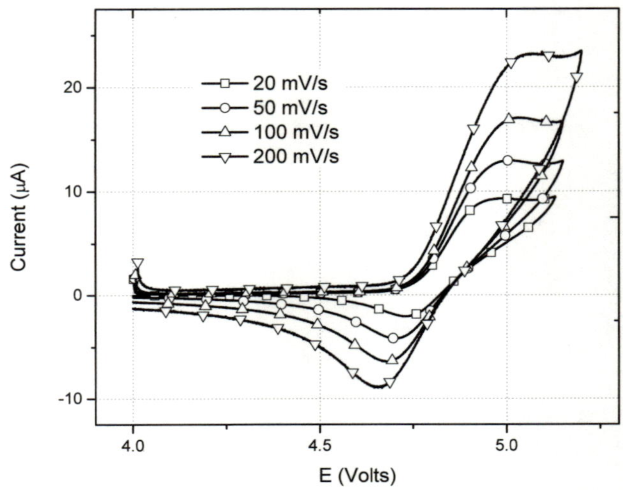

Fig. 7. Cyclic voltammogram of 1.2 M $LiPF_6$ in EC/DEC (3:7 by weight) with 0.01M TEDBPFP at various rates using a Pt/Li/Li three-electrode system.

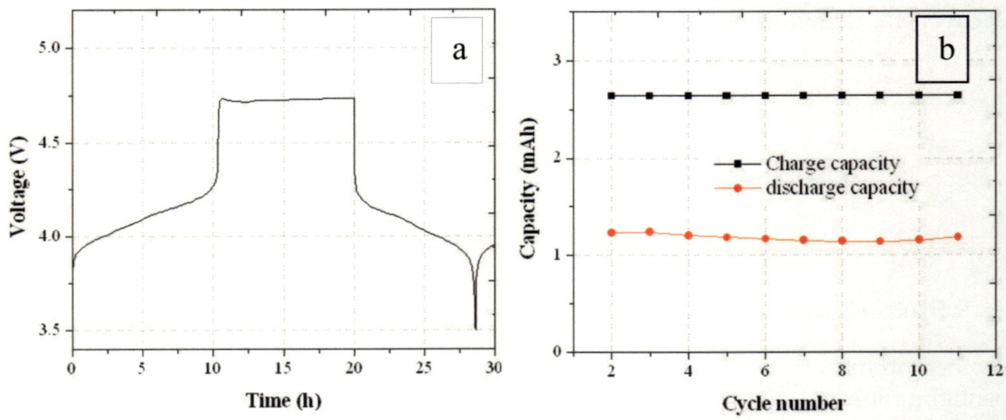

Fig. 8. a) Representative cell voltage *vs.* time for $LiMn_2O_4$/Li cell containing 5 wt% TEDBPDP; b) cell charge-discharge capacities vs. cycle number in overcharge test. Cells were charged at C/10 rate for 20 h and then discharged at C/10 rate.

For example, a new high-energy cathode material developed at Argonne National Laboratory[36,37] requires redox shuttle additives having potential higher than 4.9 V *vs.* Li/Li^+. Recently, Zhang et al. at Argonne developed a novel organophosphate functionalized redox shuttle molecule, tetraethyl-2,5-di-*tert*-butyl-1,4-phenylene diphosphate (TEDBPDP), which can provide overcharge protection for high-voltage cathode materials.[41] Figure 6 shows the reaction for the synthesis of TEDBPDP. The structure of TEDBPDP was also based upon the dimethoxybenzene ring platform. Instead of using fluorine atoms, however, organic phosphate groups were attached to the conjugated system to increase the redox potential. Belfield et al. had reported that organic phosphate groups are strong electron-withdrawing units in organic chemistry.[42] In addition, organic phosphate compounds have been studied as flame retardant additives for lithium-ion batteries,[43-47] bringing an additional benefit of this novel redox shuttle. As shown in Figure 7, the redox potential of TEDBPDP, obtained from the mean of the anodic and cathodic potentials $((E_a + E_c)/2)$, is 4.80 V *vs.* Li/Li$^+$, indicating the strong effect of the incorporated organic phosphate groups. To the best of our knowledge, 4.80 V is the highest redox shuttle potential ever reported in the literature. The oxidation potential of TEDBPDP is high enough for most >4.2 V cathode materials.

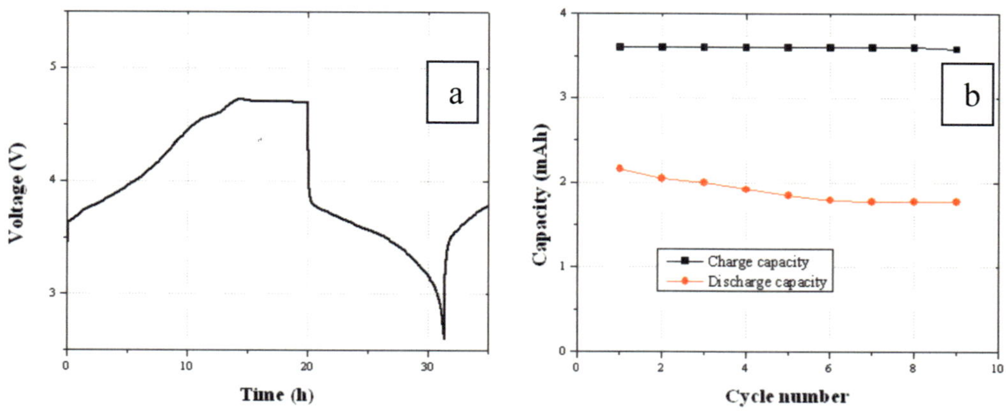

Fig. 9. a) Representative cell voltage *vs.* time for Li/$Li_{1.2}Ni_{0.15}Co_{0.1}Mn_{0.55}O_2$ cell containing 5 wt% TEDBPDP; b) cell charge-discharge capacities *vs.* cycle number in overcharge test. Cells were charged at C/10 rate for 20 h and then discharged at C/10 rate.

Figure 8 shows the overcharge performance of Li/$LiMn_2O_4$ cells containing 5 wt% TEDBPDP in the 1.2 M $LiPF_6$ electrolyte. As indicated by Figure 8(a), during the charge, lithium ion was removed from the $LiMn_2O_4$ cathode and reduced on the Li metal anode. The normal charge took place at about 3.9–4.3 V. After the cell was fully charged, the voltage climbed up quickly to 4.75 V, where the redox shuttle was activated. The flat plateau at 4.75 V indicates that the overcharge current was manipulated by the TEDBPDP molecules. At 100% overcharge, the amount of overcharge can be further increased, and TEDBPDP can provide more than 10 cycles of overcharge protection before the shuttle molecule becomes ineffective, as illustrated in Figure 8(b).

Figure 9 shows the results from an overcharge test using a Li/$Li_{1.2}Ni_{0.15}Co_{0.1}Mn_{0.55}O_2$ cell containing 5 wt% TEDBPDP in the electrolyte. This cathode required formation cycles between 2.8 V and 4.6 V before the overcharge tests. As shown in Figure 9(a), the normal charge took place starting at about 3.7 V and continued until the voltage reached 4.75 V, where the redox shuttle mechanism was activated. Even though the cell was not in an overcharge state due to the high cut-off working potential of this cathode material, the plateau at 4.75 V is a clear indication of the TEDBPDP overcharge protection. As indicated by Figure 9(b), the average capacities for charge are nearly twice those of discharge. The results demonstrate a successful example for using TEDBPDP to provide overcharge protection for a high potential cathode.

5.2 Cell compatibility

Another direction for redox shuttle improvement involves compatibility with the lithium-ion cell system. Take DDB as an example. It excels at overcharge performance in terms of suitable potential for $LiFePO_4$, length of overcharge protection, etc. However, DDB is not quite compatible with the lithium-ion cell system; specifically, it is not soluble with the conventional carbonate-based electrolytes.[48] For instance, in an electrolyte of 1.2 M $LiPF_6$ in ethylene carbonate (EC) and ethyl methyl carbonate (EMC) at a weight ratio of 3:7 (Gen 2 electrolyte), DDB can dissolve up to only 0.08 M, which is not enough in most cases. The literature studies on this molecule for overcharge protection of lithium-ion batteries mostly employ a special formulated electrolyte with reduced lithium salt concentration, such as 0.5 M lithium bis(oxalato)borate (LiBOB) in propylene carbonate (PC): dimethyl carbonate (DEC) at a volume ratio of 1:2.[34,48-50] This different formulation of the electrolyte sacrifices cell performance due to a lower conductivity, which is not acceptable for the battery industry.

Quite a few strategies have been pursued to improve the dimethoxybenzene-based redox shuttle.[28,30,51-53] For instance, to improve the solubility of DDB in carbonate-based electrolyte, Zhang et al. at Argonne[28] and others[30] have investigated an asymmetric structure design (ANL-1 in Figure 10) to create intramolecular dipole moments that facilitate the dissolution of the redox shuttle in the electrolyte. However, while successful in improving solubility, those asymmetric redox shuttles appear to be less electrochemically stable because the electronic structure of the redox center is disturbed and certain chemical bonds are weakened within the structure, which, as a result, does not provide long enough overcharge protection.

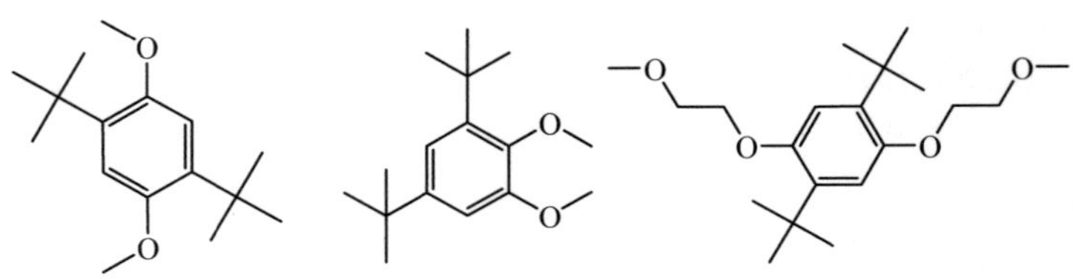

DDB ANL-1 ANL-2

Fig. 10. Chemical structures of dimethoxybenzene-based redox shuttles

Recently, a new strategy was adopted to develop a novel redox shuttle (ANL-2) that is not only soluble in carbonate-based electrolyte but also retains the excellent overcharge performance of DDB.[54-56] ANL-2 was developed by introducing oligo ether groups, such as oligo(ethylene glycol) (OEG), into the symmetric dimethoxy-di-*tert*-benzene platform. Oligo ether groups contain repeating ether units, which are well known to be soluble in polar solutions, such as water. The introduction of OEG into the redox shuttle structure significantly improved the solubility in polar carbonate electrolytes. For instance, ANL-2 redox shuttle can dissolve in Gen 2 electrolyte up to 0.4 M. More important, the modification does not break the symmetric structures of the molecules and, therefore, maintains the essential electrochemical properties.

Figure 11 shows voltage profiles of mesocarbon microbead (MCMB)/$LiFePO_4$ cells containing 0.4 M ANL-2 in Gen 2 electrolyte during 960 hours of an overcharge test. The charge rate was set to C/2, which is identical or even greater than the practical charge rate for lithium-ion batteries. Under this aggressive condition, the ANL-2 redox shuttle still worked well and provided around one-thousand hours of overcharge protection, which is more than 180 cycles (see Figure 12). This is the first redox shuttle ever reported that can work well in Gen 2 electrolyte and provide more than 180 cycles at an overcharge ratio of 100% and C/2 charge rate. Compared with DDB, the ANL-2 redox shuttle has comparable overcharge protection performance but much improved solubility in carbonate-based electrolytes, making it more applicable to lithium-ion batteries.

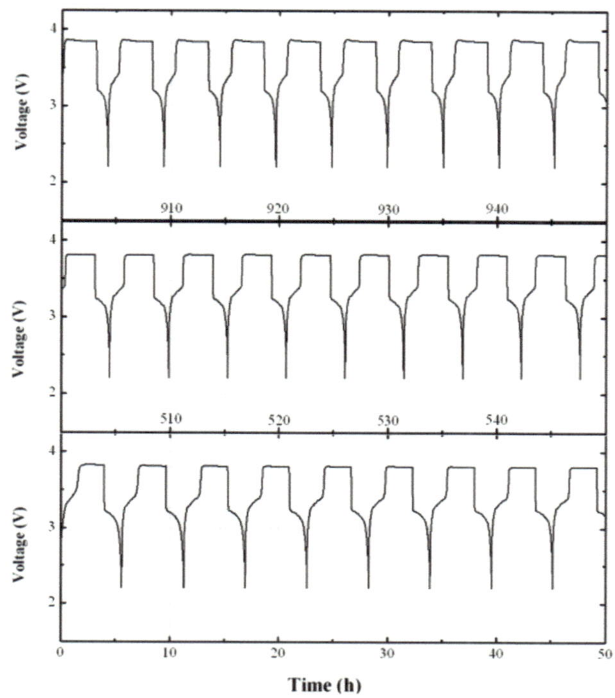

Fig. 11. Voltage profiles of MCMB/$LiFePO_4$ cells containing 0.4 M ANL-2 in Gen 2 electrolyte during 960-h overcharge test. Charging rate is C/2, and overcharge ratio is 100%.

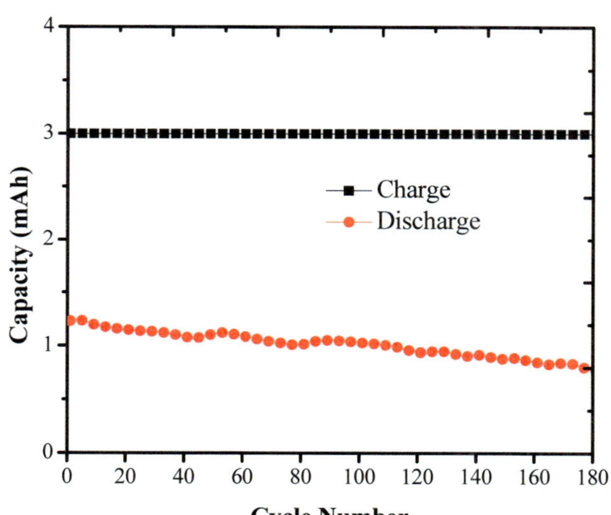

Fig. 12. Capacity retention profiles of MCMB/$LiFePO_4$ containing 0.4 M ANL-2 in Gen 2 electrolyte. Charging rate is C/2, and overcharge ratio is 100%.

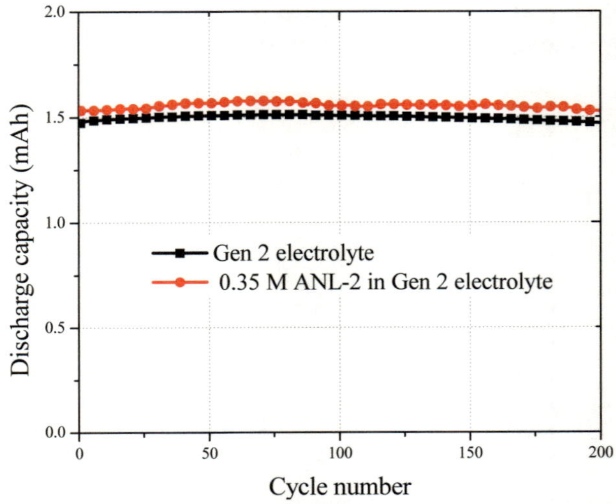

Fig. 13. Capacity retention profiles of MCMB/$LiFePO_4$ cells containing Gen 2 electrolyte with and without 0.35 M ANL-2 additive. Charging rate is C/3, and cut-off voltage is 2.3 ~3.6 V.

Figure 13 shows the discharge capacity retention of MCMB/$LiFePO_4$ cells containing Gen 2 electrolyte with and without 0.35 M ANL-2. The charge rate is set to C/3, and cut-off voltages are 2.3~3.6 V. As shown in Figure 13, the discharge capacity retention profiles for the cells with the two different electrolytes are similar, indicating that the ANL-2 additive does not degrade cell performance. This is the first redox shuttle example ever reported in the literature that not only excels at overcharge protection but also is compatible with state-of-art lithium-ion cell components. The ANL-2 redox shuttle could thus be used directly in current lithium-ion batteries.

6. Conclusion

Redox shuttle additives have been extensively studied due to their unique features and promising applications in lithium-ion batteries. This review of the literature covers their intrinsic chemical overcharge protection mechanism as determined by both theoretical and experimental studies, the correlation between chemical structure and overcharge performance, and development of novel redox shuttles with tunable properties that are related to overcharge performance and normal cycle performance.

7. References

[1] Chen, Z.; Qin, Y.; Amine, K. *Electrochimica Acta* 2009, *54*, 5605.

[2] Zhang, S. S. *Journal of Power Sources* 2006, *162*, 1379.

[3] Zhang, Z. C.; Zhang, L.; Schlueter, J. A.; Redfern, P. C.; Curtiss, L.; Amine, K. *Journal of Power Sources* 2010, *195*, 4957.

[4] Abraham, K. M.; Pasquariello, D. M.; Willstaedt, E. B. In *Journal of the Electrochemical Society* Journal of the Electrochemical Society 1990; Vol. 137, p 1856.

[5] Jonsson, M.; Lind, J.; Reitberger, T.; Eriksen, T. E.; Merenyi, G. *The Journal of Physical Chemistry* 1993, *97*, 11278.

[6] Golovin, M. N.; Wilkinson, D. P.; Dudley, J. T.; Holonko, D.; Woo, S. *Journal of The Electrochemical Society* 1992, *139*, 5.
[7] Thomas, J. R.; Philip N. Ross, Jr. *Journal of The Electrochemical Society* 1996, *143*, 3992.
[8] Demmano, G.; Selegny, E.; Vincent, J. C. *Bioelectroch Bioener* 1996, *40*, 239.
[9] Behl, W. K.; Chin, D.-T. *Journal of The Electrochemical Society* 1988, *135*, 16.
[10] Behl, W. K.; Chin, D.-T. *Journal of The Electrochemical Society* 1988, *135*, 21.
[11] Behl, W. K. *Journal of The Electrochemical Society* 1989, *136*, 2305.
[12] Abraham, K. M.; Pasquariello, D. M.; Willstaedt, E. B.; 6 ed.; ECS: 1990; Vol. 137, p 1856.
[13] Stepnicka, P. *Ferrocenes : ligands, materials and biomolecules*; J.Wiley: Chichester, England; Hoboken, NJ, 2008.
[14] Tran-Van, F.; Provencher, M.; Choquette, Y.; Delabouglise, D. *Electrochimica Acta* 1999, *44*, 2789.
[15] Adachi, M.; Tanaka, K.; Sekai, K. *Journal of the Electrochemical Society* 1999, *146*, 1256.
[16] Lee, D. Y.; Lee, H. S.; Kim, H. S.; Sun, H. Y.; Seung, D. Y. *Korean J Chem Eng* 2002, *19*, 645.
[17] Chen, J.; Buhrmester, C.; Dahn, J. R. *Electrochemical and Solid-State Letters* 2005, *8*, A59.
[18] Moshurchak, L. M.; Buhrmester, C.; Dahn, J. R. *Journal of the Electrochemical Society* 2005, *152*, A1279.
[19] Dahn, J. R.; Jiang, J. W.; Moshurchak, L. M.; Fleischauer, M. D.; Buhrmester, C.; Krause, L. J. *Journal of the Electrochemical Society* 2005, *152*, A1283.
[20] Chen, Z.; Wang, Q.; Khalil, A. *Journal of The Electrochemical Society* 2006, *153*, A2215.
[21] Wang, R. L.; Dahn, J. R. *Journal of the Electrochemical Society* 2006, *153*, A1922.
[22] Moshurchak, L. M.; Buhrmester, C.; Dahn, J. R. *Journal of the Electrochemical Society* 2008, *155*, A129.
[23] Chen, Z.; Amine, K. *Electrochemistry Communications* 2007, *9*, 703.
[24] Moshurchak, L. M.; Lamanna, W. M.; Bulinski, M.; Wang, R. L.; Garsuch, R. R.; Jiang, J. W.; Magnuson, D.; Triemert, M.; Dahn, J. R. *Journal of the Electrochemical Society* 2009, *156*, A309.
[25] Chen, Z.; Liu, J.; Jansen, N. A.; GirishKumar, G.; Casteel, B.; Amine, K. *Electrochemical and Solid-State Letters* 2010, *13*, A39.
[26] Narayanan, S. R.; Surampudi, S.; Attia, A. I.; Bankston, C. P. *Journal of The Electrochemical Society* 1991, *138*, 2224.
[27] Xu, K. *Chemical Reviews* 2004, *104*, 4303.
[28] Zhang, Z.; Zhang, L.; Schlueter, J. A.; Redfern, P. C.; Curtiss, L.; Amine, K. *Journal of Power Sources* 2010, *195*, 4957.
[29] Moshuchak, L. M.; Bulinski, M.; Lamanna, W. M.; Wang, R. L.; Dahn, J. R. *Electrochemistry Communications* 2007, *9*, 1497.
[30] Feng, J. K.; Ai, X. P.; Cao, Y. L.; Yang, H. X. *Electrochemistry Communications* 2007, *9*, 25.
[31] Wang, R. L.; Buhrmester, C.; Dahn, J. R. *Journal of the Electrochemical Society* 2006, *153*, A445.
[32] Wang, R. L.; Moshurchak, L. M.; Lamanna, W. M.; Bulinski, M.; Dahn, J. R. *Journal of The Electrochemical Society* 2008, *155*, A66.
[33] Schleyer, P. v. R. *Chemical Reviews* 2001, *101*, 1115.
[34] Claudia, B.; Jun, C.; Lee, M.; Junwei, J.; Richard Liangchen, W.; Dahn, J. R. *Journal of The Electrochemical Society* 2005, *152*, A2390.
[35] Goodenough, J. B.; Kim, Y. *Chemistry of Materials* 2009, *22*, 587.

[36] Deng, H.; Belharouak, I.; Sun, Y.-K.; Amine, K. *Journal of Materials Chemistry* 2009, *19*, 4510.

[37] Deng, H.; Belharouak, I.; Wu, H.; Dambournet, D.; Amine, K. *Journal of The Electrochemical Society* 2010, *157*, A776.

[38] Weng, W.; Zhang, Z.; Schlueter, J. A.; Redfern, P. C.; Curtiss, L. A.; Amine, K. *Journal of Power Sources* 2011, *196*, 2171.

[39] Park, S. H.; Kang, S. H.; Johnson, C. S.; Amine, K.; Thackeray, M. M. *Electrochemistry Communications* 2007, *9*, 262.

[40] Sun, Y.-K.; Myung, S.-T.; Park, B.-C.; Prakash, J.; Belharouak, I.; Amine, K. *Nat Mater* 2009, *8*, 320.

[41] Zhang, L.; Zhang, Z.; Wu, H.; Amine, K. *Energy & Environmental Science* 2011.

[42] Belfield, K. D.; Chinna, C.; Schafer, K. J. *Tetrahedron Letters* 1997, *38*, 6131.

[43] Xiang, H. F.; Xu, H. Y.; Wang, Z. Z.; Chen, C. H. *Journal of Power Sources* 2007, *173*, 562.

[44] Izquierdo-Gonzales, S.; Li, W.; Lucht, B. L. *Journal of Power Sources* 2004, *135*, 291.

[45] Xu, K.; Zhang, S.; Allen, J. L.; Jow, T. R. *Journal of The Electrochemical Society* 2002, *149*, A1079.

[46] Ma, Y.; Yin, G.; Zuo, P.; Tan, X.; Gao, Y.; Shi, P. *Electrochemical and Solid-State Letters* 2008, *11*, A129.

[47] Feng, J. K.; Cao, Y. L.; Ai, X. P.; Yang, H. X. *Electrochimica Acta* 2008, *53*, 8265.

[48] Dahn, J. R.; Junwei, J.; Moshurchak, L. M.; Fleischauer, M. D.; Buhrmester, C.; Krause, L. J. *Journal of The Electrochemical Society* 2005, *152*, A1283.

[49] Moshurchak, L. M.; Buhrmester, C.; Dahn, J. R. *Journal of The Electrochemical Society* 2005, *152*, A1279.

[50] Jun, C.; Claudia, B.; Dahn, J. R. *Electrochemical and Solid-State Letters* 2005, *8*, A59.

[51] Weng, W.; Zhang, Z.; Redfern, P. C.; Curtiss, L. A.; Amine, K. *Journal of Power Sources* 2010, *196*, 1530.

[52] Moshurchak, L. M.; Lamanna, W. M.; Mike, B.; Wang, R. L.; Rita, R. G.; Junwei, J.; Magnuson, D.; Matthew, T.; Dahn, J. R. *Journal of The Electrochemical Society* 2009, *156*, A309.

[53] Weng, W.; Zhang, Z.; Schlueter, J. A.; Redfern, P. C.; Curtiss, L. A.; Amine, K. *Journal of Power Sources* 2010, *196*, 2171.

[54] Amine, K.; Zhang, L.; Zhang, Z.; *DOE HYDROGEN and FUEL CELLS PROGRAM and VEHICLE TECHNOLOGIES PROGRAM ANNUAL MERIT REVIEW and PEER EVALUATION MEETING*. 2011.

[55] Krumdick, G. *DOE HYDROGEN and FUEL CELLS PROGRAM and VEHICLE TECHNOLOGIES PROGRAM ANNUAL MERIT REVIEW and PEER EVALUATION MEETING* 2011.

[56] Zhang, L.; Zhang, Z.; Amine, K. *unpublished* 2011.

8

Diagnosis of Electrochemical Impedance Spectroscopy in Lithium-Ion Batteries

Quan-Chao Zhuang[1], Xiang-Yun Qiu[1,2], Shou-Dong Xu[1,2],
Ying-Huai Qiang[1] and Shi-Gang Sun[3]
[1]Li-ion Batteries Lab, School of Materials Science and Engineering,
[2]School of Chemical Engineering and Technology,
China University of Mining and technology, Xuzhou
[3]State Key Laboratory of Physical Chemistry of Solid Surfaces,
Department of Chemistry, College of Chemistry and Chemical Engineering,
Xiamen University, Xiamen,
China

1. Introduction

The establishment of Electrochemical Impedance Spectroscopy (EIS), sometimes called AC impedance spectroscopy, has been initiated from 1880 to about 1900 through the extraordinary work of Oliver Heaviside. The EIS has been developed basically in the field of wet electrochemistry. In particular, the names of Sluyters, Sluyters-Rehbach[90, 91] (as well as many others) are strongly linked to this research field. Nowadays, there are no doubts that EIS has become a powerful tool for the analysis of complex processes (such as corrosion) that are influenced by many variables with regard to electrolyte, materials and interfacial geometry. The power of the technique arises from[55]: (*i*) it is a linear technique and hence the results are readily interpreted in terms of Linear Systems Theory; (*ii*) if the measurements are over an infinite frequency range, the impedance (or admittance) contains all of the information that can be gleaned from the system by linear electrical perturbation/response techniques; (*iii*) the experimental efficiency (amount of information transferred to the observer compared to the amount produced by the experiment) is particularly high; (*iv*) the validity of the data is readily determined using integral transform techniques (the Kramers–Kronig transforms) which are independent of the physical processes involved.

EIS data are often interpreted by using electrical equivalent circuits (EECs). However, the EECs are merely analogs, rather than models. But we use the EECs as a model to describe an electrochemical reaction that takes place at the electrode/electrolyte interface. Therefore they must be used with great care. The priority is to examine whether or not an EEC can indeed be appropriate as a model for the reaction at the electrified interface. It is worthwhile noting that the ultimate goal of using EIS is to characterize the mechanism of the charge transfer reaction, and thus the mere development of an analog (which is now done by various computer programs) represents an incomplete analysis of the data. For the parametric identification, the complex nonlinear least-squares (CNLS) method is used. The method is based on fitting object model parameters to the impedance spectrum[60].

In the past decades, EIS has been extensively used in the analysis of lithium battery systems, especially to predict the behavior of batteries, and to determine the factors limiting the performance of an electrode including its conductivity[75, 126], charge-transfer properties[11, 103, 117-119], properties of the passivating layer, etc. Numerous recent studies have been published on various aspects of the insertion electrodes used in lithium-ion batteries for the attempt to understand the origin of the observed capacity loss during extended cell cycling or storage[1, 30, 37, 52, 66, 78, 97].

EIS is now described in the general books on electrochemistry[9, 18, 27, 33, 56, 107], specific books[13, 59, 77] on EIS, and also by numerous articles and reviews[17, 22, 41-43, 55, 57, 87]. It became very popular in the research and applied chemistry. If a novice to EIS interested in the subject matter of this fundamental review, he/she may be strongly recommend to read the superb EIS text by Macdonald[13, 59] and Orazem et al.[77], along with a compilation of excellent research articles[24, 58, 60, 79, 121], in order to obtain a good grounding in the fundamental principles underpinning the EIS technique, as this rudimentary information is beyond the scope of this specialized review paper. The purpose of this chapter is to review recent advances and applications of EIS in the development of the kinetic model for the lithium insertion/extraction into intercalation materials and the lithiation/delithiation of simple binary transition metal compounds which can achieve reversible lithium storage through a heterogeneous conversion reaction. The typical impedance spectra and the ascription of each time constant of EIS spectra are discussed based on analyzing the potential and temperature dependence of the common EIS features. The potential and temperature dependence of the kinetic parameters, such as the charge transfer resistance, the electronic resistance of activated material, the resistance of lithium ions transferring through SEI film, the Schottky contact resistance, are also discussed based on the theoretical analysis. Moreover, the influences of non-homogeneous, multilayered porous microstructure of the intercalation electrode on the EIS feature and the kinetic parameters are discussed based on the experimental results and theoretical analysis. Finally, the inductance formation mechanism is reviewed.

2. Kinetic models for lithium ions insertion into the intercalation electrode

The intercalation electrode reaction[100, 111]:

$$xA^+ + xe^- + \langle H \rangle \Leftrightarrow A_x^+ \langle H \rangle^{x-} \tag{2-1}$$

is a special redox reaction, in which A^+ is the cation (e.g. Li^+, H^+, etc.) in the electrolyte, $<H>$ represents the host molecule (e.g. TiS_2, WO_3, etc.), $A_x<H>$ is the nonstoichiometric intercalate produced. The reactions are characterized by the reversible insertion of guest species into a host lattice without significant structural modification of the host in the course of intercalation and deintercalation. The intercalation reactions are widely used in lithium-ion batteries and in the electrochemical synthesis of intercalates.

It can be seen that the electrode composition changes during an intercalation electrode reaction, which is different from the usual first kind metal electrode reaction occurring only by electron transfer at electrode/electrolyte interfaces. The sequence of transport processes of the cations involves lithium ions transport process, the electron transport process, and the charge transfer process. Due to the differences in their time constants, EIS is a suitable technique to investigate these reactions and can allow us to separate most of these phenomena. Therefore, using EIS to analyze the kinetic parameters related to lithium ions insertion/extraction process in intercalation materials such as the SEI film resistance, charge transfer resistance, as well as the relationship between the kinetic parameters and the potential or temperature, is helpful to understand the reaction mechanism of lithium ions insert into (extract from) the intercalation materials, to study degradation effects, to facilitate further electrode optimization, and to improve the charge/discharge cycle performance as well as the rate capability of lithium-ion batteries. A major problem of impedance spectroscopy as an electroanalytical tool is the fact that, in most cases, impedance spectra cannot be simulated unambiguously by a single model. Hence, major efforts are invested by prominent electrochemical groups in the development of reliable and comprehensive models that can precisely and logically explain the impedance behavior of electrochemical systems. At present, there are two main models to expound the mechanism of intercalation electrode reaction, namely the adsorption model and the surface layer model.

The adsorption model, also called adatom model or adion model, was usually used to depict the galvanic deposit process of metal ions. It was first used to depict the lithium ions insertion/desertion process of $LiTiS_2$ by Bruce[19, 20] in the lithium-ion battery field, and then was developed by Kobayashi as well as many others[40, 71, 76, 114]. In this model, the intercalation reaction proceeds in several steps, as shown in **Figure 2-1**: a solvated cation in solution adjacent to the electrode loses part of its solvation sheath and becomes adsorbed, thus forming an adion, on the electrode surface, accompaning by the injection of an electron into the conduction band of the solid host. Subsequently, the partially solvated cation diffuses across the surface of the electrode until it reaches a site at which insertion of the ion can occur, and the ion loses therefore the remaining solvent molecules and enters the host latice. According to the adsorption model, the EIS spectra of intercalation electrode should involve three parts along with decreasing frequency: (*i*) a high-frequency dispersion because of the solvated cation losing part of its solvation sheath; (*ii*) an intermediate frequency semicircle due to the ion losing the remaining solvent molecules and enterring the host latice; (*iii*) a low-frequency spike of the ionic diffusion. Although the adsorption model can expose some results, it is not widely accepted because of the involvement of solid electrolyte interphase (abbreviated as SEI) with some consequence during the delithiation/lithiation process[106].

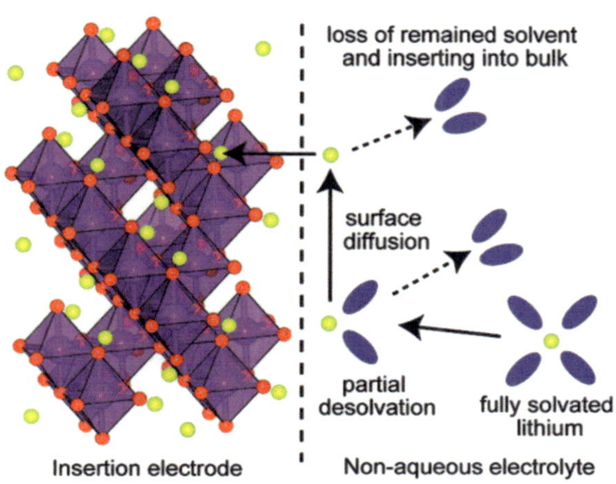

Fig. 2-1. Schematic representation of the adion mechanism of intercalation[76].

The surface layer model (also called SEI model) with approbation was first used to depict the lithium-ion insertion/desertion process of $LiCoO_2$ by Thomas[102], and developed mainly by Aurbach and co-workers[3-6, 31, 47, 50, 124]. The surface layer model was based on the assumption that electronic conductivity of the intercalation electrode was high and that the power was compacted sufficiently to ensure that each particle was in contact with the aggregate across a solid-solid interface making an ohmic contact of low resistance to electron flow. Under these conditions, the intercalation electrode formed a rough, but continuously interconnected, porous solid of low bulk resistance $R_b \ll R_{el}$ (the bulk electrolyte resistance). According to the surface layer model based on the above assumption, lithium insertion results in a series of complex phenomena. This model reflects the steps involved during lithium-ion insertion: (*i*) lithium-ion transport in an electrolyte; (*ii*) lithium-ion migration through the SEI film; (*iii*) charge-transfer through the electrode/electrolyte interface; (*iv*) lithium-ion diffusion in an electrode; (*v*) accumulation-consumption of Li in the electrode, which accompanies phase transition between intercalation stages; (*vi*) electron transport in an electrode and at an electrode/current collector interface. Among these steps of lithium-ion insertion, the electron transport (*vi*) and the lithium-ion transport in the electrolyte solution (*i*) usually do not present semicircle in the frequency range 10^5~10^{-2} Hz range due to their high characteristic frequencies, and these components of resistance appear as a Z′ intercept in the Nyquist plot. The process (*iv*) gives the Warburg impedance, which is observed as a straight line with an angle of 45° from the Z′ axis. The process (*v*) accumulation-consumption of Li in the electrode yields the bulk intercalation capacitance, which is observed as a vertical line, and the other two processes (*ii*) and (*iii*) generate their own semicircles at each characteristic frequency, respectively. In the past decades, the modified Voigt-FMG EEC for surface layer model, as shown in **Figure 2-2** and suggested by Aurbach et al.[47], is considered to provide the best account of the lithium-ion insertion process in intercalation electrode.

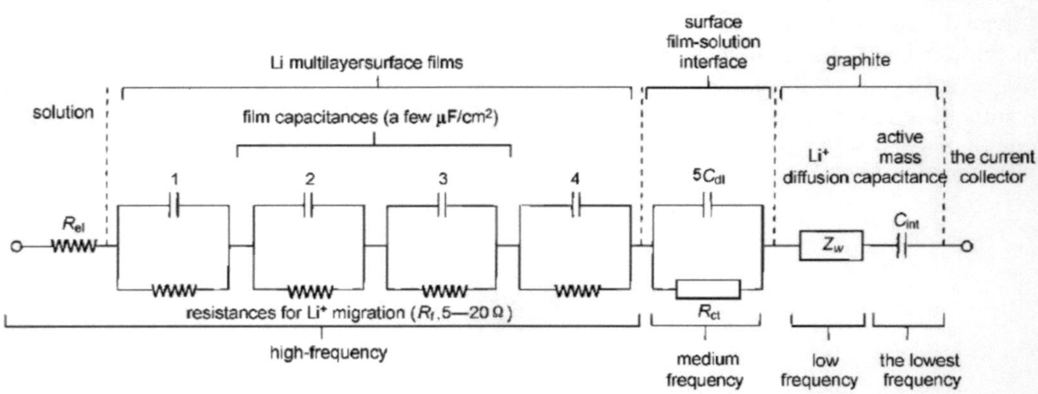

Fig. 2-2. EEC evolved by Aurbach used for analysis of impedance spectra of the lithium-ion insertion/desertion in the intercalation electrode[47].

Yet the above model is a simplification of real situation that the electrode is assumed to be built up of spherical particles of uniform size, and no change of the particle structure or new phase formation in the lithium insertion process. Afterwards, Barsoukov et al.[11, 12] proposed a new model based on single particles for commercial composite electrode, as shown in **Figure 2-3**. They supposed that the electrochemical kinetics characteristic of battery materials was represented by several common steps, as shown in **Figure 2-4**: (*i*) ionic charge conduction through electrolyte in the pores of the active layer and electronic charge conduction through the conductive part of the active layer; (*ii*) lithium-ion diffusion through the surface insulating layer of the active material; (*iii*) electrochemical reaction on the interface of active material particles including electron transfer; (*iv*) lithium-ion diffusion in the solid phase and (*v*) phase-transfer in cases where several phases are present and a capacitive behavior that is related to the occupation of lithium ions, which give a semicircle and straight line perpendicular to Z′ axis in the Nyquist plot (commonly below 10^{-2} Hz), respectively.

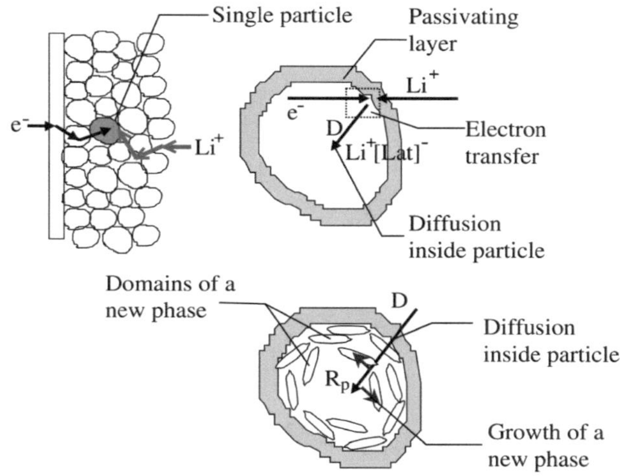

Fig. 2-3. Pictorial representation model for lithium-ion insertion/deinsertion into the intercalation electrode proposed by Barsoukov et al[12].

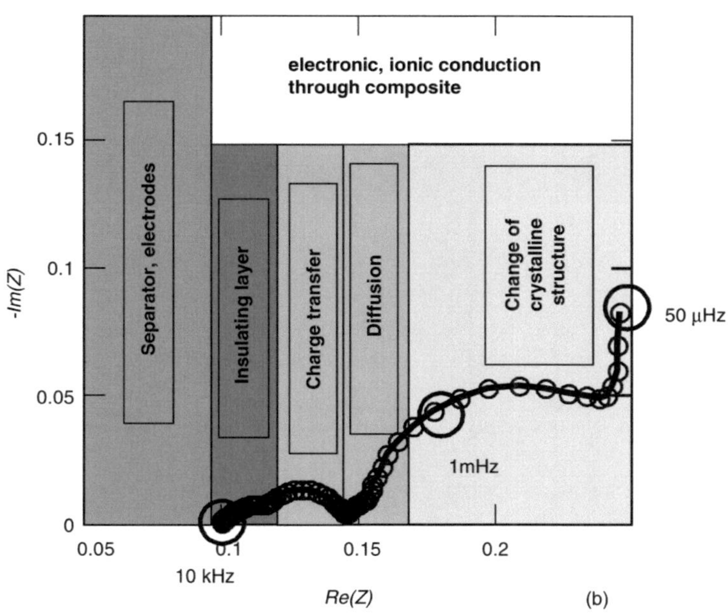

Fig. 2-4. Typical impedance spectra of intercalation electrode proposed by Barsoukov et al.[13].

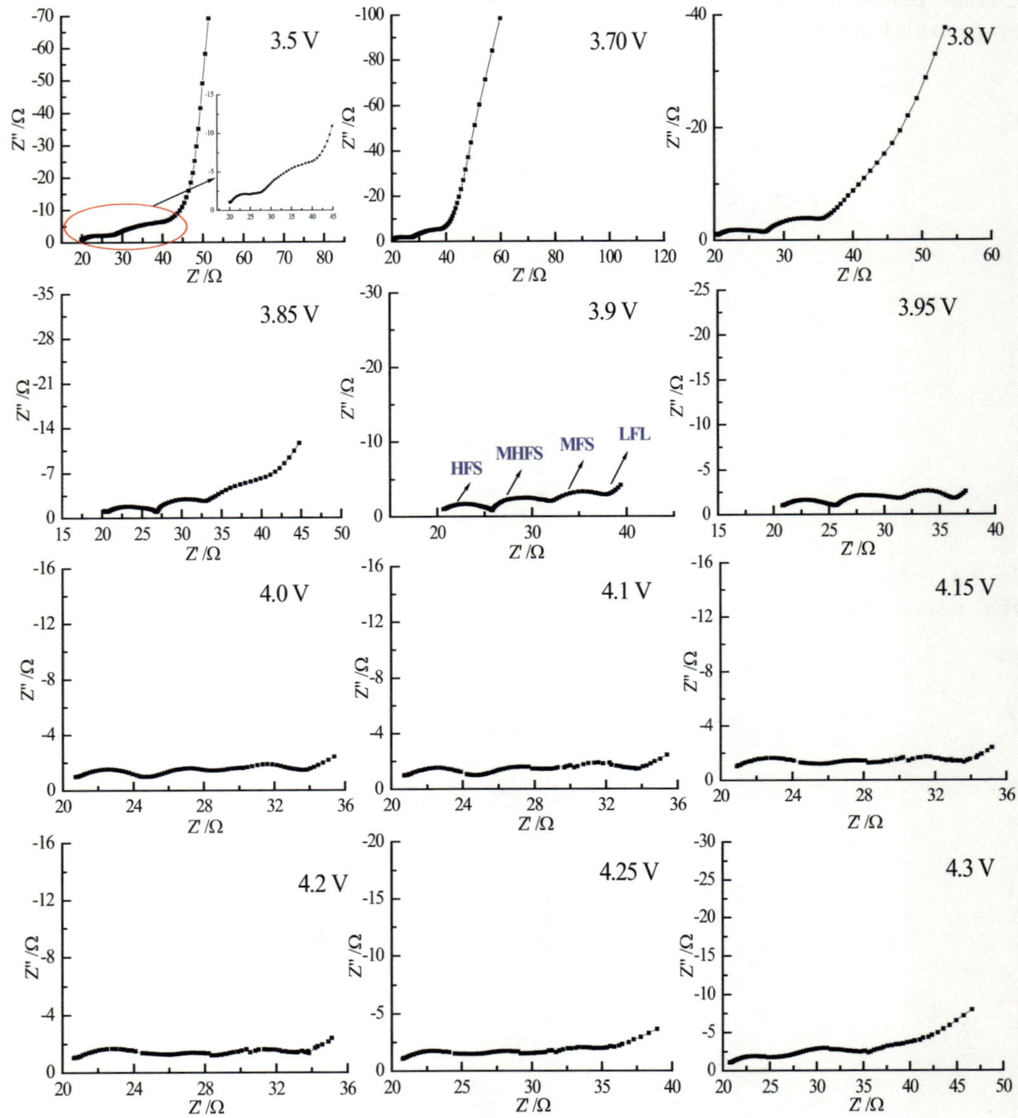

Fig. 2-5. Nyquist plots of the spinel $LiMn_2O_4$ electrode at various potentials from 3.5 to 4.3 V during the first delithiation[126].

However, many electrochemically active materials, such as $LiFePO_4$, $LiMn_2O_4$ and $LiCoO_2$, are not good electronic conductors. As a concequence, it is necessary to add an electronically conductive material such as carbon black. To physically hold the electrode together, a binder is also added. In these cases the electrochemical reaction can only occur at the points where the active material, the conductive diluent and electrolyte meet[110]. So the electrical conductivity of the component materials is one of the most important issues in connection with the intercalation electrode reaction. Nobili and co-workers[25, 61, 72-75] suggested that the Nyquist plots of $LiCoO_2$ in the delithiated state should involve a third semicircle relating to the electronic propetries of the material, however, the three suggested semicircles could not be observed in their experimental results. In our previous work[126], as illustrated in **Figure 2-5,** at intermediate degrees of the first delithiation process in spinel $LiMn_2O_4$ electrode, three semicircles are observed in the Nyquist diagram, and it was demonstrated that the semicircle in the middle to high frequency range (MHFS) should be attributed to the electronic properties of the material. Therefore, a modified model is put forward, as shown in **Figure 2-6**, and the EEC is illustrated in **Figure 2-7.** In this EEC, R_s represents the ohmic resistance, R_{SEI} and R_{ct} are resistances of the SEI film and the charge transfer reaction. The capacitance of the SEI film and the capacitance of the double layer are represented by the constant phase elements (CPE) Q_{SEI} and Q_{dl}, respectively. The very low frequency region, however, cannot be modeled properly by a finite Warburg element; therefore, it is chosen to replace the finite diffusion by a CPE, i.e., Q_D. The electronic resistance of the material and the associated capacitance used to characterize the electronic properties of the material are represented by R_e and the constant phase elements Q_e. The EIS spectra in the frequency range 10^5~10^{-2} Hz are interpreted in terms of the following physical phenomena in an order of decreasing frequency: (*i*) a high frequency semicircle (HFS) because of the presence of a surface layer; (*ii*) a middle to high frequency semicircle (MHFS) related to the electronic

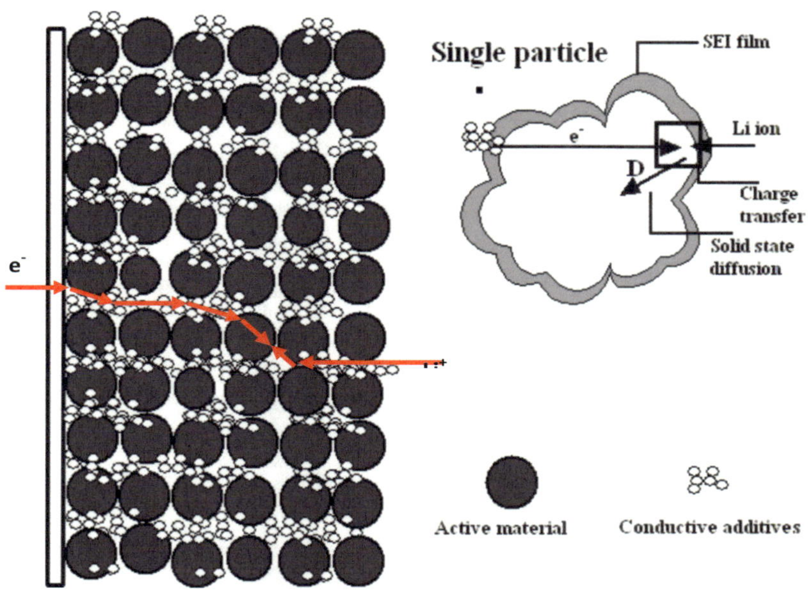

Fig. 2-6. Pictorial representation model for lithium ion insertion/de-insertion into the intercalation electrode[126].

properties of the material; (*iii*) a middle frequency semicircle (MFS) associated with charge transfer, and finally, and (*iv*) the very low frequency incline line attributed to the solid state diffusion.

Although the adsorption model or surface layer model both can fit partially some results, a debate is still open on that whether or not the solvated/lost solvated or the migration of lithium ions through the SEI film is the rate determining step in the transport of lithium-ion? Namely, what should be the origin of the semicircle in the high frequency range (HFS)? On the base of the adsorption model, the HFS is attributed to the solvated/lost solvated of lithium ions; while according to the surface layer model, the HFS is ascribed to the migration of lithium ions through the SEI film. So, it needs to be further investigated.

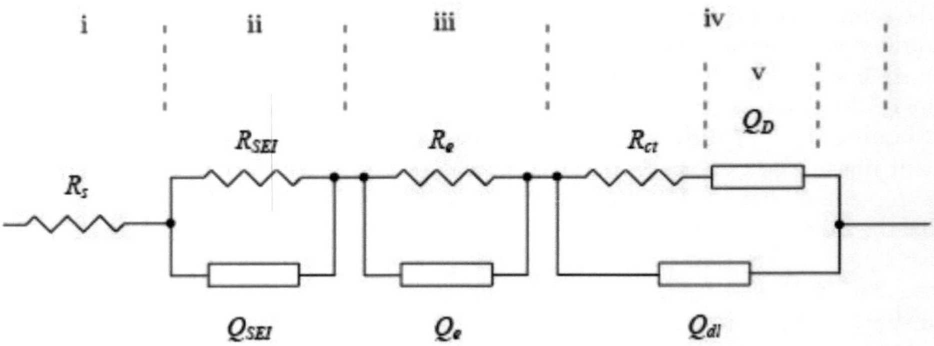

Fig. 2-7. Equivalent circuit proposed for analysis of the intercalation electrode during the charge-discharge processes[126].

3. Kinetic models for the lithiation and delithiation of simple binary transition metal compounds

In 2000, Poizot et al.[84] reported for the first time that lithium could be stored reversibly in some simple binary transition metal (TM) oxides MO_x (M=Fe, Co, Ni, etc.) through a heterogeneous conversion reaction, i.e. $MO_x + 2xLi \rightarrow M^0 + xLi_2O$, which is different from the intercalation/deintercalation mechanism. Later, reversible lithium storage was also observed in transition metal fluorides as well as sulfides, nitrides, selenides and phosphides[7, 8, 26, 63, 81, 82, 89, 94, 116]. As opposed to intercalation reactions, the reversible conversion process enables the full redox utilization of the transition metal and has 2-4 times the specific capacity of intercalation compounds as high as 600-1000 $mAh.g^{-1}$[54]. Therefore, as potential alternatives for intercalation compounds, transition metal oxides as well as other transition metal compounds attract lots of attention in recent years.

As a new type of reaction, the kinetic behavior of the conversion reaction is not very clear and remains to be revealed. **Figure 3-1** and **3-2** show Nyquist plots of the NiF_2/C composites electrode prepared through high energy mechanical milling during the first discharge and charge process, respectively. It can be seen that the typical EIS characteristics appeared with three semicircles in the Nyquist diagram at 1.6 V, similar to that of intercalation electrode. However, most metal fluorides such as NiF_2 are insulators with large band gap, the fitted values of the semicircle in the middle frequency region (MFS), with the

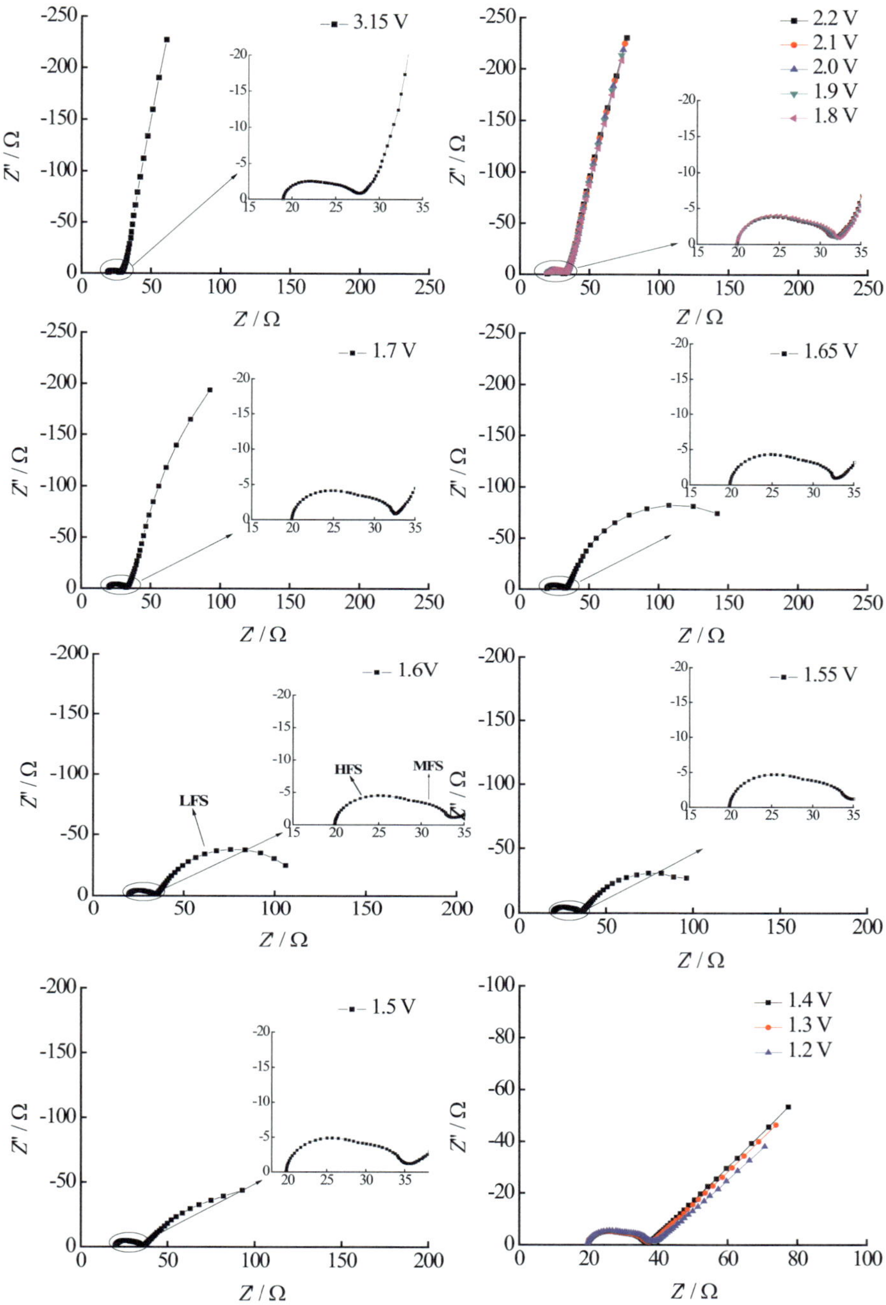

Fig. 3-1. Nyquist plots of the NiF_2/C composites electrode during the first discharge process[88].

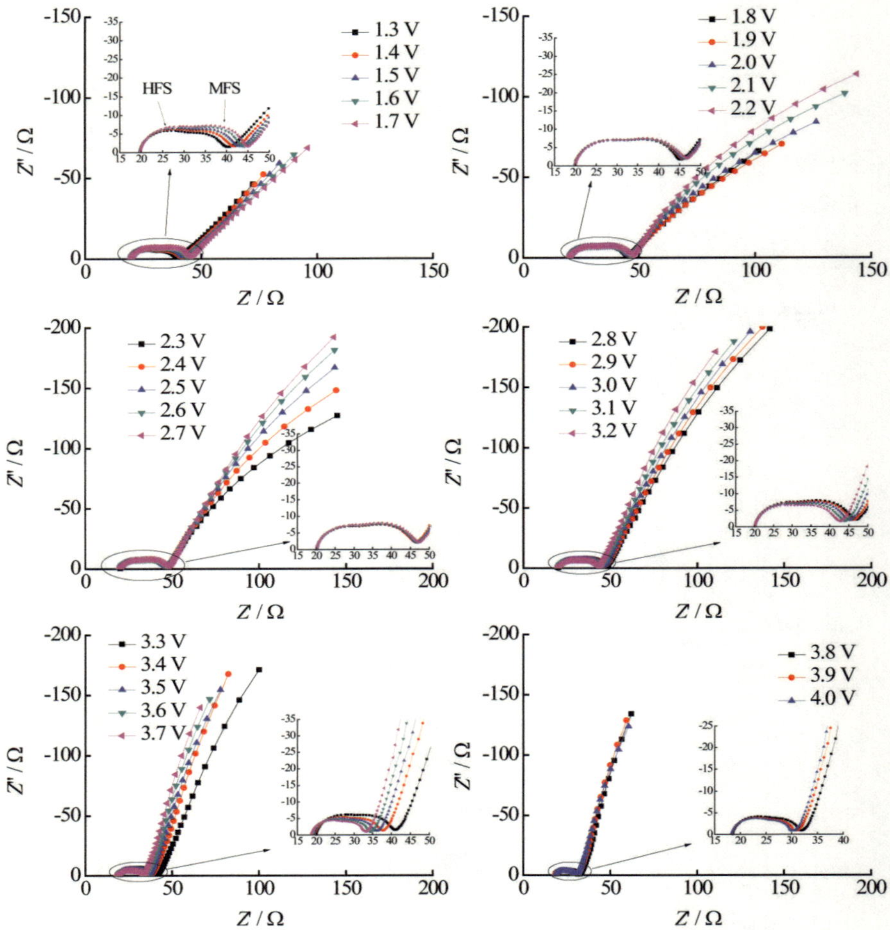

Fig. 3-2. Nyquist plots of the NiF_2/C composites electrode during the first charge process[88].

value of only scores of ohms, could not be the reflection of the electronic resistance of NiF_2, and it has been demonstrated that the MFS should be related to the contact between conductive agents and active materials. Based on the above experimental results, a new model (**Figure 3-3**) is put forward in our recent studies[88]. The EIS is interpreted in terms of the following physical phenomena in an order of decreasing frequency: firstly ionic charges conduct through electrolyte and lithium ions diffuse through the SEI film to the surface of the NiF_2; To maintain electrical neutrality of particles, electrons conduction along the external circuit, conductive agents, and the contact points between NiF_2 and C, then hop on the surface of NiF_2. Later the charge transfer process takes place close to the carbon particles. But for the diffusion process, because F^- could migrate in the NiF_2 particles[2] and compete favorably with Li^+[120], there may exist two ways: (*i*) Li^+ diffuses to contact points to form LiF along the particle surface; (*ii*) F^- migrates further from the carbon, through the NiF_2 phase, or along the newly formed Ni-NiF_2 interfaces, so as to the surface of NiF_2 and form LiF with Li^+.

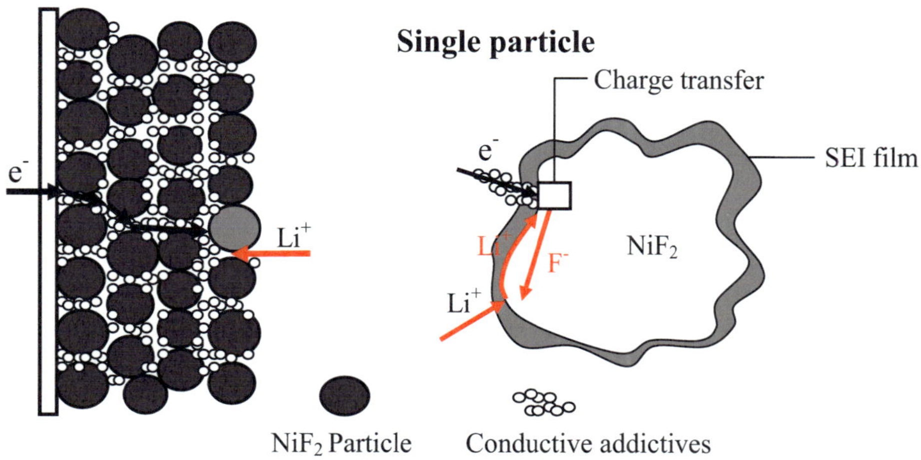

Fig. 3-3. Pictorial representation model for lithium-ion insertion/de-insertion into NiF_2/C composite electrode based on conversion reaction[88].

4. The potential and temperature dependence of the kinetic parameters

The main problem remaining in the practical application of EIS is its ambiguity: many physically different procedures or separate stages of a complicated process show similar features in terms of impedance spectroscopy. A classical electrochemical system contains processes such as electron-transfer, diffusion and absorption of reacting species, and eluciadting these processes could be easy if all absorption sites have the same energy, i.e. the electrode surface is energetically homogeneous. However, the processes occurring in practical ion insertion electrodes are much more complicated compared to the classical systems. The practical ion insertion electrodes are composite materials, in which the active mass particles are bound to a current collector with a polymeric binder such as polyvinylidene fluoride (PVDF). In addition, the composite electrodes have to contain a conductive additive, usually carbon black. The electrodes are regularly prepared from slurry of the particles and the binder in an organic solvent, which is spread on the current collector, followed by drying. The final shape of the electrode is obtained by applying some pressure to the electrode. So the shape and value of the resistance in the impedance spectrum should be affected by the amount of conductive additive in the composite materials, the contact between the electrode film and the current collector, the solvent, particle size, thickness of electrode, and stack pressure[23, 48, 93]. Without systematic and integrated comparison of EIS spectra of these processes, incorrect interpretation may arise. So it is improtant to determine the physical processes corresponding to time constants ranging via theoretical analysis and contrast tests in detail.

4.1 Kinetic parameters for lithium-ion migration through the SEI film

The SEI film is a key element of traditional Li-ion batteries and plays an important role in the electrochemical performance of the electrode material and a battery[113]. The SEI film prevents lithium from being intercalated in the solvated state, which leads to the exfoliation (swelling) of the carbon. In addition, the SEI film inhibits also the further reduction of electrolyte by active lithium and thus limits the degradation of the electrolyte. The semicircle in the Nyquist plots in the high frequency range is commonly attributed to lithium-ion migration through the SEI film, and its characteristic parameters are the resistance due to Li ions migration, R_{SEI}, and geometric capacitance of the surface films, C_{SEI}.

The resistance and capacitance values corresponding to the migration of Li ions through the SEI film can be described by Eq. (4-1) and (4-2)

$$R_{SEI} = \rho l / S \tag{4-1}$$

$$C_{SEI} = \varepsilon S / l \tag{4-2}$$

where l is the thickness of the SEI film, S the electrode surface area, ρ the resistivity, and ε the permittivity of the SEI film. If we assume that the surface S, permittivity ε, and resistivity ρ remain constant, it is clear that a thickness increase will lead to a resistance increase, and to a decrease in capacitance.

4.1.1 The growth rate of SEI film

A simple model for SEI film growth can be formulated similarly to the growth of oxides on metals. A metal exposed to air reacts to form an oxide which helps passivate the surface and slows further reaction. Similarly, lithiated graphite exposed to electrolyte reacts to form an SEI film which helps passivate the surface and slows further reactions. Lawless reviewed numerous models of oxide growth on metals which he showed followed a huge variety of rate laws[44]. The simplest of these is the "parabolic growth law" which assumes that the rate of increase in the thickness of the passivating layer, l, is inversely proportional to the thickness of the layer

$$dl/dt = k/l \tag{4-3}$$

where k is a proportionally constant. This can be rewritten as

$$ldl = kdt \tag{4-4}$$

and the integration yields

$$1/2l^2 = kt + C \tag{4-5}$$

If the thickness is taken x=0 at t=0, then the constant, C is zero and one can write

$$l = (2kt)^{1/2} \tag{4-6}$$

The rate of change of the thickness of the passivating layer is formulated by combining Eqs. (4-3) and (4-6), i.e.

$$dl/dt = (1/2k)^{1/2} t^{-1/2} \tag{4-7}$$

In a Li/graphite cell, it is believed that the SEI film begins to form as lithium is transferred electrochemically to the graphite electrode. With continueously cycling more and more Li is irreversibly consumed as the SEI film becomes more and more thick. The total amount of Li consumed is directly proportional to the SEI film thickness, which might be described by Eq.

(4-6). The amount of lithium consumed by SEI film growth is directly proportional to the irreversible capacity in each cycle, which might be described by Eq. (4-7). Many researchers studying the failure of Li-ion batteries have identified this capacity loss vs $t^{1/2}$ (or equivalent) relationship and have demonstrated that their models fit experimental data very well[16, 85, 92, 95, 122].

4.1.2 The potential dependence of the SEI film resistance

According to the SEI model mentioned, it is possible to use Eq. (4-8) describing the migratin of ions in a solid crystal under external field[80]. It is assumed that the thickness (l) is larger than the space charge lengths in the SEI, and there is no change in the concentration or molility of the mobile lattice defects through the SEI film.

$$i = 4zFacv\exp(-W / RT)\sinh(azFE / RT) \tag{4-8}$$

in Eq. (4-8) a = half-jump distance, v = vibration frequency of the ion in the crystal, c = concentration of the lattice cationic defect, z = the valance of the mobile ion, W = a barrier energy for jumping, E = electricfield, F = Faraday constant. The assumption that the migration of the cation in the SEI is the rate determining step (rds) means the largest fraction of the electrode overpotential (η) will develop on the SEI, i.e.

$$\eta = \eta_{SEI} = El \tag{4-9}$$

For low field conditions, Eq. (4-8) can be linearization into Eq. (4-10)

$$i = (4z^2F^2a^2cv / RTl)\exp(-W / RT)\eta \tag{4-10}$$

thus

$$R_{SEI} = (RTl / 4z^2F^2a^2cv)\exp(W / RT) \tag{4-11}$$

or

$$\ln R_{SEI} = \ln(RTl / 4z^2F^2a^2cv) + W / RT \tag{4-12}$$

The relation between lnR_{SEI} and $1/T$ is linear given by Eq. (4-12), and W can be obtained from the line gradient.

According to Eq. (4-12), the energy barriers for the ion jump relating to migration of lithium ions through the SEI film of the spinel $LiMn_2O_4$ electrode in 1 mol/L $LiPF_6$-EC (ethylene carbonate): DEC (diethyl carbonate) was determined to be 15.49 kJ/mol[109], and that of $LiCoO_2$ electrode in 1 mol/L $LiPF_6$-EC:DEC:DMC (dimethyl carbonate) and 1 mol/L $LiPF_6$-PC (propylene carbonate):DMC+5% VC (vinylene carbonate) electrolyte solutions were calculated to be 37.74 and 26.55 kJ/mol[125], respectively, in our previous studies.

4.2 Kinetic parameters for electron charge conduction

4.2.1 The electronic properties of intercalation materials

Among the characteristics of insertion materials, the electrical conductivity of the component materials is one of the most important issues in connection with the rate performance of batteries. In addition to such practical importance, conductivity measurements during the lithium insertion (and extraction) reaction would be an attractive approach for the study of the variation in the electronic structure of the materials as a function of lithium content[68, 69]. The semicircle in the Nyquist plots in the middle to high frequency range is commonly ascribed to the electronic properties of intercalation materials, and its characteristic parameters are the electronic resistance of the insertion materials R_e and geometric capacitance of the insertion materials C_e.

According to the Ohm's law, the electrical conductivity can be obtained through the equality (4-13):

$$R_e = l / (\sigma S) \tag{4-13}$$

where σ is the conductivity, l is the thickness of the electrode film, and S is the area of the electrode film.

4.2.1.1 The temperature dependence of R_e

All cathode materials for lithium-ion batteries have semiconductor features. Electrical conduction in a semiconductor is a thermally activated phenomenon and usually follows an Arrhenius type relationship as shown below[99]:

$$\sigma T = \sigma_0 \exp(-E_a / k_B T) \tag{4-14}$$

where σ is the electrical conductivity, σ_0 is the pre-exponential factor, E_a is the activation energy, k_B is the Boltzmann constant and T is the temperature.

On the substitution of Eq. (4-14) in Eq. (4-13), one gets

$$R_e = \frac{lT}{S\sigma_0} \exp(E_a / k_B T) \tag{4-15}$$

ln1/T is acted by Taylor series expansion and ignored the high-order component, Eq. (4-15) can become into

$$\ln R_e = \ln \frac{l}{S\sigma_0} + \frac{(E_a - k_B)}{k_B T} + 1 \tag{4-16}$$

Eq. (4-16) presents explicitly that when $1/T \rightarrow 0$ and constant electrode potential, the relation between lnR_e and $1/T$ is linear, and E_a can be obtain from the line gradient.

4.2.1.2 The potential dependence of R_e

As noted above, all cathode materials for lithium-ion batteries have semiconductor features, and can be to be divided into *n*-type semiconductor and *p*-type semiconductor, which have different potential dependence of R_e.

It is well known [93, 108, 123] that $LiCoO_2$ is a *p*-type semiconductor (band-gap E_g =2.7 eV)[105], while Li_xCoO_2 exhibits a metal-like behavior for x<0.75. Li_xCoO_2 is predicted to have partially filled valence bands for x lower than 1.0[104]. For each Li removed from $LiCoO_2$ lattice, an electron hole is created within the valence band. Namely,

$$p = 1 - x \tag{4-17}$$

where p is the concentration of electron hole. We may expect that there will be sufficient holes, when x is below 0.75, to allow for a significant degree of screening. And in this regime, the hole states in the valence bands are likely to be delocalised so that Li_xCoO_2 exhibits metallic-like electronic properties, namely, the existence of a drastic change of the electronic conductivity occurs at early stage of lithium deintercalation, which may be caused by a transition from insulator to metal (or so called metal-insulator transition). This behavior is clearly observed in infrared absorption spectra where a strong absorption by holes occurs at low wavenumbers[38]. Accordingly, the variation with potential of electronic conductivity of $LiCoO_2$ in the charge-discharge process may be divided into three regions: (1) the region in which Li_xCoO_2 has a semiconductor-like behavior; (2) the region in which the hole states in the valence bands are likely to be delocalised; (3) the region in which Li_xCoO_2 has a metal-like behavior.

The general expression for the electronic conductivity for p-type semiconductor is given by Eq. (4-18).

$$\sigma = pq\mu \tag{4-18}$$

where μ is carrier hole mobility, and q is electron charge.

The Langmuir insertion isotherm could be used for lithium-ion deintercalation from $LiCoO_2$ hosts by assuming that the interaction between the intercalated species and the host material as well as the interaction between the intercalated are absent. Thus, the intercalation level, x, is given by[45]:

$$x/(1-x) = \exp\left[f(E-E_0)\right] \tag{4-19}$$

where f=F/RT (F and R, Faraday and gas constant respectively, T, absolute temperature), E and E_0 define the electrode's real and standard potentials in the equilibrium.

On the substitution of Eq. (4-17) in Eq. (4-19), one gets

$$p = 1/\{1+\exp[f(E-E_0)]\} \tag{4-20}$$

While on the substitution of Eq. (4-13) and Eq. (4-18) in Eq. (4-20), one has

$$\ln R_e = \ln(S/q\mu l) + \ln\{1+\exp[f(E-E_0)]\} \tag{4-21}$$

$\ln\{1+\exp[f(E-E_0)]\}$ is acted by Taylor series expansion. If the high-order component is ignored, we have

$$\ln R_e = \ln(2S/q\mu l) + \frac{1}{2}f(E-E_0) \tag{4-22}$$

It can be seen from Eq. (4-22) that, the value of ln$R$ shows a linear dependence on electrode potential. Therefore the variation of the electronic conductivity of $LiCoO_2$ in the charge-discharge process with potential may be divided into three different parts: (*i*) when Li_xCoO_2 has a semiconductor-like behavior, the value of lnR_e shows a linear dependence on electrode potential; (*ii*) when the hole states in the valence bands are likely to be delocalised, the value of lnR_e increases or decreases drastically with electrode potential; (*iii*) when Li_xCoO_2 has a metal-like behavior, the value of lnR_e also exhibits a linear dependence on electrode potential.

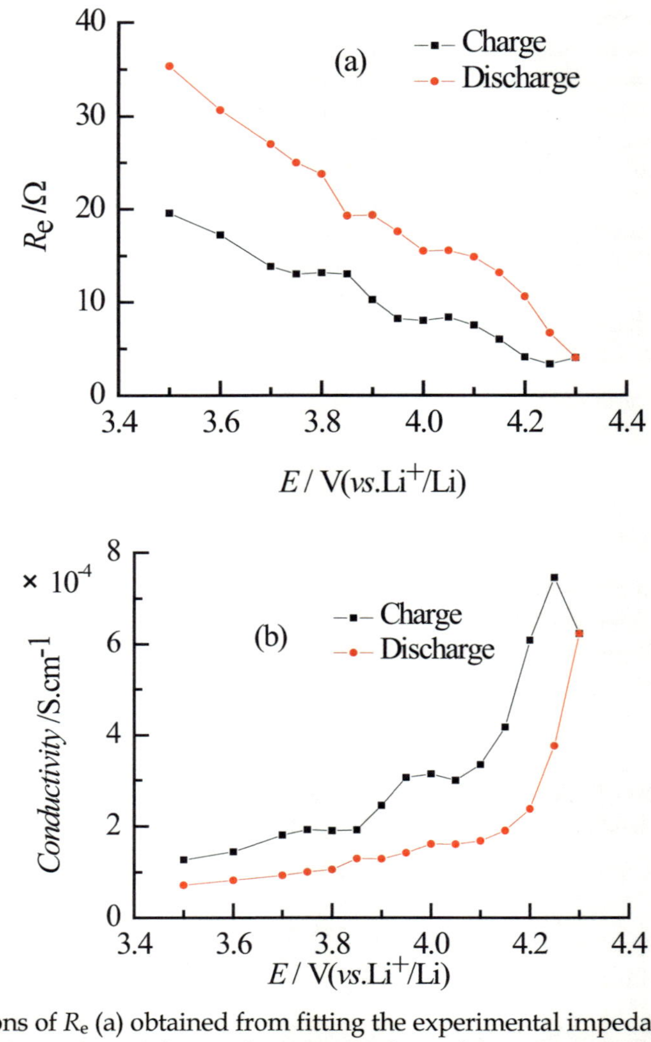

Fig. 4-1. Variations of R_e (a) obtained from fitting the experimental impedance spectra of the spinel $LiMn_2O_4$ electrode and the conductivity (b) derived from R_e with electrode potential during the first charge-discharge cycle[126].

While spinel $LiMn_2O_4$ is an *n*-type semiconductor, which is a mixed-valence (Mn^{3+}/Mn^{4+}) compound and its electronic conduction takes place by electron-hopping between high-valence (Mn^{4+}) and low-valence (Mn^{3+}) cations[62, 83]. Conductivity of this type would be governed by the concentration of carriers (electrons from Mn^{3+}) and the hopping length (Mn-Mn interatomic distance). The number of electron carriers decreases proportionally with the degree of de-lithiation coupled with oxidation of Mn^{3+} to Mn^{4+}. On the other hand, the Mn-Mn distance in the spinel structure is reduced by delithiation. Variations of R_e obtained from fitting the experimental impedance spectra of the spinel $LiMn_2O_4$ electrode and the conductivity derived from R_e with electrode potential during the first charge-discharge cycle are shown in **Figure 4-1**. The conductivity derived from R_e is in the range of 10^{-4} $S{\cdot}cm^{-1}$, being roughly in agreement with previous reports[39, 70]. The electronic resistance of the material in the charge and discharge process has the same change relationship with potential, namely, R_e decreases with the increase of the electrode polarization potential in the charge process, and increases with the decrease of the electrode polarization potential, indicating that the effect of the contraction of hopping length predominated over the decrease of the numbers of electron carriers.

4.2.2 The potential dependence of Schottky contact resistance

The semicircle in the Nyquist plots of binary transition metal (TM) compounds in the middle frequency range is commonly ascribed to the Schottky contact and its characteristic parameters are the resistance R_c and capacitance of the Schottky contact. According to thermionic emission diffusion theory, a Schottky contact behaviour can be described by the equation which takes into account the defects of lattice, electric field, tunneling effects, the presence of an interfacial layer, and carrier recombination in the space charge region of the metal-semiconductor contact as given by[21, 86, 98]:

$$I = I_0 \exp(\frac{qE}{nk_B T})[1 - \exp(\frac{-qE}{k_B T})] \tag{4-23}$$

where I_0 is the saturation current, q the electronic charge, k Boltzmann constant, T the absolute temperature, E the applied bias voltage, n is the ideal factor. The expression for the saturation current, I_0 is:

$$I_0 = AA^* T^2 \exp(-\frac{q\Phi_B}{k_B T}) \tag{4-24}$$

where A is the Schottky contact area, A^* the effective Richardson constant, and $q\Phi_B$ is the Schottky barrier height. When $E > 3kT / q$, Eq. (4-24) can be simplified as,

$$I = I_0 \exp(\frac{qE}{nk_B T}) \tag{4-25}$$

According to the Ohm's Laws, the Schottky contact resistance R_c can be presented in the following form:

$$R_c = (\frac{dI}{dE})^{-1} \tag{4-26}$$

substitute Eq. (4-25) into Eq. (4-26), R can be expresse as following.

$$R_c = (I_0 \frac{q}{nk_B T})^{-1} \exp(-\frac{qE}{nk_B T}) \tag{4-27}$$

When the contact media do not change, A^* and Φ_B could be considered to be the same, Eq. (4-27) could be written as:

$$R_c = C \exp(-\frac{qE}{nk_B T}) \tag{4-28}$$

where C is constant. Therefore change Eq. (4-28) to linear equation by logarithm, we can obtain finally the following expression:

$$\ln R_c = \ln C - \frac{qE}{nk_B T} \tag{4-29}$$

The plot of lnR_c versus E should give a straight line with the slope = $-q/nk_BT$ and y-intercept at ln$C$ on condition that no change occur with contact media.

4.3 Kinetic parameters for the charge transfer process

Among several processes of lithium ion and electron transport, the charge transfer at an electrode/electrolyte interface is an essential process of the charge-discharge reaction of lithium ion batteries. The semicircle in the Nyquist plots in the low frequency range is commonly attributed to the charge transfer process, and its characteristic parameters are the charge-transfer resistance, R_{ct}, and the double layer capacitance, C_{dl}.

4.3.1 The potential dependence of R_{ct}

The electrochemical reaction of intercalation electrode is given by Eq. (4-30)

$$(1-x)Li^+ + (1-x)e + Li_xMO_y \Leftrightarrow LiMO_y \tag{4-30}$$

where M is the transition metal.

Suppose that the velocity of the forward reaction r_f (lithiation in intercalation electrode) is proportional to $c_{\max}(1-x)$ and the concentration (M_{Li^+}) of lithium-ion in the electrolyte near the electrode. $c_{\max}(1-x)$ is the insertion sites on the intercalation electrode surface not occupied by lithium ions, x is the insertion level, $c_{\max}$ (mol/cm^3) is the maximum concentration of lithium ion in intercalation electrode. Then, the velocity of the backward reaction rate r_b is proportional to $c_{\max}x$, $c_{\max}x$ is the sites already occupied by lithium ions. r_f and r_b can be written as[10, 128]

$$r_f = k_f c_{\max}(1-x)M_{Li^+} \tag{4-31}$$

$$r_b = k_b c_{\max} x \tag{4-32}$$

Therefore:

$$i = r_f - r_b = n_e F c_{\max}\left[k_f(1-x)M_{Li^+} - k_b x\right] \tag{4-33}$$

where n_e is the number of electron exchange in the processes of lithium ion insertion and extraction, F is the Faraday constant; k_f and k_b are the velocity constants of the forward and backward reactions.

The molar intercalation energy ΔG_{int} of the intercalation electrode can be expressed as

$$\Delta G_{int} = a + mx \tag{4-34}$$

where a is the constant about the interaction energy between an intercalated ion and a host lattice near it, and m is the constant about the interaction energy between two intercalated ions in different sites.

According to the activated complex theory, k_f and k_b can be expressed from[53]

$$k_{\mathrm{f}} = k_{\mathrm{f}}^{0} \exp\left[\frac{-\alpha\left(n_{\mathrm{e}}FE + \Delta G_{\mathrm{int}}\right)}{RT}\right] \tag{4-35}$$

$$k_{\mathrm{b}} = k_{\mathrm{b}}^{0} \exp\left[\frac{(1-\alpha)\left(n_{\mathrm{e}}FE + \Delta G_{\mathrm{int}}\right)}{RT}\right] \tag{4-36}$$

with a representing symmetry factor for the electrochemical reaction, and k_f^0 and k_b^0 can be written in the Arrhenius form

$$k_{\mathrm{f}}^{0} = A_{\mathrm{f}} \exp\left(\frac{-\Delta G_{0\mathrm{f}}}{RT}\right) \tag{4-37}$$

$$k_{\mathrm{b}}^{0} = A_{\mathrm{b}} \exp\left(\frac{-\Delta G_{0\mathrm{b}}}{RT}\right) \tag{4-38}$$

Substitute Eqs. (4-35) and (4-36) into Eq. (4-33), the current i can be obtained in the following expression

$$i = n_{\mathrm{e}}Fc_{\max}k_{\mathrm{f}}^{0}(1-x)M_{\mathrm{Li}^{+}} \exp\left[\frac{-\alpha\left(n_{\mathrm{e}}FE + \Delta G_{\mathrm{int}}\right)}{RT}\right] - n_{\mathrm{e}}Fc_{\max}k_{\mathrm{b}}^{0}x\exp\left[\frac{(1-\alpha)\left(n_{\mathrm{e}}FE + \Delta G_{\mathrm{int}}\right)}{RT}\right] \tag{4-39}$$

when the forward reaction r_f equal the backward reaction r_b, that is to say i=0, so the exchange current density i_0 can be obtained

$$i_0 = n_{\mathrm{e}}Fc_{\max}k_{\mathrm{f}}^{0}(1-x)M_{\mathrm{Li}^{+}} \exp\left[\frac{-\alpha\left(n_{\mathrm{e}}FE + \Delta G_{\mathrm{int}}\right)}{RT}\right] = n_{\mathrm{e}}Fc_{\max}k_{\mathrm{b}}^{0}x\exp\left[\frac{(1-\alpha)\left(n_{\mathrm{e}}FE + \Delta G_{\mathrm{int}}\right)}{RT}\right] \tag{4-40}$$

Thus

$$i_0 = n_{\mathrm{e}}Fc_{\max}k_0\left(M_{Li^{+}}\right)^{(1-\alpha)}(1-x)^{(1-\alpha)}x^{\alpha} \tag{4-41}$$

where k_0 is the standard reaction speed constant and k_0 can be expressed as

$$k_0 = k_{\mathrm{f}}^{0} \exp\left[\frac{-\alpha\left(n_{\mathrm{e}}FE_0 + \Delta G_{\mathrm{int}}\right)}{RT}\right] = k_{\mathrm{b}}^{0} \exp\left[\frac{(1-\alpha)\left(n_{\mathrm{e}}FE_0 + \Delta G_{\mathrm{int}}\right)}{RT}\right] \tag{4-42}$$

The charge transfer resistance can be defined as

$$R_{\mathrm{ct}} = RT/n_{\mathrm{e}}Fi_0 \tag{4-43}$$

Substitute Eq. (4-41) into Eq. (4-43), we get

$$R_{ct} = \frac{RT}{n_e{}^2F^2c_{max}k_0\left(M_{Li^+}\right)^{(1-\alpha)}(1-x)^{(1-\alpha)}x^{\alpha}} \tag{4-44}$$

Suppose that the delithiation/lithiation process in intercalation electrode is invertible, namely, $a = 0.5$, the Eq. (4-44) can be expressed as

$$R_{ct} = \frac{RT}{n_e{}^2F^2c_{max}k_0\left(M_{Li^+}\right)^{0.5}(1-x)^{0.5}x^{0.5}} \tag{4-45}$$

Eq. (4-45) predicts clearly a rapid increase in R_{ct} with the decrease of x as $x<0.5$, a rapid decrease in R_{ct} with the increase of x as $x>0.5$, and the minimum R_{ct} will be reached at $x=0.5$.

At the beginning of lithium ion inserts into or the very end of lithium ion extracts from the active mass during the electrochemical processes, that is to say, for very low insertion level ($x \to 0$). Eq. (4-19) takes the form

$$x = \exp[f(E - E_0)] \tag{4-46}$$

Substitute Eq. (4-46) into Eq. (4-44), we get

$$R_{ct} = \frac{RT}{n_e{}^2F^2c_{max}k_0\left(M_{Li^+}\right)^{(1-\alpha)}}\exp\left[\frac{-\alpha F(E-E_0)}{RT}\right] \tag{4-47}$$

changing Eq. (4-47) to linear equation by logarithm, we can obtain finally the following expression:

$$\ln R_{ct} = \ln\frac{RT}{n_e{}^2F^2c_{max}k_0\left(M_{Li^+}\right)^{(1-\alpha)}} - \frac{\alpha F(E-E_0)}{RT} \tag{4-48}$$

Eq. (4-48) presents explicitly that when $x\to 0$, $\ln R_{ct} \sim E$ shows a linear variation and the symmetry factors of charge transfer in electrochemical kinetics, α, can be calculated from the straight line slope. By using Eq. (4-48), the symmetry factor of charge transfer of lithium-ion insertion-desertion in $LiCoO_2$ is determined as 0.5 in our previous studies[128], and that of lithium-ion insertion-desertion in graphite materials is determined as 0.56 by Holzapfel et al.[36].

4.3.2 The temperature dependence of R_{ct}

According to Eq. (4-37), Eq. (4-42) and Eq. (4-44), we can get

$$R_{ct} = \frac{RT}{n_e{}^2F^2c_{max}A_f\left(M_{Li^+}\right)^{(1-\alpha)}(1-x)^{(1-\alpha)}x^{\alpha}} \times \exp\left[\frac{\Delta G_{0c} + \alpha\left(nFE_0 + \Delta G_{int}\right)}{RT}\right] \tag{4-49}$$

the intercalation-deintercalation reaction active energies ΔG can be expressed from

$$\Delta G = \Delta G_{0c} + \alpha(nFE_0 + \Delta G_{int}) = \Delta G_{0c} + \alpha(nFE_0 + a + gx) \tag{4-50}$$

Substitute Eq. (4-50) into Eq. (4-49), we get

$$R_{\rm ct} = \frac{RT}{n_{\rm e}^2 F^2 c_{\max} A_f \left(M_{\rm Li^+}\right)^{(1-\alpha)} (1-x)^{(1-\alpha)} x^{\alpha}} \times \exp(\frac{\Delta G}{RT}) \tag{4-51}$$

and Eq. (4-51) can become as

$$\ln R_{\rm ct} = \ln \frac{R}{n_{\rm e}^2 F^2 c_{\max} A_f \left(M_{\rm Li^+}\right)^{(1-\alpha)} (1-x)^{(1-\alpha)} x^{\alpha}} + \frac{\Delta G}{RT} - \ln\frac{1}{T} \tag{4-52}$$

ln$1/T$ is acted by Taylor series expansion and ignored the high-order component, one gets

$$\ln R_{\rm ct} = \ln \frac{R}{n_{\rm e}^2 F^2 c_{\max} A_f \left(M_{\rm Li^+}\right)^{(1-\alpha)} (1-x)^{(1-\alpha)} x^{\alpha}} + \frac{(\Delta G - R)}{RT} + 1 \tag{4-53}$$

Eq. (4-53) presents explicitly that when $1/T \to 0$ and at constant electrode potential, the relation between $\ln R_{ct}$ and $1/T$ is linear and ΔG can be obtain from the line gradient.

According to Eq. (4-53), the intercalation-deintercalation reaction active energies of $LiCoO_2$ are calculated in our previous studies[128] to be 68.97 and 73.73 kJ/mol, respectively, in 1 mol/L $LiPF_6$-EC: DEC: DMC and 1 mol/L $LiPF_6$-PC: DMC+5%VC electrolyte solutions, and that of $Li_{4/3}Ti_{5/3}O_4$ was determined at 48.6±0.3 kJ·mol^{-1} in PC-based electrolyte and 44.0±1.2 kJ·mol^{-1} in EC +DEC-based electrolyte by Doi et al.[29].

4.4 Diffusion coefficients

Normally, the concentration of the lithium ions in solution is nearly 1mol·L^{-1} and the diffusion coefficient, D, is about 10^{-5}cm^2·s^{-1}, and the values of these two parameters are both bigger than that in intercalation compounds (concentration ~ 10^{-2}mol·L^{-1}, D ~ 10^{-10}cm^2·s^{-1}). Therefore, the charge transfer in the liquid phase could be neglected. The feature of EIS of intercalation compounds electrode in the low-frequency region shows an inclined line. It represents the Warburg impedance (Z_w), which is associated with lithium ions diffusion in the intercalation particles. Ho and Huggins et al.[35] first applied EIS to investigate the intercalation reaction of the intercalation electrode, and they did some theoretical derivations in order to get the expression of Z_w. Thus, the diffusion coefficient D can be calculated form Z_w.

For the case of semi-infinite diffusion

$$Z_w = \frac{\mathrm{B}}{\sqrt{\omega}} - j\frac{\mathrm{B}}{\sqrt{\omega}} \tag{4-54}$$

here ω is the radial frequency, $j = \sqrt{-1}$, and B is a constant which contains a concentration independent diffusion coefficient.

For a thin film electrode, the diffusion coefficients were obtained by Ho[35]and Wu[112] et al. Here, two extreme cases are considered.

First, if $\omega >> \frac{2D}{L^2}$, D is the diffusion coefficient and L is the thickness of the thin film electrode. The phase angle is equal to π/4[35, 112], thus, Z_w has the form

$$Z_w = |Z|\cos(\frac{\pi}{4}) - j|Z|\sin(\frac{\pi}{4}) \quad (4\text{-}55)$$

where

$$|Z| = \left|\frac{V_m}{FAD}\frac{dE}{dx}\omega^{-1/2}\right| \quad (4\text{-}56)$$

V_m is the molar volume of the materials, A is the electrode's surface area, D is the diffusion coefficient, and $\frac{dE}{dx}$ is the slope of coulometric titration curve.

Comparison of Eq. (4-56) with Eq. (4-54), the diffusion coefficient D can be calculated from the slope of $Z' \sim \frac{1}{\sqrt{\omega}}$ or $Z'' \sim \frac{1}{\sqrt{\omega}}$, B can be written as[35]

$$\mathrm{B} = -\frac{V_m}{\sqrt{2}FAD}\frac{dE}{dx} \quad (4\text{-}57)$$

The second condition consists in that, if $\omega << \frac{2D}{L^2}$, then the phase angle is equal to π/2. Such phase angle indicates that the current is 90° out of phase with the voltage and is independent of the diffusion coefficient.

Thus, D can be calculated from the limiting low frequency resistance R_L and capacitance C_L. The R_L and C_L can be written as

$$R_L = -\frac{V_m L}{3FAD}\frac{dE}{dx} \quad (4\text{-}58)$$

$$\frac{1}{\omega C_L} = -\frac{V_m}{FA\omega L}\frac{dE}{dx} \quad (4\text{-}59)$$

From Eq. (4-58) and (4-59), we can get

$$D = \frac{L^2}{3R_L C_L} \quad (4\text{-}60)$$

4.5 Phase transformation and intercalation capacitance

At very-low-frequencies, the impedance spectrum starts to deviate from the shape expected due to diffusion alone. The features of EIS of intercalation electrode in the very-low-frequency region are composed of a semicircle and a vertical line, which can be attributed to the phase transformation of the intercalation particles or the growth of a new phase and the accumulation of lithium ions in the intercalation particles, respectively. These processes could be modeled using a resistance (R_b)-capacitance (C_b) parallel element with a capacitance (C_{int}) in series. Here, R_b and C_b represent the resistance and the capacitance of the phase change of the intercalation particles, C_{int} is the intercalation capacitance of the electrodes.

4.5.1 Phase transformation

Normally, the measured frequency range of EIS in lithium ion batteries system is from 10^5 to 10^{-2} Hz, and the volume changes of the practical intercalation compounds, such as $LiCoO_2$, $LiMn_2O_4$ and graphite materials etc. are not obvious. Moreover, there is little difference between the physical/chemical properties of the two phases which are the new phase and the origin phase. Hence, the semicircle related to the phase transformation is seldom found in EIS in the low-frequency region (≥0.01Hz) for the intercalation compounds. However, for the Cu_6Sn_5[32] alloy or Si anodes for lithium-ion batteries, a new arc is detected which can be attributed to the phase formation process in the low-frequency region (**Figure 4-2**). For the Cu_6Sn_5 alloy, there is a phase transformation process form Cu_6Sn_5 to $Li_xCu_6Sn_5$ ($0<x<13$) under 0.4 V (Li/Li^+) with large volume expansion. Owing to large differences between the physical/chemical properties of the Cu_6Sn_5 phase and that of the $Li_xCu_6Sn_5$ ($0<x<13$) phase, two independent time constants for lithium ions transference in the two phases appear and, as a result, a new arc is formed. For Si anodes, the same phenomenon can be found in the potential region 0.1-0.01 V (Li/Li^+), and there exists a phase transformation process from Si to Li_xSi ($0<x\leq4.4$).

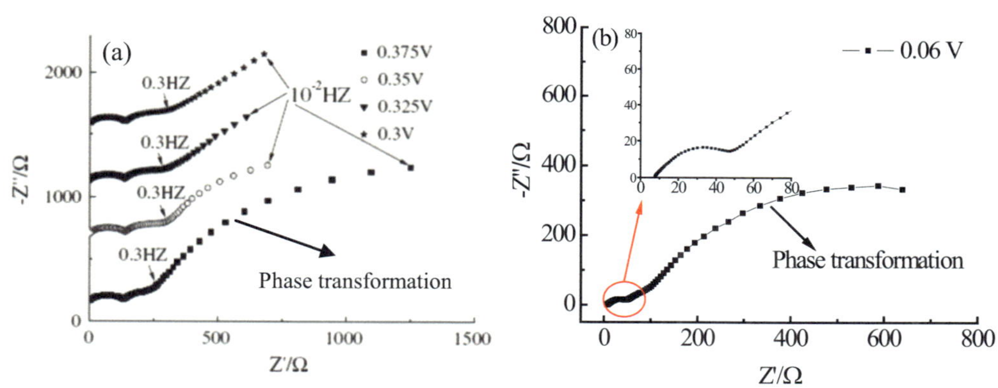

Fig. 4-2. Nyquist plots of Cu_6Sn_5 alloy electrode (a) [32] and Si electrode (b).

4.5.2 Intercalation capacitance

The composition of the intercalation electrode is described in terms of an occupancy fraction $x=n/N$, where n is the number density of intercalated atoms, and N is some number density proportional to the size of the host, such as the number density of one type of site or one type of host atom (i.e. Li in Li_xCoO_2). The steady state voltage of the working electrode, V, is related to the chemical potential of the intercalated atom in the working electrode, μ, and that in the reference electrode, μ_{ref}, by the following equation[64]

$$V = -\frac{1}{e}(\mu - \mu_{ref}) \tag{4-61}$$

where e stands for the positive elementary charge. Note that z=1 for Li. The term $-zeV$ is the work done on the cell per ion intercalated under a voltage difference of V. Therefore, the voltage (V) is determined by the variation of the chemical potential of the cathode as a function of composition, $\mu(x)$. Note that measuring the cell voltage at equilibrium versus the charge passed between the electrodes is equivalent to measuring the chemical potential as a function of x, i.e. the Li content of the Li_xCoO_2 compound. The thermodynamics requires that μ increases with the concentration of guest ions, so V decreases as ions are added to the positive electrode. The voltage V is determined by the variation of the chemical potential of the cathode with the composition, μ(x). The equilibrium intercalation capacitance can be defined in the following way[14]:

$$C_0 = -\frac{dQ}{dV} = LAe^2N\frac{dx}{d\mu} \tag{4-62}$$

where Q=$LAne$ is the total extent of charge passed, L is the film thickness, and A is the area. For noninteracting particles that occupy identical sites of energy E_0 (lattice gas model), the probability of occupancy at equilibrium is determined by Fermi-Dirac statistics

$$F(E_0,\mu) = \frac{1}{1+\exp[(E_0-\mu)/k_BT]} \tag{4-63}$$

Generally, one complication of the lattice-gas model applied to intercalation compounds is related to the dissociation of the intercalated atom into ions and electrons. In the present situation, the electronic contributions to the chemical potential (μ) will be neglected so that μ_e (the chemical potential of electrons) is considered as constant, and only μ will be considered to vary as a function of Li insertion, that is, x. Therefore, it is possible to focus attention on the contributions of intercalated ions to the intercalation capacitance. Because x=n/N =$F(E_0, \mu)$, the chemical potential function of ions distributed in the sites of a solid matrix takes the form

$$\mu(x) = E_0 + k_BT\ln(\frac{x}{1-x}) \tag{4-64}$$

In this case, the intercalation capacitance, C_{int}, is given by:

$$C_0 = \frac{Le^2N}{k_BT}x(1-x) \tag{4-65}$$

At the beginning of lithium ion inserts into or the very end of lithium ion extracts from the active mass during the electrochemical processes, that is to say, for very low insertion level ($x \to 0$). Eq. (4-65) takes the form

$$C_{int} = \frac{Le^2N}{k_BT}x \tag{4-66}$$

Substituting $x = \exp\left[f\left(E - E_0\right)\right]$ into Eq. (4-66) and changing the obtained equation to a linear equation by a logarithm, we can obtain the following expression of C_{int}

$$\ln C_{\text{int}} = \ln \frac{Le^2 N}{k_B T} + f\left(E - E_0\right) \tag{4-67}$$

Eq. (4-67) presents explicitly that, when $x \to 0$, $lnC_{int} \sim E$ shows a linear variation.

A more complete description of intercalation electrodes considers, in addition to the entropic term in Eq. (4-67), other significant effects such as the interactions among ions and the lattice distortions induced by the guest species. In many cases, these effects are described by effective potential terms that are a function of x. An example is the following chemical potential function,

$$\mu(x) = E_0 + k_B T \ln(\frac{x}{1-x}) + g k_B T x \tag{4-68}$$

where g is the dimensionless interaction parameter that includes both the interactions between intercalated ions and strain fields caused by expansion or contraction of the lattice [45, 96]. The critical value of the interaction parameter is g = -4. The capacitance related to Eq. (4-66) is[14, 15]

$$C_{\text{int}} = \frac{Le^2 N}{k_B T} \left[g + \frac{1}{x\left(1-x\right)} \right]^{-1} \tag{4-69}$$

From the Eq. (4-69), it can be seen that when the insertion lever x=0.5, C_{int} has the maximum value. Meanwhile, when $x \to 0$, that means there are few intercalated lithium ions in the materials' lattice.

C_{int} can be obtained from the imaginary part of the impedance in the limit of very low frequencies, $\omega \to 0$, thus, C_{int} can be written as the form[46, 51]

$$C_{\text{int}} = -\frac{1}{\omega Z''} \tag{4-70}$$

where ω is the angular frequency of a small-amplitude ac voltage.

5. Impedance spectra of nonhomogeneous, multilayered porous composite intercalation electrode

The electrochemical response of Li insertion electrodes is usually very complicated. In this part, graphite electrodes are discussed as a typical example, since graphite is an excellent electronic conductor, and there is no semicircle related to the electronic properties of graphite materials in the Nyquist plots. The practical ion intercalation electrodes are composite materials, in which the active mass particles are bound to a current collector with PVDF binder, and the electrodes are usually prepared from slurry of the particles and the binder in an organic solvent that spreads out on the current collector, and followed by drying. Therefore, the preparation of composite intercalation electrode coatings, especially with manual preparation, may result in a nonhomogeneous distribution of the mass of porous electrodes. To analyze the consequence of such a thickness distribution on the electrode's impedance, as shown in **Figure 5-1**, we assume a two-thickness (L_1 and L_2) layer distribution, with a related contribution to the total current of θ_{L1} and (1-θ_{L1}). Every layer composes of only two spherical particles with different radii, r_1 and r_2. The impedance behavior of the non-homogeneous, multilayered porous composite electrode will be analyzed in the following discussion.

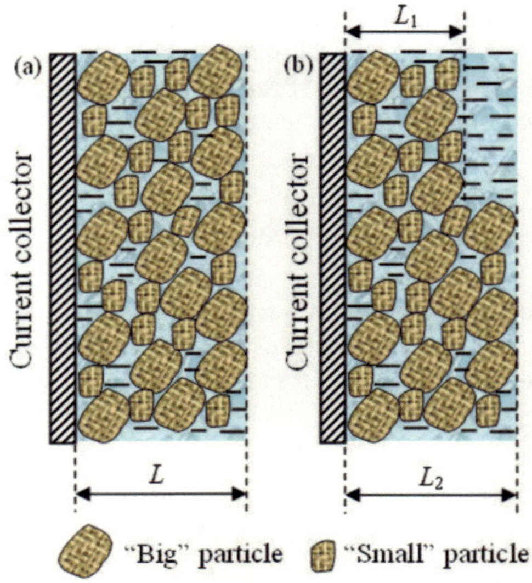

Fig. 5-1. Schematic views of two general models of the porous electrode[115]
Both of the two models of the porous electrode contain two different radii:
(a) homogeneous, porous electrode and (b) nonhomogeneous, multilayered porous electrode.

The total admittance of an electrode which contains two types of spherical particles with different radii ($1/Z_{mix}$) can be regarded as an averaged sum of the individual admittances $1/Z_{part,1}$ and $1/Z_{part,2}$[46, 49, 65].

$$\frac{1}{Z_{mix}}=\frac{\theta_1}{Z_{part,1}}+\frac{1-\theta_1}{Z_{part,2}} \tag{5-1}$$

where θ_1 is the fraction of the total capacity due to a contribution of the "small" particles. $Z_{part,1}$ and $Z_{part,2}$ here represent the impedance of the "small" particles and "big" particles, respectively.

The impedance of an individual insertion/extraction particle can be written from[46]

$$Z_{part,i}=\frac{R_{SEI,i}}{1+j\omega R_{SEI,i}C_{SEI,i}}+\frac{R_{ct,i}+Z_{W,i}}{1+j\omega C_{dl,i}[R_{ct,i}+Z_{W,i}]} \tag{5-2}$$

where i denotes particles with different radii, i=1, 2; R_{SEI} and C_{SEI} stand for the resistance of ions' migration through the SEI films and the films' capacitance, respectively. Z_W is the Warburg resistance, and Z_W can be presented in the following form[46]

$$Z_w = \frac{R_{part,\,i}}{Y_{s,\,i}} \tag{5-3}$$

with the finite-space diffusion resistive element, $R_{part,i}$ and $(1/Y_{s,i})$ of the form[46, 49, 65]

$$R_{part,\,i} = \frac{r_{s,\,i}^2}{3D_s C_{part,\,i}} = \frac{\tau_i}{3C_{part,\,i}} \tag{5-4}$$

$$\frac{1}{Y_{s,\,i}} = \frac{\tanh(\sqrt{j\omega\tau_i})}{\left(\sqrt{j\omega\tau_i} - \tanh(\sqrt{j\omega\tau_i}\,)\right)} \tag{5-5}$$

where $C_{part,i}$ is the limiting low-frequency capacitance of a spherical particle, $r_{s,i}$ represents the diameter of the spherical particle, τ_i is the diffusion time constant, and ω is the angular frequency.

Meyers et al.[65] described an important theoretical model for porous intercalation electrodes that consisted of individual spherical particles with different kinds of size distribution (see Fig. 5-1 (a)). For a porous electrode composed of only the same sized spherical particles, the characteristics of EIS are affected by four major parameters, (*i*) the electronic conductivities of the intercalation particles, ε; (*ii*) the solution in the pore space, γ; (*iii*) the ratio of the true surface area of the particles to their volume δ; and (*iv*) the sharpness of the particles' distribution. The impedance of the porous electrode, Z_{porous}, is relative to the impedance of the mixed particle electrode, and Z_{porous} can be written as the following form[23, 46, 49, 65]

$$Z_{porous} = \frac{L}{\varepsilon + \gamma}[1 + \frac{2 + ((\varepsilon/\gamma) + (\gamma/\varepsilon))\cosh\nu}{\nu\sinh\nu}] \tag{5-6}$$

with the parameter ν of the form

$$\nu = L(\frac{\varepsilon + \gamma}{\gamma\varepsilon})^{1/2}(\frac{\delta}{Z_{mix}})^{1/2} \tag{5-7}$$

here, L is the thickness of the porous electrode.

From the Equation (5-1), the impedance of a parallel combination of two porous layers, Z_{L1+L2}, can be expressed below

$$\frac{1}{Z_{L1+L2}} = \frac{\theta_{L1}}{Z_{mix,\,L1}} + \frac{1-\theta_{L1}}{Z_{mix,\,L2}} \tag{5-8}$$

where $Z_{mix,L1}$, $Z_{mix,L2}$ and L_1, L_2 are the impedance and the thickness of two layers, respectively.

According to the model presented in this study and Eqs. (5-2)-(5-8), a typical Nyquist plot of the electrode that contains three semicircles and a slope line is simulated, which is shown in **Figure 5-2**. As can been seen, the calculated impedance spectroscopy has the similar characteristic with the experimental results, and the impedance plots calculated according to

Eqs. (5-2)-(5-8) are in good qualitative agreement with the experimental impedance spectroscopy of composite, porous graphite electrode. The parameters involved in the calculation are indicated in **Table 5-1** and **5-2**. The calculated thicknesses of the two sublayers are 0.06 and 0.12 cm, which are both thicker than the practical, industrial intercalation electrodes.

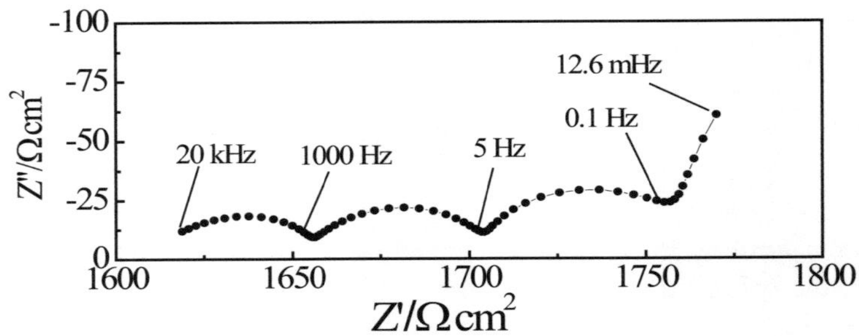

Fig. 5-2. Nyquist plot simulated by computer according to the non-homogeneous, multilayered porous electrode model[115].

	$r_{s,i}/\mu m$	$R_{part,i}/\Omega\ cm^2$	$C_{dl,i}/\mu F\ cm^{-2}$	$R_{ct,i}/\Omega\ cm^2$	$C_{SEI,i}/\mu F\ cm^{-2}$	$R_{SEI,i}/\Omega\ cm^2$
Big particles	2	200	200	300	5	100
Small particles	0.3	200	100	200	4	80

Table 5-1. Parameters of the Graphite Particles with Different Radii

Parameters	$D_s/cm^2\ s^{-1}$	L_1/cm	L_2/cm	δ/cm^{-1}	$\gamma/\Omega^{-1}\ cm^{-1}$	$\varepsilon/\Omega^{-1}\ cm^{-1}$	θ_{L1}	θ_{L2}
Model data	3×10^{-10}	0.06	0.12	5×10^{3}	5.5×10^{-5}	1×10^{-5}	0.15	0.85

Table 5-2. Porous Structure Parameters

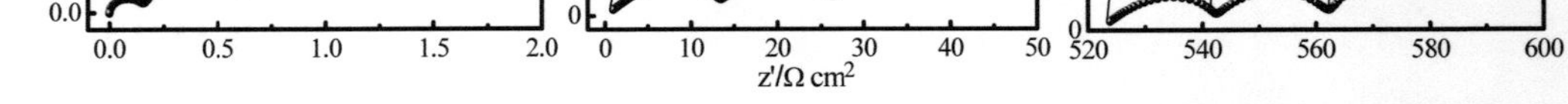

Fig. 5-3. Nyquist plots for a porous electrode composed of mixture of "small" and "big" particles with the different values for ε and γ according to the non-homogeneous, multilayered porous electrode model[115].

To further make clear the limitations and implications of our model, some simulations were carried out to discuss the effect of the porosity of the porous electrode (the electronic conductivities of the insertion particles, ε, and the solution in the pore space, γ), particle size, and the layer distribution on the impedance response of porous electrodes. **Figure 5-3** illustrates the effect of different ε and γ on the impedance structure of a porous electrode composed of mixture of "small" and "big" particles according to the nonhomogeneous, multilayered porous electrode model. Other parameters for the porous electrode are as same as indicated in the **Table 5-1**. It can be seen that, whether the values of the two conductivities are high or low, the feature of the impedance of a porous electrode does not change. When the values of ε and γ are increased, it causes a decrease in resistance and capacitance for the whole impedance, together with a drastic shift of the curve along the real impedance axis.

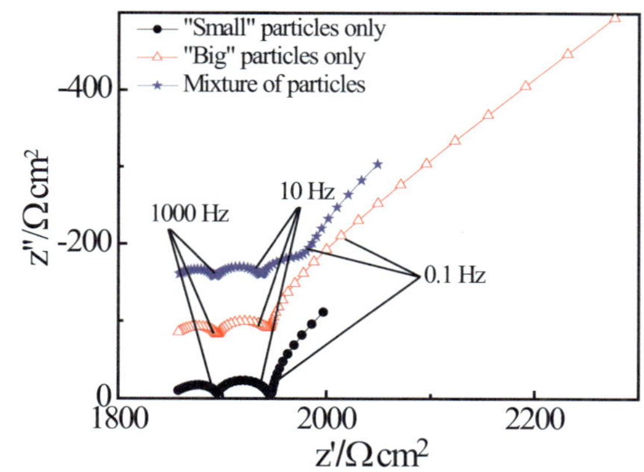

Fig. 5-4. Simulated impedance spectra of a porous electrode composed of "small", "big" and mixture particles according to the homogeneous, porous electrode model[115].

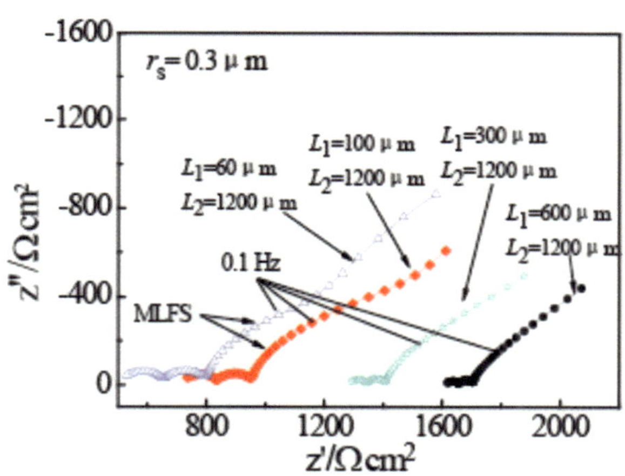

Fig. 5-5. Family of Nyquist plots for a porous electrode composed only of "small" particles with the set of different thickness of the "thinner" layer[115].

The homogeneous, porous electrode model was used to investigate the impedance spectra for a porous electrode composed of "small", "big", and mixture of particles. The results are shown in **Figure 5-4**, and the radii of the particles are 0.3 and 1 μm. As can be seen, the shapes of the impedance for the porous electrode with a single particle distribution ("big" or "small" particles) are the same. When the porous electrode contains a mixture of particles,

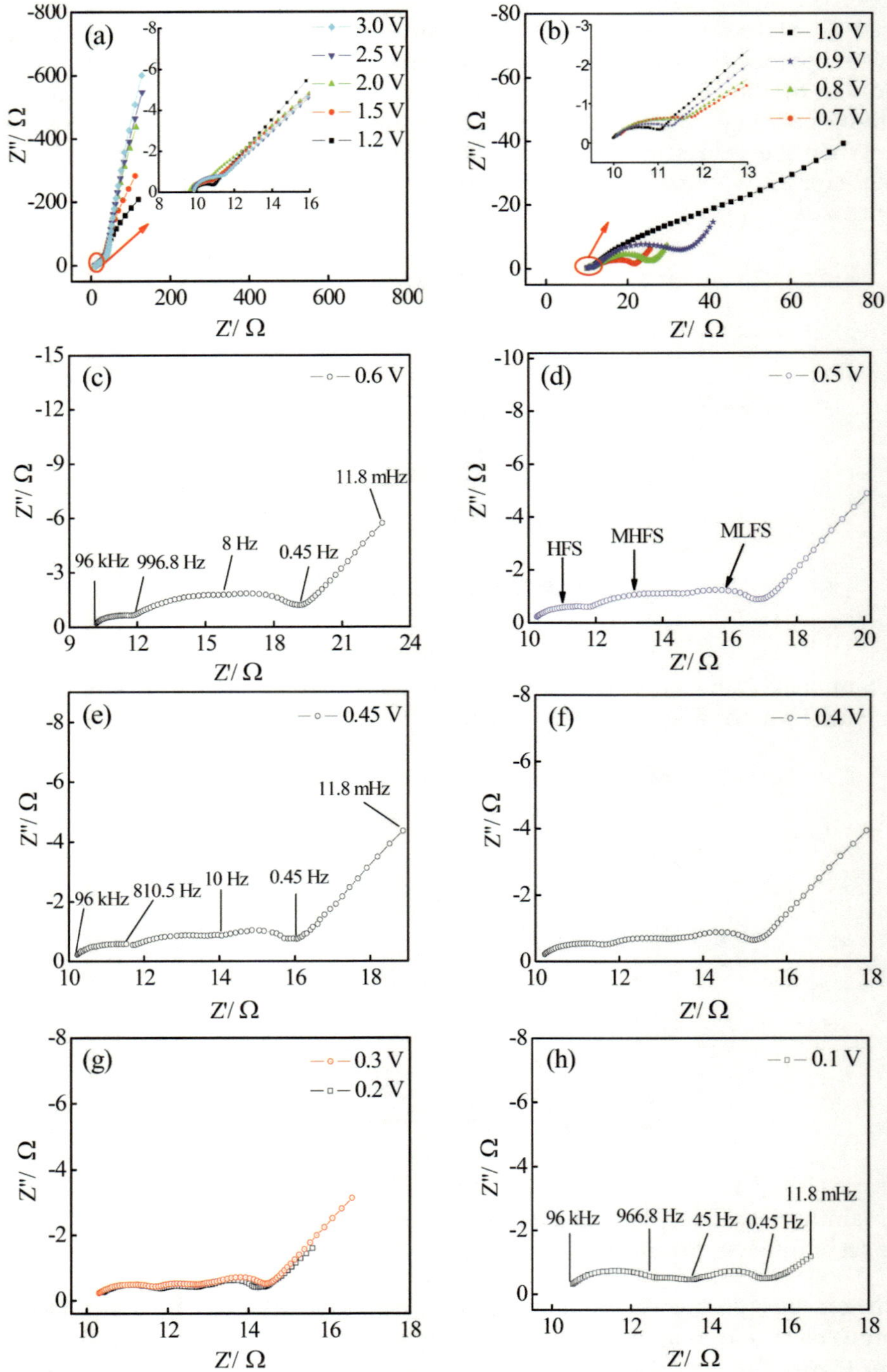

Fig. 5-6. Nyquist plots of the graphite electrode at various potentials from 3.0 to 0.1 V during the first lithium-ion insertion[115].

this yields the impedance spectral feature where an arc is gradually formed in the frequency domain 10-0.1 Hz. This phenomenon demonstrates that the particle size distributions may result in a certain inclination towards an arc in the middle-to-low frequency domain, corresponding to the result reported by Diard et al.[28].

The impedance spectra of a porous electrode composed only of "small" particles with the set of different thickness of the "thinner" layer were obtained in **Figure 5-5**. When the thickness of the "thinner" layer is low, this causes a shift of the curve along the real impedance axis and an arc in the middle-to-low frequency domain is formed gradually.

On the basis of the above analysis, it can be concluded that the particle size and layer distributions may result in a certain inclination toward an arc in the middle-to-low frequency domain. It is worthy to note that an MLFS will be well-developed even if all of the particles have the same radius, but the electrodes are also composed of two parts (thin and thick parts). That is to say, the appearance of the middle-to-low frequency semicircle could be achieved by adopting a nonhomogeneous, layered distribution of the electrode with a different particle size distribution. The nonhomogeneous, multilayered porous electrode model prediction is in good qualitative agreement with the impedance spectra of nonhomogeneous, multilayered porous graphite electrode measured in our previous study[115], as shown in **Figure 5-6**.

6. The inductance formation mechanism

An inductive loop (IL) is often observed in the impedance spectra, but the source for this element in the electrode system remains unclear. An inductance is defined as the properties of an electric circuit that causes an electromotive force to be generated in it as a result of a change in the current flowing through the circuit. Gnanaraj et al.[34] suggested that a plausible explanation for the IL is the formation of an electromotive force superimposed on

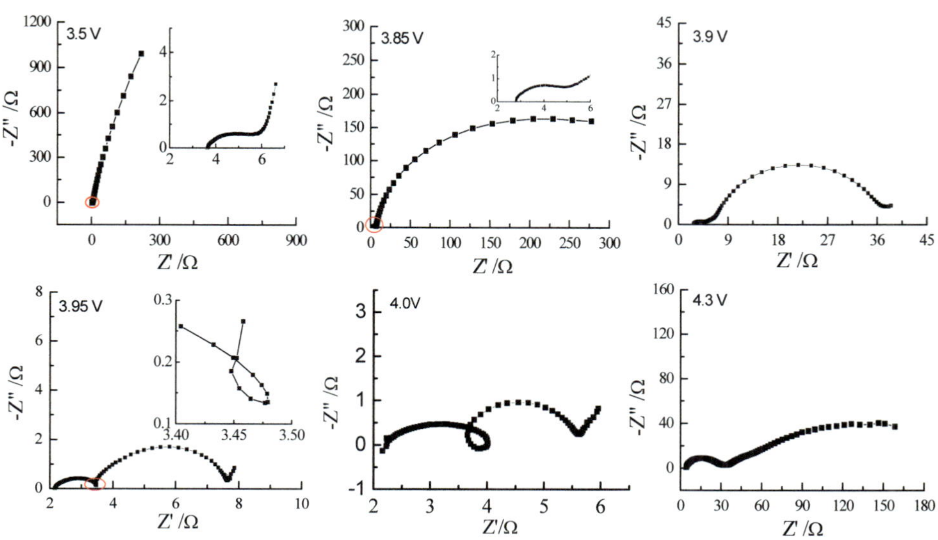

Fig. 6-1. Variations of impedance spectra of $LiCoO_2$ electrode with the polarization potential in the first delithiation[128].

the lithium ion extraction. The inductive loop has been found in EIS in our previous studies about layered $LiCoO_2$[128] and spinel $LiMn_2O_4$[127] for lithium ion batteries (see **Figure 6-1**). For layered $LiCoO_2$ electrode, when lithium ions are extracted from the $LiCoO_2$ electrode, isolation of Li-rich and Li-poor (deficient) regions in the electrode may be created by the SEI film due to disequilibrium with respect to electronic continuity, thus a concentration cell is established between $LiCoO_2$ and delithiation $LiCoO_2$ ($Li_{1-x}CoO_2$, $0<x<0.5$) separated by the SEI film. Because the SEI film is imperfect in the first delithiation of $LiCoO_2$, a current flows within the concentration cell, which generates a field that opposes the filed due to the process of lithium ions extraction. The discharge of a concentration cell involves current flow opposed to charging the $LiCoO_2$ electrode, such a situation meets well with the requirements for the formation of an inductive loop.

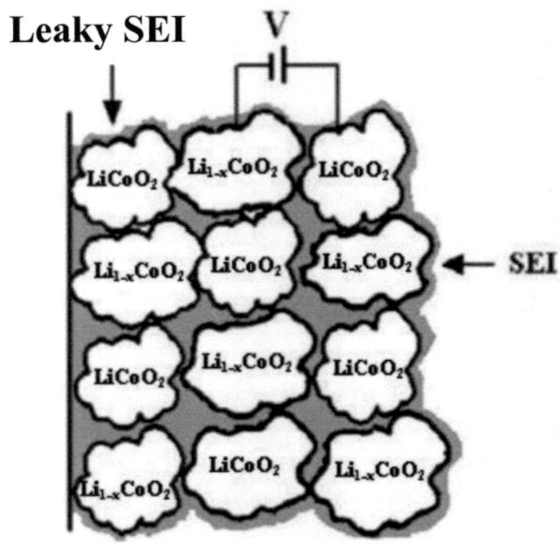

Fig. 6-2. Pictorial representation model for $LiCoO_2/Li_{1-x}CoO_2$ concentration cell[128].

To well understand the inductive loop in the $Li/LiCoO_2$ cell systems, a pictorial model representation of the SEI film growth and the concentration cell is presented in **Figure 6-2**. With lithium ion extraction from the $LiCoO_2$ electrode in charge-discharge processed, the $LiCoO_2/Li_{1-x}CoO_2$ concentration cell continues to leak current until the electrode fully deintercalated (corresponding to $LiCoO_2$ fully converted to $Li_{0.5}CoO_2$), i.e. the lithium ion concentration differences in the electrode are removed. Thus, the SEI film on the $LiCoO_2$ electrode surface may be termed a "leak SEI film".

For the $Li/LiMn_2O_4$ cell system, an IL also appeared in the potential regions of partial lithium ions deintercalated, that implies the mechanism of the inductance formation is as same as the $Li/LiCoO_2$ cell system. But interestingly, another IL was turning up along with the former inductive loop at the potential of 3.975 V (see **Figure 6-3**), that means there are two different mechanism of the inductance formation in the $Li/LiMn_2O_4$ cell system, corresponding to two concentration cells exist. When the $LiMn_2O_4$ fully converted to Mn_2O_4, the inductive loop was disappeared. As reported in the literature[67, 101], there has a two-step reversible (de)intercalation reaction, in which lithium ions occupy two different tetragonal 8a sites in spinel $Li_{1-x}Mn_2O_4$ ($0<x<1$), so the two concentration cells could be signified as $LiMn_2O_4/Li_{1-x}Mn_2O_4$ and $Li_{0.5}Mn_2O_4/Li_{0.5-x}Mn_2O_4$ ($0<x<0.5$) concentration cells, and the schematic presentation of models for $LiMn_2O_4/Li_{1-x}Mn_2O_4$ and $Li_{0.5}Mn_2O_4/Li_{0.5-x}Mn_2O_4$ ($0<x<0.5$) concentration cells is presented in **Figure 6-4**. The IL caused by the $LiMn_2O_4/Li_{1-x}Mn_2O_4$ concentration cell appears in the middle-to-high frequency region (~100 Hz) in the Nyquist plots, and the middle-to-low frequency region (~1 Hz) for $Li_{0.5}Mn_2O_4/Li_{0.5-x}Mn_2O_4$ concentration cell.

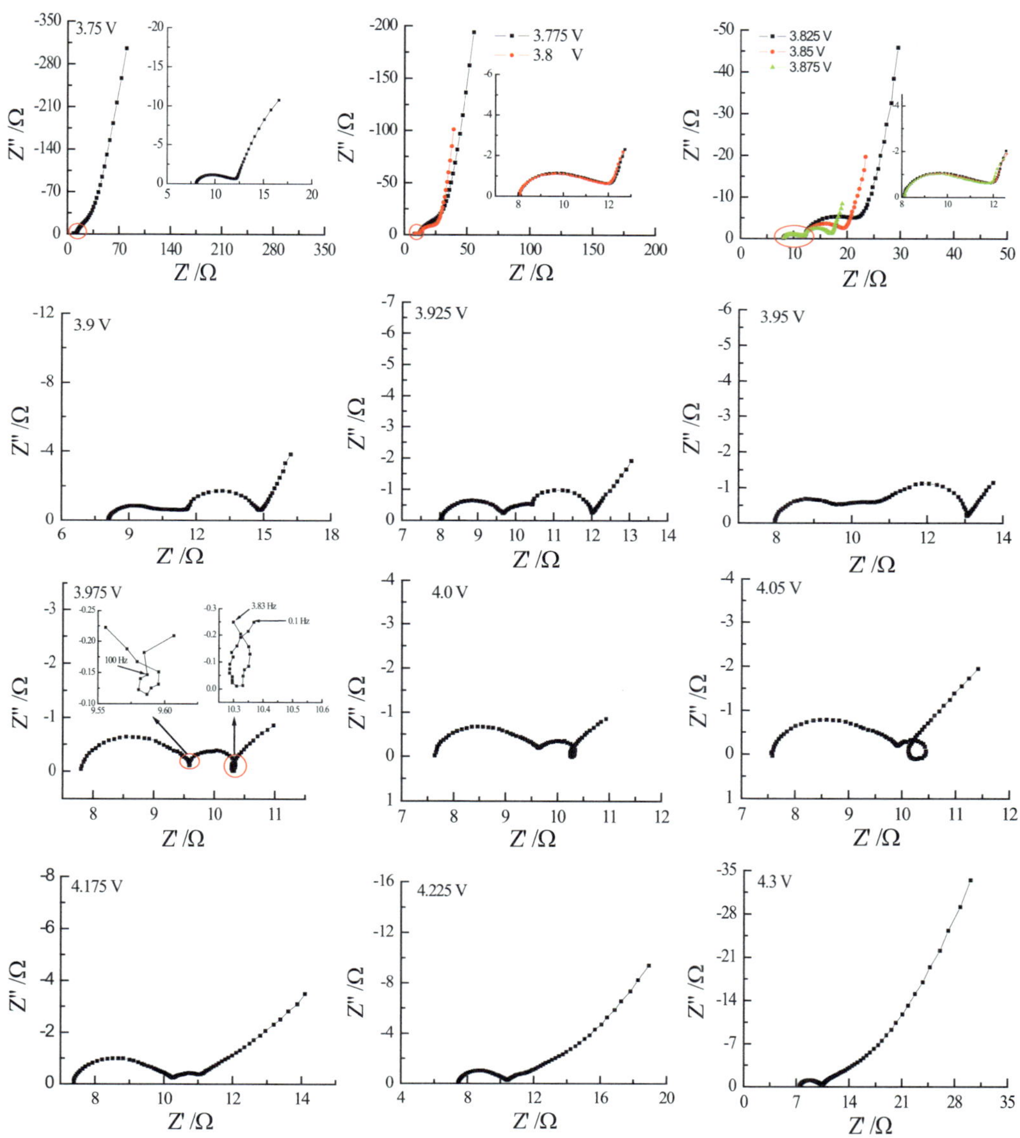

Fig. 6-3. Variations of impedance spectra of $LiMn_2O_4$ electrode with the polarization potential in the first delithiation[127].

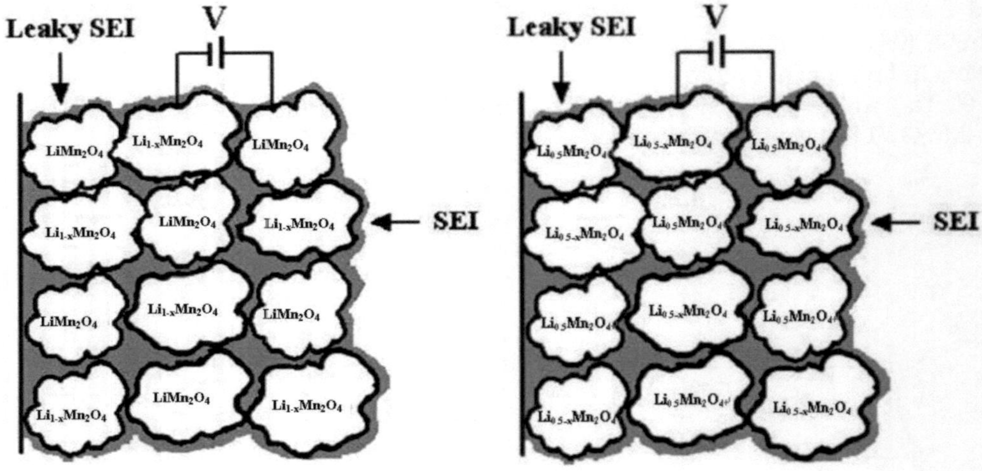

Fig. 6-4. Pictorial representation model for
$LiMn_2O_4/Li_{1-x}Mn_2O_4$ and $Li_{0.5}Mn_2O_4/Li_{0.5-x}Mn_2O_4$ ($0<x<0.5$) concentration cells[127].

7. Conclusions

Impedance measurements have changed the way electrochemists interpret the electrode/electrolyte interface. The technique offers the most powerful analysis on the status of electrodes during charging/discharging processes of lithium-ion batteries. More importantly, this technique may be used to monitor and probe many different processes that occur during electrochemical experiments, including adsorption of reactants and products as well as various reactions that either precede or follow the experiments, thereby changing the electrical characteristics of electrode/electrolyte interfaces. However, there are also some technical matters in application: (*i*) Lots of factors such as the electrolytic cell system, the composition of intercalation electrode, the content and particle size of active materials, as well as the thickness and preparation of electrode film, can affect the shape and value of the resistance in the impedance spectrum. Therefore, it is improtant to determine the physical processes corresponding to time constants ranging via theoretical analysis and contrast tests in detail; (*ii*) A debate is still open on the attribution of time constants relating to different physical processes in lithium insertion, and it needs to be further investigated; (*iii*) At present, it takes a long time to make impedance measurements in a full frequency region at times more than a few hours, particularly when the frequencies of as low as milli- or submillihertz are included, even for a stable system. Thus, the challenge of the future is to devise the fast impedance technique to shorten its measurement time. At last, correlation of impedance measurements and other experiments (frequently employing non-electrochemical techniques, e.g. in-situ FTIR, in-situ XRD, etc.) would make the analysis of EIS more persuasive.

8. Acknowledgement

This work was supported by the Fundamental Research Funds for the Central Universities (2010LKHX03) and Major State Basic Research Development Program of China (2009CB220102).

9. References

[1] Abraham, D. P. ; Reynolds, E. M. ; Schultz, P. L. ; Jansen, A. N. ; Dees, D. W. *J. Electrochem. Soc.* 2006, *153*, A1610-A1616.

[2] Atkinson, K. J. W.; Grimes, R. W.; Owens, S. L. *Solid State Ionics* 2002, *150*, 443-448.

[3] Aurbach, D.; Ein-Eli, Y.; Markovsky, B. *J. Electrochem. Soc.* 1995, *142*, 2882-2890.

[4] Aurbach, D.; Gamolsky, K,; Markovsky, B. *J. Electrochem. Soc.* 2000, *147*, 1322-1331.

[5] Aurbach, D.; Levi, M. D.; Levi, E. *J. Electrochem. Soc.* 1998, *145*, 3024-3034.

[6] Aurbach, D.; Zaban, A.; Schechter, A. *J. Electrochem. Soc.* 1995, *142*, 2873-2882.

[7] Badway, F.; Cosandey, F.; Pereira, N.; Amatucci, G. G. *J. Electrochem. Soc.* 2003, *150(10)*, A1318-A1327.

[8] Badway, F.; Pereira, N.; Cosandey, F.; Amatucci, G. G. *J. Electrochem. Soc.* 2003, *150(9)*, A1209-A1218.

[9] Bard, A. J.; Faulkner, L. R. Electrochemical Methods Fundamentals and Applications. John Wiley & Sons, Inc. New York, 2001.

[10] Barral, G.; Diard, J. P.; Montella, C. *Electrochim. Acta* 1984, *29*, 239-246.

[11] Barsoukov, E.; Kim, D. H.; Lee, H. S.; Lee, H.; Yakovleva, M.; Gao, Y.; Engel, J. F. *Solid State Ionics* 2003, *161*, 19-29.

[12] Barsoukov, E.; Kim, J. H.; Kim, J. H.; Yoon, C. O.; Lee, H. *Solid State Ionics* 1999, *116*, 249-261.

[13] Barsoukov, E.; Macdonald, J. R. Impedance Spectroscopy Theory, Experiment, and Applications. John Wiley & Sons, Inc. Hoboken, New Jersey, 2005.

[14] Bisquert, J. *Electrochim. Acta* 2002, *47*, 2435-2449.

[15] Bisquert, J.; Vikhrenko, V. S. *Electrochim. Acta* 2002, *47*, 3977-3988.

[16] Blomgren, G. E. *J. Power Sources* 1999, *81–82*, 112.

[17] Boukamp, B. A. *Solid State Ionics* 2004, *169*, 65–73.

[18] Brett, C. M. A.; Oliveira, B. A. M. Electrochemistry, Principles, Methods, and Applications, Oxford University Press, 1993.

[19] Bruce, P. G.; Saidi, M. Y. *J. Electrochem. Chem.* 1992, *322*, 93-105.

[20] Bruce, P. G.; Saidi, M. Y. *Solid State Ionices* 1992, *51*, 187-190.

[21] Card, H. C.; Rhoderick, E. H.; *J. Phys. D* 1971, *4*, 1589-1601.

[22] Chang, B. Y.; Park, S. M. *Annu. Rev. Anal. Chem.* 2010, *3*, 207–29.

[23] Chang, Y. C.; Sohn, H. J. *J. Electrochem. Soc.* 2000, *147*, 50-58.

[24] Chiodelli, G.; Lupotto, P. *J. Electrochem. Soc.* 1991, *138(9)*, 2073-2711.

[25] Croce, F.; Nobili, F.; Deptula, A.; Lada, W.; Tossici, R.; D´Epifanio, A.; Scrosati, B.; Marassi, R. *Electrochem. Commun.* 1999, *1*, 605-608.

[26] Débart, A.; Dupont, L.; Patrice, R.; Tarascon, J. M. *Solid State Sci.* 2006, *8(6)*, 640-651.

[27] Delahay, P. New Instrumental Methods in Electrochemistry, Interscience, New York, 1954.

[28] Diard, J. P.; Le Gorrec, B.; Montella, C. *J. Electroanal. Chem.* 2001, *499*, 67-77.

[29] Doi, T.; Iriyama, Y.; Abe, T.; Ogumi, Z. *Anal. Chem.* 2005, *77*, 1696-1700.

[30] Dokko, K.; Mohamedi, M.; Fujita, Y.; Itoh, T.; Nishizawa, M.; Umeda, M.; Uchida, I. *J. Electrochem. Soc.* 2001, *148*, A422-A426.

[31] Ely, Y. E.; Aurbach, D. *Langmuir* 1992, *8*, 1845-1850.

[32] Fan, X. Y.; Zhuang, Q. C.; Wei, G. Z.; Huang, L.; Dong, Q. F.; Sun, S. G. *J. Appl. Electrochem.* 2009, *39*, 1323-1330.

[33] Gileadi, E. Electrode Kinetics for Chemists, Engineers, and Material Scientists, VCH, New York, 1993.

[34] Gnanaraj, J. S.; Thompson, R. W.; Iaconatti, S. N. *Electrochem. Solid-State Lett.* 2005, *8(2)*, A128-132.

[35] Ho, C.; Raistrick, I. D.; Huggins, R. A. *J. Electrochem. Soc.*, 1980, *127*, 343-350.

[36] Holzapfel, M.; Martinent, A.; Alloin, F.; Gorrec, B. L.; Yazami, R.; Montella, C. *J. Electroanal. Chem.* 2003, *546*, 41-50.

[37] Itagaki, M.; Kobari, N.; Yotsuda, S.; Watanabe, K.; Kinoshita, S.; Ue, M. *J. Power Sources* 2005, *148*, 78-84.

[38] Julien, C. M. *Mater. Sci. Eng. R* 2003, *40*, 47-102.

[39] Kanoh, H.; Feng, Q.; Hirotsu, T.; Ooi, K. *J. Electrochem. Soc.* 1996, *143*, 2610-2615.

[40] Kobayashi, S.; Uchimoto, Y. *J. Phys. Chem. B* 2005, 109, 13322-13326.

[41] Kurzweil, P. AC Impedance Spectroscopy – A Powerful Tool for the Characterization of Materials and Electrochemical Power Sources. In Proceedings: the 14th International Seminar on Double Layer Capacitors, Deerfield Beach, FL., U.S.A., 2004, December 6-8, p1-16.

[42] Lasia, A. Applications of the Electrochemical Impedance Spectroscopy to HydrogenAdsorption, Evolution and Absorption into Metals, In: Modern Aspects of Electrochemistry, Conway, B. E.; R. E. White, Edts, Kluwer/Plenum, New York, 2002, vol. 35, p. 1-49.

[43] Lasia, A. Electrochemical Impedance Spectroscopy and Its Applications. In: Modern Aspects of Electrochemistry, Conway, B. E.; Bockris, J.; White, R. E. Edts., Kluwer Academic/Plenum Publishers, New York, 1999, Vol. 32, p. 143-248.

[44] Lawless, K. R. *Rep. Prog. Phys.* 1974, *37*, 231.

[45] Levi, M. D.; Aurbach, D. *Electrochim. Acta* 1999, *45*, 167-185.

[46] Levi, M. D.; Aurbach, D. I *J. Phys. Chem. B* 2004, *108*, 11693-11703.

[47] Levi, M. D.; Aurbach, D. *J. Phys. Chem. B* 1997, *101*, 4630-4640.

[48] Levi, M. D.; Aurbach, D. *J. Phys. Chem. B* 2005, *109*, 2763-2773.

[49] Levi, M. D.; Aurbach, D. *J. Power Sources* 2005, *146*, 727-731.

[50] Levi, M. D.; Gamolsky, K.; Aurbach, D. *Electrochim. Acta* 2000, *45*, 1781-1789.

[51] Levi, M. D.; Salitra, G.; Markovsky, B.; et al. *J. Electrochem. Soc.* 1999, *146*, 1279-1289.

[52] Levi, M. D.; Wang, C.; Aurbach, D. *J. Electrochem. Soc.* 2004, *151*, A781-A790.

[53] Li, Y.; Wu, H. *Electrochim. Acta* 1989, *34*, 157-159.

[54] Liao, P.; MacDonald, B. L.; Dunlap, R. A.; Dahn, J. R. *Chem. Mater.* 2008, *20*, 454-461.

[55] Macdonald, D. D. *Electrochim. Acta* 2006, *51*, 1376-1388.

[56] Macdonald, D.D. Transient Techniques in Electrochemistry, Plenum Press, New York, 1977.

[57] Macdonald, J. R. *Ann. Biomed. Eng.* 1992, *20*, 289-305.

[58] Macdonald, J. R. *Electrochim. Acta* 1990, *35(10)*, 1483-1492.

[59] Macdonald, J. R. Impedance Spectroscopy Emphasizing Solid Materials and Systems, Wiley, New York, 1987.

[60] Macdonald, J. R. *Solid State Ionics* 2005, *176*, 1961–1969.

[61] Marassi, R.; Nobili, F.; Croce, F.; Scrosati, B. *Chem. Mater.* 2001, *13*, 1642-1646.

[62] Marzec, J.; Świerczek, K.; Przewoźnik, J.; Molenda, J.; R Simon, D.; M Kelder, E.; Schoonman, J. *Solid State Ionics* 2002, *146*, 225-237.

[63] Mauvernay, B.; Doublet, M. L.; Monconduit, L. *J. Phys. Chem. Solids* 2006, *67*, 1252-1257.

[64] McKinnon, W. R.; Haering, R. R. in: White R. E.; Bockris J. O. M.; Conway, B. E. (Eds.), Modern Aspects of Electrochemistry, vol.15, Plenum Press, New York, 1983, p. 235.

[65] Meyers, J. P.; Doyle, M.; Darling, R. M.; Newman, J. *J. Electrochem. Soc.* 2000, *147*, 2930-2940.
[66] Mirzaeian, M.; Hall, P. J. *J. Power Sources* 2010, *195*, 6817-6824.
[67] Miura, K.; Yamada, A.; Tanaka, M. *Electrochim. Acta* 1996, *41*, 249-256.
[68] Molenda, J. *Solid State Ionics* 2004, *175*, 203.
[69] Molenda, J. *Solid State Ionics* 2005, *176*, 1687.
[70] Molenda, J.; Swierczek, K.; Kucza, W.; Marzec, J.; Stoklosa, A. *Solid State Ionics* 1999, *123*, 155-163.
[71] Nakayama, M.; Ikuta, H.; Uchimoto, Y.; Wakihara, M.; Kawamura, K. *J. Phys. Chem. B* 2003, *107*, 10603-10607.
[72] Nobili, F.; Dsoke, S.; Corce, F.; Marassi, R. *Electrochimica Acta* 2005, *50*, 2307-2313.
[73] Nobili, F.; Dsoke, S.; Minicucci, M.; Croce, F.; Marassi, R. *J. Phys. Chem. B* 2006, *110 (23)*, 11310-11313.
[74] Nobili, F.; Tossici, R.; Croce, F.; Scrosati, B.; Marassi, R. *J. Power Sources* 2001, *94*, 238-241.
[75] Nobili, F.; Tossici, R.; Marassi, R.; Croce, F.; Scrosati, R. *J. Phys. Chem. B* 2002, *106*, 3909-3915.
[76] Okumura, T.; Fukutsuka, T.; Matsumoto, K.; Orikasa, Y.; Arai, H.; Ogumi, Z.; Uchimoto, Y. *J. Phys. Chem. C* 2011, *115(26)*, 12990-12994.
[77] Orazem, M. E.; Tribollet, B. Electrochemical Impedance Spectroscopy. John Wiley & Sons, Inc. Hoboken, New Jersey, 2008.
[78] Osaka, T.; Nakade, S.; Rajamäki, M.; Momma, T. *J. Power Sources* 2003, *119-121*, 929-933.
[79] Park, S. M.; Yoo, J. S.; Chang, B. Y.; Ahn, E. S. *Pure Appl. Chem.* 2006, *78(5)*, 1069–1080.
[80] Peled, E. *J. Electrochem. Soc.* 1979, *126*, 2047-2051.
[81] Pereira, N.; Dupont, L.; Tarascon, J. M.; Klein, L. C.; Amatucci, G. G. *J. Electrochem. Soc.* 2003, *150(9)*, A1273-1286.
[82] Pereira, N.; Klein, L. C.; Amatucci, G. G. *J. Electrochem. Soc.* 2002, *149(3)*, A262-A271.
[83] Pistoia, G.; Zane, D.; Zhang, Y. *J Electrochem Soc.* 1995, *142(8)*, 2551-2557.
[84] Poizot, P.; Laruelle, S.; Grugeon, S.; Dupont, L.; Tarascon, J. M. *Nature* 2000, *407*, 496-499.
[85] Ramadass, P.; Haran, B.; White, R.; Popov, B. N. *J. Power Sources* 2003, *123*, 230.
[86] Rhoderick, E. H.; Williams, R. H. *Metal-Semiconductor Contacts;* Clarendon: Oxford, 1988.
[87] Rodrigues, S.; Munichandraiah, N.; Shukla, A. K. *J. Power Sources* 2000, *87*, 12–20.
[88] Shi, Y. L.; Shen, M. F.; Xu, S. D.; Qiu, X. Y.; Jiang, L.; Qiang, Y. H.; Zhuang, Q. C.; Sun, S. G. *Int. J. Electrochem. Sci.* 2011, *6*, 3399-3415.
[89] Silva, D. C. C.; Crosnier, O.; Ouvrard, G.; Greedan, J.; Safa-Sefat, A.; Nazar, L. F. *Electrochem. Solid-State Lett.* 2003, *6(8)*, A162-A165.
[90] Sluyters-Rehbach, M.; Sluyters, J. H. in E. Yaeger, J. O' M. Bockris, B. E. Conway, S. Sarangapini (Eds.), Comprehensive Treatise of Electrochemistry, Plenum, New York, 1984, Vol. 9, p. 177.
[91] Sluyters-Rehbach, M.; Sluyters, J. H. in Electroanalytical Chemistry, Bard, A. J. Ed. Marcel Dekker, New York, 1977, Vol. 4, p. 1-127.
[92] Smith, A. J.; Burns, J. C.; Zhao, X. M.; Xiong, D. J.; Dahn, J. R. *J. Electrochem. Soc.* 2011, *158 (5)*, A447-A452.
[93] Song, J. Y.; Lee, H. H.; Wang, Y. Y. et al. *J. Power Sources* 2002, *111*, 255-267.
[94] Souza, D. C. S.; Pralong, V.; Jacobson, A. J.; Nazar, L. F. *Science* 2002, *296(5575)*, 2012-2015.

[95] Spontiz, R. *J. Power Sources* 2003, *113*, 72.
[96] Stromme, M. *Phys. Rev. B* 1998, *58*, 11015.
[97] Sun, X. G.; Dai, S. *J. Power Sources* 2010, *195*, 4266-4271.
[98] Sze, S. M. *Physics of Semiconductor Devices*; Wiley: New Jersey, 1981.
[99] Takamura, T.; Endo, K.; Fu, J.; Wu, Y.; Lee, K. J.; Matsumoto, T. *Electrochim. Acta* 2007, *53*, 1055–1061.
[100] Thackeray, M. M. *Prog. Solid State Chem.* 1997, *25*, 1.
[101] Thackeray, M. M.; David, W. I. F.; Bruce, P. G. et al. *Mater. Res. Bull.* 1983, *18*, 461-472.
[102] Thomas, M. G. S. R.; Bruce, P. G.; Goodenough, J. B. *J. Electrochem. Soc.* 1985, *132*, 1521-1528.
[103] Umeda, M.; Dokko, K.; Fujita, Y.; Mohamedi, M.; Uchida, I.; Selman, J. R. *Electrochim. Acta* 2001, *47*, 885-890.
[104] Van der Ven, A.; Aydinol, M. K.; Ceder, G. *Mater. Res. Soc. Symp. Proc.* 1998, *496*, 121.
[105] Van, E. J.; Wieland, J. L.; Eskes, H.; Kuiper, P.; Sawatzky, G. A.; De Groot, F. M. F.; Turner, T. S.; *Phys. Rev. B* 1992, *44*, 6090.
[106] Verma, P.; Maire, P.; Novák P. *Electrochim. Acta* 2010, *55*, 6332-6341.
[107] Vetter, K. J. Electrochemical Kinetics, Academic Press, New York, 1967.
[108] Wang, Y.; Fu, Z. W.; Yue, X. L. et al. *J. Electrochem. Soc.* 2004, *151(4)*, E162-E167.
[109] Wei, T.; Zhuang, Q. C.; Wu, C.; Cui, Y. L.; Fang, L.; Sun, S. G. *Acta Chim. Sinica* 2010, *68*, 1481-1486.
[110] Whittingham, M. S. *Chem. Rev.* 2004, *104*, 4271.
[111] Whittingham, M. S. *Prog. Solid State Chem.* 1978, *12*, 41.
[112] Wu, H. Q.; Li, Y. F. Electrochemical kinetics. Beijing: China Higher Educarion Press, 1998, 217-225.
[113] Xu, K. Chem. *Rev.* 2004, *104*, 4303-4417.
[114] Xu, K.; Cresce, A. V.; Lee, U. *Langmuir* 2010, *26(13)*, 11538-11543.
[115] Xu, S. D.; Zhuang, Q. C.; Tian, L. L.; et al. *J. Phys. Chem. C* 2011, *115*, 9210-9219.
[116] Xue, M. Z.; Fu, Z. W. *Electrochem. Commun.* 2006, *8(12)*: 1855-1862.
[117] Yamada, I.; Abe, T.; Iriyama, Y.; Ogumi, Z. *Electrochem. Commun.* 2003, *5*, 502-505.
[118] Yamada, I.; Iriyama, Y.; Abe, T.; Ogumi, Z. *J. Power Sources* 2007, *172*, 933-937.
[119] Yamada, Y.; Iriyama, Y.; Abe, T.; Ogumi, Z. *Langmuir* 2009, *25(21)*, 12766-12770.
[120] Yamakawa, N.; Jiang, M.; Grey, C. P. *Chem. Mater.* 2009, *21*, 3162-3176.
[121] Yoo, J. S.; Park, S. M. *Anal. Chem.* 2000, *72*, 2035-2041.
[122] Yoshida, T.; Takahashi, M.; Morikawa, S.; Ihara, C.; Katsukawa, H.; Shiratsuchi, T.; Yamakic, J. *J. Electrochem. Soc.* 2006, *153*, A576.
[123] Yu, F.; Zhang, J. J.; Wang, C. Y. *Progress in Chemistry* 2010, *22(1)*, 9-18.
[124] Zaban, A.; Zinigrad, E.; Aurbach, D. *J. Phys. Chem.* 1996, *100*, 3089-3101.
[125] Zhuang, Q. C.; Wei, G. Z.; Xu, J. M.; Fan, X. Y.; Dong, Q. F.; Sun, S. G. *Acta Chimica. Sinica.* 2008, *66*, 722-728.
[126] Zhuang, Q. C.; Wei, T.; Du, L. L.; Cui, Y. L.; Fang, L.; Sun, S. G. *J. Phys. Chem. C* 2010, *114*, 8614-8621.
[127] Zhuang, Q. C.; Wei, T.; Wei, G. Z. et al. *Acta Chim. Sinica* 2009, *67(19)*, 2184-2192.
[128] Zhuang, Q. C.; Xu, J. M.; Fan, X. Y.; Wei, G. Z.; Dong, Q. F.; Jiang, Y. X.; Huang, L.; Sun, S. G. *Sci. China, Ser. B* 2007, *50(6)*, 776-783.